# Helicopter Instructor's Handbook

U.S. Department
of Transportation

Federal Aviation
Administration

MW00963768

# Preface

The Helicopter Instructor's Handbook is designed as a technical manual for applicants who are preparing for their flight instructor pilot certificate with a helicopter class rating. This handbook contains detailed coverage of aerodynamics, flight controls, systems, performance, flight maneuvers, emergencies, and aeronautical decision-making. Topics such as weather, navigation, radio navigation and communications, use of flight information publications, and regulations are available in other Federal Aviation Administration (FAA) publications.

This handbook conforms to flight instructor pilot training and certification concepts established by the FAA. There are different ways of teaching, as well as performing flight procedures and maneuvers, and many variations in the explanations of aerodynamic theories and principles. Occasionally the word "must" or similar language is used where the desired action is deemed critical. The use of such language is not intended to add to, interpret, or relieve a duty imposed by Title 14 of the Code of Federal Regulations (14 CFR).

This handbook is available for download, in PDF format, from www.faa.gov.

This handbook is published by the United States Department of Transportation, Federal Aviation Administration, Airman Testing Standards Branch, AFS-630, P.O. Box 25082, Oklahoma City, OK 73125.

Comments regarding this publication should be sent, in email form, to the following address:

AFS630comments@faa.gov

# Acknowledgments

The Helicopter Instructor's Handbook was produced by the Federal Aviation Administration (FAA) with the assistance of Safety Research Corporation of America (SRCA). The FAA wishes to acknowledge the following contributors:

NZ Civil Aviation Authority for image of safety procedures for approaching a helicopter (Chapter 1)

David Park (www.freedigitalphotos.net) for image used in Chapter 2

Paul Whetstone (www.meyersaircraft.com) for image used in Chapter 5

Burkhard Domke (www.b-domke.de) for images used in Chapter 5

Bishop Equipment Mfg. Inc (www.bishopequipment.com) for image used in Chapter 6

Terry Simpkins of FLYIT Simulators (www.flyit.com) for image used in Chapter 13

Additional appreciation is extended to W.A. (Dub) Blessing, HAI Outstanding CFI award 1985; Donovan L. Harvey (1928 to 2000), HAI Outstanding CFI award 1987; Neil Jones, CFI; Chin Tu, CFI; the Helicopter Association International (HAI), Aircraft Owners and Pilots Association (AOPA), and the AOPA Air Safety Foundation for their technical support and input.

# Table of Contents

## Chapter 17
## Single-Pilot Resource Management, Aeronautical Decision-Making, and Risk Management

# Chapter 1
# Introduction to Flight Training

## Purpose of Flight Training

It is the helicopter instructor's responsibility to discuss the overall purpose of flight training with the student. Explain that the goal of flight training is the acquisition and honing of basic airmanship skills that provide the student with:

- An understanding of the principles of flight.

- The ability to safely operate a helicopter with competence and precision both on the ground and in the air.

- The knowledge required to exercise sound judgment when making decisions affecting operational safety and efficiency.

Ensure the student understands that a helicopter operates in a three-dimensional environment and requires specific skills to control the aircraft:

- Coordination—the ability to use the hands and feet together subconsciously and in the proper relationship to produce desired results in the helicopter control.

- Control touch—to develop the ability to sense and evaluate the varying pressures and resistance of the control surfaces and/or the instructor's input transmitted through the cockpit flight controls and apply inputs in response to those pressures.

- Timing—the application of muscular coordination at the proper instant to make maneuvering flight a constant smooth process.

- Mental comprehension of aerodynamic state, power required versus power available, and hazards present.

Keep in mind that an accomplished pilot demonstrates the ability to assess a situation quickly and accurately and to determine the correct procedure to be followed under the circumstance; to analyze accurately the probable results of a given set of circumstances or of a proposed procedure; to exercise care and due regard for safety; to gauge accurately the performance of the aircraft; and to recognize personal limitations and limitations of the aircraft and avoid approaching the critical points of each. The development of airmanship skills requires effort and dedication on the part of both the student and the flight instructor. It begins with the first training flight when the instructor encourages proper habit formation by introducing and modeling safe operating practices.

While every aircraft has its own particular flight characteristics, the purpose of primary and intermediate flight training is not to learn how to fly a particular make and model of helicopter; it is to develop skills and safe habits that are transferable to any helicopter. *[Figure 1-1]* Basic airmanship skills serve as a firm foundation for this. Acquiring necessary airmanship skills during training and demonstrating these skills by flying with precision and safe flying habits allows the pilot to transition easily to more complex helicopters. Remember, the goal of flight training is to become a safe and competent pilot, and that passing required tests for pilot certification is only the first step toward this goal.

## Practical Flight Instructor Strategies

As discussed in Chapter 8 of the Aviation Instructor Handbook, certified flight instructors (CFIs) should remember they are a role model for the student. The flight instructor should demonstrate good aviation air sense and practices at all times.

**Figure 1-1.** *As part of flight training, a pilot instructs a student on proper techniques for landing at an airport.*

For the helicopter CFI, this means:

- Before the flight—discuss the procedures for the exchange of controls, establish scan areas for clearing the aircraft, and establish who is responsible for initiating immediate action in an emergency.

- During flight—prioritize the tasks of aviating, navigating, and communicating. Instill the importance of "see and avoid" and utilizing aircraft lighting to be more visible in certain flight conditions.

- During landing—conduct stabilized approaches, maintain proper angle and desired rate of closure on final. Use aeronautical decision-making (ADM) to demonstrate good judgment for go-arounds, wake turbulence avoidance, traffic, and terrain avoidance.

- Always—remember that safety is paramount.

Flight instructors have the responsibility of producing the safest pilots possible. For that reason, CFIs should tirelessly encourage each student to learn as much as he or she is capable of and keep raising the bar toward the ultimate goal. When introducing lesson tasks, flight instructors should introduce the student to the Practical Test Standards (PTS) and discuss that the minimum acceptable standards for passing a given maneuver are stated therein. The CFI must stress to the student that these are only the minimum standards and that he or she should strive for much higher performance.

The PTS is not a teaching tool. It is a testing tool. The overall focus of flight training should be on learning, which includes gaining an understanding of why the standards exist and how they were determined. *[Figure 1-2]* Use the PTS as a training aid. Title 14 of the Code of Federal Regulations (14 CFR) does require specific training for the PTS endorsements, but this should not be presented to the student at the end of the training. The CFI should take into consideration all of the

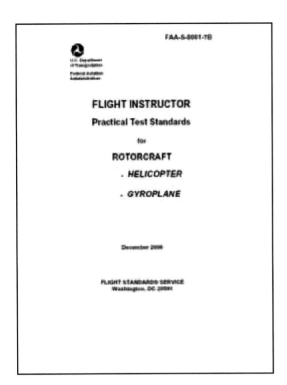

**Figure 1-2.** *Practical Test Standards.*

necessary training and strategically plan that training so the student has time to practice and prepare. It is the ultimate goal of the CFI to produce the safest, most competent pilot from his or her course of instruction and take pride in knowing that the student not only passed the test standards but exceeds those standards when conducting any and all helicopter procedures, on the ground or in the air.

# The Federal Aviation Administration (FAA)

## Role

It is imperative that a new student be introduced and become familiar with the role of the Federal Aviation Administration (FAA) in aviation. For the new student, this includes introducing him or her to the parts and subparts of 14 CFR that relate to flight training and pilot certification. To be included are pertinent handbooks, the PTS, and any references the CFI determines to be valuable to the student pilot learning experience. For transitioning pilots, the PTS for the helicopter is a key reference. The student should also be introduced to the Knowledge Test Guides that can be found at www.faa.gov.

An online session at the FAA website provides the CFI with an opportunity to introduce the new student and/or transitioning pilot to the many resources now available around the clock. The site has easy-to-access handbooks, regulations, standards, manuals, references, and even online courses. With the advent of the Integrated Airman Certificate and/or Rating Application (IACRA), the FAA can process airman certification documents via the Internet, interfacing with multiple FAA national databases to validate data

and verify specific fields. IACRA automatically ensures applicants meet regulatory and policy requirements and forwards the FAA Form 8710-1 application and test results to the FAA Airmen Certification Branch. *[Figure 1-3]* While many younger students interface easily with the Internet, a CFI trains pilots of all ages. Ensuring the student is comfortable using the FAA's Internet resources is part of a good training program.

**Figure 1-3.** *IACRA processes applications for airman certification via the Internet and automatically ensures applicants meet regulatory and policy requirements through programming rules and data validation.*

## FAA Reference Material

The reference materials described below, as revised, can be used by the CFI to assemble a handout for the student. An example of such a handout can be found in Appendix A.

- Pilot's Handbook of Aeronautical Knowledge (FAA-H-8083-25)—provides essential knowledge for pilots as they progress through pilot training. Useful to beginning pilots, as well as those pursuing more advanced certificates.

- Helicopter Flying Handbook (FAA-H-8083-21)— designed as a technical manual for applicants who are preparing for their private, commercial, or flight instructor pilot certificates with a helicopter class rating. The handbook contains detailed coverage of aerodynamics, flight controls, systems, performance, flight maneuvers, emergencies, and ADM specific to helicopter flight, which makes it a valuable training aid. Helicopters are rotorcraft as are gyroplanes.

Gyroplanes and helicopters are the two classes of aircraft in the rotorcraft category. Therefore, to differentiate between the classes of aircraft with different skill requirements, the FAA issues rotorcraft helicopter ratings or rotorcraft gyroplane ratings.

- Instrument Flying Handbook (FAA-H-8083-15)—designed for use by instrument flight instructors and pilots preparing for instrument rating tests, this handbook is a valuable training aid for CFIs as it includes basic reference material for knowledge testing and instrument flight training. *[Figure 1-4]*

**Figure 1-4.** *The Instrument Flying Handbook is one of many training aids provided by the FAA Airman Testing Standards Branch.*

- Risk Management Handbook (FAA-H-8083-2)—provides tools to help pilots determine and assess each situation for the safest possible flight with the least amount of risk. This handbook presents methods pilots can use to manage the workloads associated with each phase of flight, resulting in a safer, more enjoyable, and less stressful experience for both themselves and their passengers.

- Advanced Avionics Handbook (FAA-H-8083-6)—provides general aviation users with comprehensive information on the advanced avionics equipment available in technically advanced aircraft.

- Aeronautical Information Manual (AIM)—Chapter 10 of the AIM includes items that specifically pertain to helicopter operations. The AIM also provides the aviation community with basic flight information and Air Traffic Control (ATC) procedures for use in the National Air Space (NAS) of the United States. It also contains items of interest to pilots concerning health/medical facts, factors affecting flight safety, etc.

- Airport/Facility Directory—containing information on public and joint use airports, communications, navigation aids, instrument landing systems, very high frequency (VHF) omnirange navigation system (VOR) receiver checkpoints, preferred routes, Automated Flight Service Station (AFSS)/Weather Service telephone numbers, Air Route Traffic Control Center (ARTCC) frequencies, part-time surface areas, and various other pertinent special notices essential to air navigation, the directory is now available in digital format at www.faa.gov.

- Practical Test Standards—the Rotorcraft (Helicopter and Gyroplane) PTS establishes the standards for pilot certification practical tests for the rotorcraft category, helicopter, and gyroplane classes. FAA inspectors and designated pilot examiners (DPEs) conduct practical tests in compliance with these standards. Flight instructors and applicants should find these standards helpful during training and when preparing for the practical test. More detailed information can be found at www.faa.gov. Refer the new student to page 3 of the PTS which provides a list of references used to compile the standards under which he or she is tested. This list identifies the publications that describe the various tasks that need to be mastered prior to the test. While explaining the PTS, be sure to review the Rotorcraft Practical Test Prerequisites.

An applicant for the Rotorcraft Practical Test is required by 14 CFR part 61 to:

1. Be able to read, speak, write, and understand the English language. (If there is a doubt, use Advisory Circular (AC) 60-28, English Language Skill Standards.)

2. Have passed the appropriate pilot knowledge test since the beginning of the 24th month before the month in which the practical test is completed.

3. Have satisfactorily accomplished the required training and obtained the aeronautical experience prescribed.

4. Possess a current Medical Certificate.

5. Have an endorsement from an authorized instructor certifying that the applicant has received and logged training time within 60 days preceding the date of application.

6. Also have an endorsement certifying that the applicant has demonstrated satisfactory knowledge of the subject areas in which the applicant was deficient on the airman knowledge test.

## Role of the Examiner

The subject of the PTS also offers the CFI an opportunity to discuss the role of the examiner who plays an important role in the FAA's mission of promoting aviation safety by administering FAA practical tests for pilot and flight instructor certificates and associated ratings. To satisfy the need for pilot testing and certification services, the FAA delegates certain of these responsibilities to private individuals who are not FAA employees.

Appointed in accordance with 14 CFR section 183.23, a designated pilot examiner (DPE) is an individual who meets the qualification requirements of the Pilot Examiner's Handbook, Order 8710.3, and who:

- Is technically qualified.

- Holds all pertinent category, class, and type ratings for each aircraft related to their designation.

- Meets requirements of 14 CFR part 61, sections 61.56, 61.57, and 61.58, as appropriate.

- Is current and qualified to act as pilot in command (PIC) of each aircraft for which they are authorized.

- Maintains at least a third-class medical certificate if required.

- Maintains a current flight instructor certificate, if required.

Designated to perform specific pilot certification tasks on behalf of the FAA, a DPE may charge a reasonable fee. Generally, a DPE's authority is limited to accepting applications and conducting practical tests leading to the issuance of specific pilot certificates and/or ratings. The majority of FAA practical tests at the private and commercial pilot level are administered by DPEs, following FAA provided test standards.

DPE candidates must have good industry reputations for professionalism, integrity, a demonstrated willingness to serve the public, and adhere to FAA policies and procedures in certification matters. The FAA expects the DPE to administer practical tests with the same degree of professionalism, using the same methods, procedures, and standards as an FAA aviation safety inspector (ASI).

## Role of the Certificated Flight Instructor (CFI)

The FAA places full responsibility for student flight training on the shoulders of the CFI, who is the cornerstone of aviation safety. It is the job of the flight instructor to train the student in all the knowledge areas and teach the skills necessary for the student to operate safely and competently as a certificated pilot in the NAS. The training is not limited to airmanship skills, but includes pilot judgment and decision-making and good operating practices.

A pilot training program depends on the quality of the ground and flight instruction the student receives. A competent instructor must possess a thorough understanding of the learning process, knowledge of the fundamentals of teaching, and the ability to communicate effectively with the student. He or she uses a syllabus and teaching style that embodies the "building block" method of instruction. In this method, the student progresses from the unknown to the known via a course of instruction laid out in such a way that each new maneuver embodies the principles involved in the performance of maneuvers previously learned. Thus, with the introduction of each new subject, the student not only learns a new principle or technique, but also broadens his or her application of those principles or techniques previously learned.

Insistence on correct techniques and procedures from the beginning of training by the CFI ensures the student develops proper habit patterns. Any deficiencies in maneuvers or techniques must immediately be emphasized and corrected. A CFI serves as a role model for the student who observes the flying habits of his or her flight instructor during flight instruction, as well as when the instructor conducts other pilot operations. Thus, the CFI becomes a model of flying proficiency for the student who, consciously or unconsciously, attempts to imitate the instructor. The CFI's advocacy and description of safety practices mean little to a student if the instructor does not demonstrate them consistently. For this reason, CFIs must observe recognized safety practices, as well as regulations during all flight operations.

An appropriately rated CFI is responsible for training the pilot applicant to acceptable standards in all subject matter areas, procedures, and maneuvers included in the tasks within the appropriate PTS. Because of the impact of their teaching activities in developing safe, proficient pilots, flight instructors should exhibit a high level of knowledge, skill, and the ability to impart that knowledge and skill to students.

Additionally, the flight instructor must certify that the applicant is able to perform safely as a pilot and is competent to pass the required practical test. Throughout the applicant's training, the CFI is responsible for emphasizing the performance of effective visual scanning, collision avoidance, and runway incursion avoidance procedures.

Anyone who enrolls in a pilot training program commits considerable time, effort, and expense to earn a pilot certificate. Many times an individual judges the effectiveness of the flight instructor and the success of the pilot training program based on his or her ability to pass the requisite FAA practical test. A truly professional flight instructor stresses to the student that practical tests are a sampling of pilot ability compressed into a short period of time. The goal of a CFI is to train the "total" pilot.

## Flight Safety Practices

A major component of the FAA's mission is to improve the nation's aviation safety record by conveying safety principles and practices through training, outreach, and education. The goal to reduce the number of accidents in the ever increasingly populated airways means safe flight practices are an important element of flight instruction. It is the CFI's responsibility to incorporate flight safety into the program of training.

Do not become complacent about safety while instructing. The CFI must always be vigilant about safety and must instill a safety-first attitude in the student. According to statistics from Helicopter Association International's (HAI) Five-Year Comparative U.S. Civil Helicopter Safety Trends, the ratio of instructional/training-related accidents to total accidents in the United States has increased more than 18 percent between January 1, 2002, and December 31, 2006. Interestingly enough though, the total number of helicopter flight hours has increased by 37 percent, while the accident rate per 100,000 flight hours has drastically decreased—by 42 percent in the same time period. The entire U.S. Civil Helicopter Safety Statistic - Summary Report can be found at www.rotor.com.

Accidents happen quickly during flight instruction, as this recent National Transportation Safety Board (NTSB) accident report reveals:

During a training flight, a helicopter collided with terrain. Weather was visual flight rules (VFR) with no flight plan filed. This was the CFI's first instructional flight with this student. They conducted the preflight inspection of the helicopter together, started up, and departed for the practice area.

Once the student had a general understanding of the controls, they did an approach that terminated in a hover. The CFI set up the helicopter for a slight right quartering headwind to compensate for translating tendencies, then allowed the student to manipulate the controls. During hover, the helicopter exhibited pendulum action that is common for new students learning to hover. During one of the right lateral oscillations, the helicopter unexpectedly lost altitude. The right skid contacted the ground, and the helicopter rolled over onto its right side. Within seconds, it ignited. Both pilots exited immediately.

Since the helicopter and engine had no mechanical failures or malfunctions during the flight, the accident might have been prevented by:

- Maintaining a proper skid height during instruction at all times.

- Stopping the lateral and aft movement sooner.

- Restricting hovering flight to later lessons after the student has gained some insight and appreciation of the control responsiveness and sensitivity of the helicopter.

The CFI also should have stayed on the controls longer to give the student more time to become familiar with them. The CFI violated the building block principle of simple to complex. The student had no experience to build upon. Helicopter students learn best by beginning in the air where there is a greater margin of error and then learning to fly closer to the ground.

Accident data at the NTSB offer CFIs excellent scenario material for safety discussions. Updated daily and located at www.ntsb.gov, descriptions of more than 140,000 aviation accidents can be searched by a variety of factors, such as date or aircraft category.

## Helicopter Hazards

During the entire training program, CFIs should emphasize safe operation of the aircraft. The student must be introduced to and completely understand the flight characteristics of the type helicopter being flown. Loss of tail rotor effectiveness (LTE), dynamic rollover (DRO), and the meaning of and how to interpret the height velocity diagram are three topics of discussion for continuous review. By virtue of its many moving parts, the helicopter presents numerous hazards. *[Figure 1-5]* It is the responsibility of the CFI to teach safe operating practices in and around the aircraft.

A CFI should draw to the attention of the student the hazards that include, but are not limited to the following:

- For single rotor helicopters, students should be taught from the beginning that it is preferred to approach

# SAFETY AROUND HELICOPTERS

## Approaching or Leaving a Helicopter

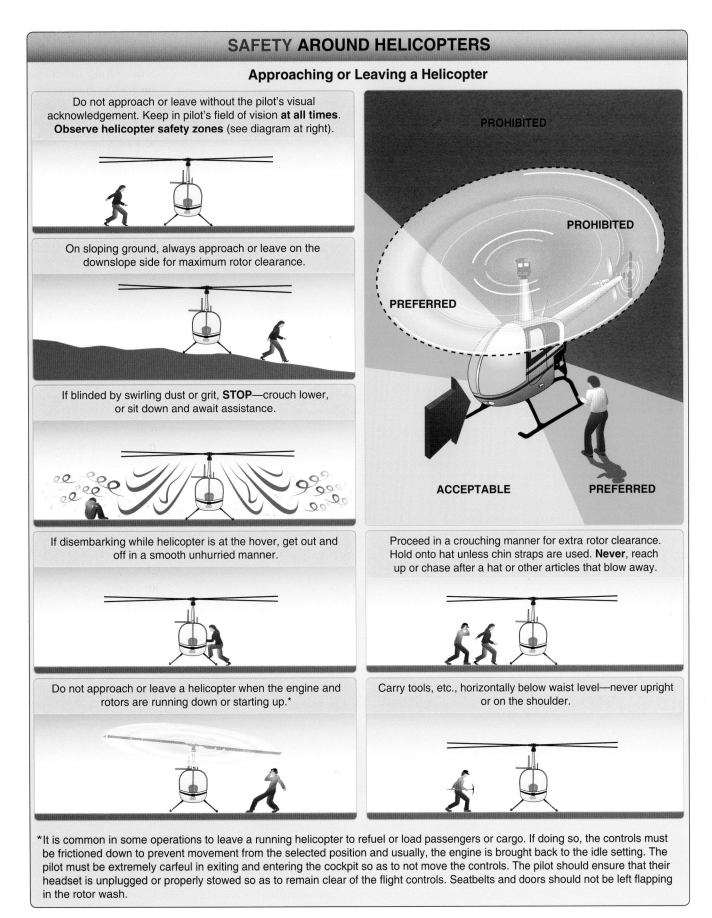

Do not approach or leave without the pilot's visual acknowledgement. Keep in pilot's field of vision **at all times**. **Observe helicopter safety zones** (see diagram at right).

On sloping ground, always approach or leave on the downslope side for maximum rotor clearance.

If blinded by swirling dust or grit, **STOP**—crouch lower, or sit down and await assistance.

If disembarking while helicopter is at the hover, get out and off in a smooth unhurried manner.

Do not approach or leave a helicopter when the engine and rotors are running down or starting up.*

PROHIBITED

PROHIBITED

PREFERRED

ACCEPTABLE

PREFERRED

Proceed in a crouching manner for extra rotor clearance. Hold onto hat unless chin straps are used. **Never**, reach up or chase after a hat or other articles that blow away.

Carry tools, etc., horizontally below waist level—never upright or on the shoulder.

*It is common in some operations to leave a running helicopter to refuel or load passengers or cargo. If doing so, the controls must be frictioned down to prevent movement from the selected position and usually, the engine is brought back to the idle setting. The pilot must be extremely carfeul in exiting and entering the cockpit so as to not move the controls. The pilot should ensure that their headset is unplugged or properly stowed so as to remain clear of the flight controls. Seatbelts and doors should not be left flapping in the rotor wash.

**Figure 1-5.** *Safe operating procedures in and around the helicopter.*

and exit the helicopter from the sides but that the forward quarter is acceptable. If approaching or exiting a helicopter that is on a slope, always exit on the downward side to avoid contact with the rotor blades. Limited access to the near aft portion of the fuselage is acceptable for some helicopters, such as the BO-105 and BK-117, in which the tail rotor has been elevated and loading is in the rear of the fuselage. CFIs should advise students to always consult with the pilot or trained personnel before going aft of the cockpit doors. This instills in the students the preferred direction to enter and exit the rotor disk area so the pilot can maintain eye contact with personnel around the aircraft. During preflight, the CFI should teach students to do a proper walk-around before moving any control surfaces to ensure that nothing is in the way of the main or tail rotor blades.

- Always avoid the tail rotor by approaching from the sides. The rotor disk should be tipped so the students understand just how low the main rotor blades may dip in winds and as a result of exaggerated control movements.

- Hands and fingers can be pinched by rotor hubs and hinges during preflight and postflight inspections.

- Main and tail rotor blades pose significant hazards for those unaccustomed to being around helicopters during ground operations.

- Any moving blade is dangerous and can cause injury or damage while under power or during the start up and coast down periods after engine power has been removed.

- Wind or a control input can easily cause slow moving blades to droop or flex, reducing clearance for people standing underneath the rotor disk.

- If the helicopter must be moved from the hangar, students should be cautioned on the hazards of having a piece of machinery raised off the surface and the correct methods of raising and lowering the aircraft. Since helicopters may be taller than an equal size airplane; the student should be taught to ensure plenty of vertical clearance for the aircraft as it is moved. Trip hazards, such as ground wires, should be explained as to requirements, storage, and attachment at end of flights.

- The movement of the helicopter for flight should include preplanning to prevent the hangar from filling with grass, dirt, and excessive wind in the facility. The direction of the wind and airflow around the building should be considered before selecting a takeoff point for the helicopter.

- Jewelry, especially rings, should be removed before preflight and postflight to ensure that they will not be caught on any fasteners or sharp objects. Loose clothing should be secured, and objects in pockets should be removed if the pockets cannot be fastened.

In hover flight, the CFI should emphasize the hazards that rotor wash presents to persons or light aircraft nearby. Dust and debris cause eye injuries and vortices damage light aircraft. A tail rotor is another source of significant hazard because it is out of sight of the pilot. Instructors should ensure the student is aware of the requirement to keep the tail rotor area cleared. Hazards such as those listed above are but a few of the hazards unique to the helicopter. The observant CFI identifies potential hazards during the lesson, corrects the deficiency immediately with an explanation, and develops them as teaching points.

## Instructional Hazards

Flying a helicopter offers a different set of physical and mental challenges for a student. The stress of learning how to fly is coupled with the physical demands of flying the helicopter. The constant vibration of the aircraft, as well as the continually need to make control inputs to "fly" the aircraft, make helicopter flight a more physically and mentally strenuous type of flying. The vibration, noise, and stress can lead to fatigue, which can have a detrimental effect upon the ability of the student not only to fly a helicopter but to absorb instruction. To combat this hazard, limit the length of the lesson to less than an hour until the student becomes accustomed to the demands of this type of flying. For further discussion of medical factors associated with flying, refer to the Pilot's Handbook of Aeronautical Knowledge.

As shown in *Figure 1-6,* the CFI must remain vigilant when the student has control of the helicopter because the student's knee may get in the way of the cyclic movement. The student's size cannot be changed, but it is the CFI's responsibility to teach the student to be aware of how their size may affect the flight controls input.

The instructor should always ensure that all of the flight controls are unencumbered. Students are so focused on the task at hand when learning to fly and often times will unknowingly obstruct the flight controls. For example, water bottles, clothing and cameras can get stuck under the collective levers preventing movement, or anti-torque pedals can get blocked from movement by the students boot or shoe.

Another potential instructional hazard stems from the ability of helicopter rotor blades to strike the terrain or objects in a 360° arc. This unique capability of the helicopter must be stressed when teaching a student who is transitioning from fixed-wing aircraft. A fixed-wing pilot is accustomed only to the idea that one wing will hit if the aircraft is banked too

**Figure 1-6.** *Robinson Helicopter R-22.*

far. If teaching someone who is transitioning from airplanes, the CFI needs to stress to the student the speed of the rotor and its close proximity to the ground.

## Collision Avoidance

While pilots often believe that having a CFI on board minimizes the possibility of a midair collision (MAC), FAA research reveals that flight instructors were on board the aircraft in 37 percent of the accidents studied. From a collision perspective, flight training is one of the most dangerous missions—an especially frightening fact, considering that flight instructors comprise less than 10 percent of the pilot population.

### See and Avoid

As discussed in the Aviation Instructor's Handbook, the CFI must ensure from the start of flight training that the student develops the habit of maintaining airspace surveillance at all times. *[Figure 1-7]* If a student believes the instructor assumes all responsibility for scanning and collision avoidance procedures, he or she will not develop the habit of maintaining the constant vigilance essential to safety. Establish scan areas and communication practices for keeping the aircraft cleared as outlined in the AIM, paragraphs 4-4-15 and 8-8-6c. For example, "Clear left? Cleared left. Turning left." should be verbalized in conjunction with the actual scanning. In addition to clearing left and right, a helicopter pilot must also clear directly above and below since the helicopter has the ability of climbing and descending

**Figure 1-7.** *Collision avoidance, both in the air and on the ground, is one of the most basic responsibilities of a pilot flying in visual conditions.*

vertically. This ability has resulted in helicopters climbing directly into overhead hangar doors and power lines. Any observed tendency of a student to enter flight maneuvers without first making a careful check for other air traffic must be corrected immediately. In addition to the statistic quoted above, recent studies of midair collisions determined that:

- Most of the aircraft involved in collisions are engaged in recreational flying, and not on any type of flight plan.

- Most midair collisions occur in VFR weather conditions during weekend daylight hours.

- The vast majority of accidents occurred at or near nontowered airports and at altitudes below 1,000 feet.

- Pilots of all experience levels were involved in midair collisions, from pilots on their first solo ride to 20,000-hour veterans.

- Most collisions occur in daylight with visibility greater than three miles.

It is imperative to introduce 14 CFR section 91.113, Right-of-Way Rules: Except Water Operations," for the "see and avoid" concept immediately to the student. Practice the "see and avoid" concept at all times regardless of whether the training is conducted under VFR or instrument flight rules (IFR). A CFI and student can review the FAA's suggestions for how to contribute to professional flying and reduce the odds of being involved in a midair collision, at www.faa.gov. Other references that contain collision avoidance information for both the CFI and student are AC 90-48, Pilot's Role in Collision Avoidance; FAA-H-8083-25, Pilot's Handbook of Aeronautical Knowledge; and the Aeronautical Information Manual (AIM) (all as revised) located online at www.faa.gov.

## Positive Exchange of Flight Controls

Incident/accident statistics indicate a need to place additional emphasis on the exchange of control of an aircraft by pilots. Numerous accidents have occurred due to a lack of communication or misunderstanding as to who actually had control of the aircraft, particularly between students and flight instructors. Establishing the following procedure during initial training ensures the formation of a habit pattern that should stay with students throughout their flying careers. They are more likely to relinquish control willingly and promptly when instructed to do so during flight training.

During flight training, there must always be a clear understanding between the student and the flight instructor of who has control of the aircraft. *[Figure 1-8]* Prior to flight, a briefing should be conducted that includes the procedure for the exchange of flight controls. A positive three-step process in the exchange of flight controls between pilots is

**Figure 1-8.** *There should never be any doubt about who is flying the helicopter.*

a proven procedure and one that is strongly recommended. During this procedure, a visual check is recommended to see that the other person actually has the flight controls. When returning the controls to the instructor, the student should follow the same procedure the instructor used when giving control to the student. There should never be any doubt as to who is flying the aircraft.

CFIs should always guard the controls and be prepared to take control of the aircraft. When necessary, the instructor should take the controls and calmly announce, "I have the flight controls." If an instructor allows a student to remain on the controls, the instructor may not have full and effective control of the aircraft. Anxious students can be incredibly strong and usually exhibit reactions inappropriate to the situation. If a recovery is necessary, there is absolutely nothing to be gained by having the student on the controls and needing to fight for control of the aircraft. Students should never be allowed to exceed the flight instructor's limits. Flight instructors should not exceed their own ability to perceive a problem, decide upon a course of action, and physically react within their ability to fly the aircraft.

## Single-Pilot Resource Management (SRM)

According to data presented at the 2005 International Helicopter Safety Symposium, the helicopter accident rate is 30 percent higher than the general aviation (GA) accident rate. Reducing this rate is an industry wide goal and the CFI plays an important role in reaching it by stressing single-pilot resource management (SRM) and risk management during flight training.

As discussed in the Aviation Instructor's Handbook and the Pilot's Handbook of Aeronautical Knowledge, SRM is the art and science of managing all resources (both onboard the aircraft and from outside sources) available to a single pilot (prior and during flight) to ensure the successful outcome of the flight. SRM grew out of the airline industry's crew resource management (CRM) training for flight crews that was launched in an effort to reduce human factors-related aircraft accidents. SRM is the effective use of all available resources: human, hardware, and information to ensure a safe flight. The CFI must keep in mind that SRM is not a single task; it is a set of skill competencies that must be evident in all tasks. Aviation resource management charges the flight instructor with the responsibility of teaching the student a safety mindset that enhances his or her decision-making skills.

SRM depends upon teaching the student higher order thinking skills (HOTS) as discussed in Chapter 2 of the Aviation Instructor's Handbook. HOTS are taught from simple to complex and from concrete to abstract. To teach HOTS effectively involves strategies and methods that include:

- Using problem-based learning (PBL) instruction,
- Authentic problems,
- Student-centered learning,
- Active learning,
- Cooperative learning, and
- Customized instruction to meet the individual learner's needs.

These strategies engage the student in some form of mental activity, have the student examine that mental activity and select the best solution, and challenge the student to explore other ways to accomplish the task or the problem.

Student understanding of risk management and judgment is enhanced when the instructor includes the student in all preflight practices and procedures, as the instructor shares the logic behind decisions whether to fly or not to fly. If the instructor uses the performance charts every time before flying to ensure sufficient power, control authority, and lift is available, then the student will probably acquire that habit. If the instructor always prompts the student to call for a weather, NOTAMS, and TFR briefing, then the student will learn proper preflight planning techniques. If the instructor determines what the student wants to be able to do with the helicopter, then the instructor can makes plans to ensure that the hazards inherent to those operations are covered completely and emphasized during training.

## Risk Management

The FAA is committed to reducing the number of helicopter accidents and promoting risk management as an important component of flight training. The objective of risk management is to provide a proper balance between risk and opportunity. Two elements define risk management: hazard and risk. Hazard is a real or perceived condition, event, or circumstance that a pilot encounters. Risk is how the pilot views the potential impact of the hazard.

Risk management is the method used to control, eliminate, or reduce the hazard to an acceptable level. The individual pilot is unique to risk management. An acceptable level of risk to one pilot may not necessarily be the same to another pilot. Unfortunately in many cases, the pilot perceives that his or her level of risk acceptability is actually greater than their capability, thereby taking on risk that is dangerous.

For example, prior to entering a helicopter, the CFI must establish his or her own limitations. How far is the CFI willing to allow the student to drift during a hover? Once personal limitations are established, the CFI must fly within them. The CFI should always ensure that the helicopter is never allowed to depart the instructor's comfort zone and maneuvering limitations. In reality, the instructor is observing the maneuvering of the helicopter and monitoring the control movements by sight or feel. The helicopter instructor has to be very familiar with that particular helicopter and it's responses to control inputs and winds, especially at a hover with a wing with an airspeed of 400+ knots flying while at 3 feet landing gear height above the surface. A split second delay in correcting an errant control input can be disastrous.

References and resources for risk management include:

- Pilot's Handbook of Aeronautical Knowledge, FAA-H-8083-25
- Pilot risk management brochures located at www.faa.gov (brochures include tips for teaching practical risk management) [Figure 1-9]
- Risk Management Handbook, FAA-H-8083-2

Since the DPE evaluates the applicant's ability to use good ADM procedures in order to evaluate risks throughout the practical test, it is important the CFI incorporates risk management into the flight lessons as soon as possible. The scenarios should be realistic and within the capabilities of the helicopter used for the practical test.

To teach risk management, CFIs must understand system safety flight training occurs in three phases. First, there are the traditional aircraft control maneuvers. In order to apply critical thinking skills, the student must first have a high degree of confidence in their ability to fly the aircraft. Basic airmanship skill is the priority during this phase of flight training. The CFI accepts the responsibility of risk management until the student is able to accept more tasking.

**Figure 1-9.** *Brochure available from the FAA website for teaching practical risk management.*

In the second phase, the CFI teaches the student how to identify hazards, manage risk, and use all available resources to make each flight as safe as possible. This can be accomplished through scenarios that emphasize the skill sets being taught. For example, the CFI could inform the student that they were going to do some photography in the mountains for a survey. The instructor could give the student two temperatures and one elevation for the areas. Then, the instructor would assist the student in reviewing the performance charts for the two temperatures and have the student determine the differences in helicopter performance with those temperatures and how to determine any maneuvering restrictions from those temperatures. "Does the lack of OGE hover restrict anything?" could be one question. Then hopefully, the CFI and student would fly up to some point for the student to have a safe and real-life experience of the difference in aircraft performance in higher temperatures and higher density altitudes.

In the third phase, as the student is completing the course of training, the instructor should begin exposing the student to practical scenarios of helicopter flight and enable the student to discern the hazards associated with each profile. Using the "simple to complex" method at all times, the student is introduced to scenarios demanding focus on several safety-of-flight issues. *[Figure 1-10]*

The CFI must present the subject of risk management as it relates to helicopter operations for the level of instruction being presented. For example, a new helicopter student has different requirements from those of a prospective commercial Emergency Medical Services (EMS) pilot.

## Chapter Summary

This chapter introduced the purpose of flight training, resources available to the CFI, a brief overview of flight safety practices, and hazards unique to helicopter flying.

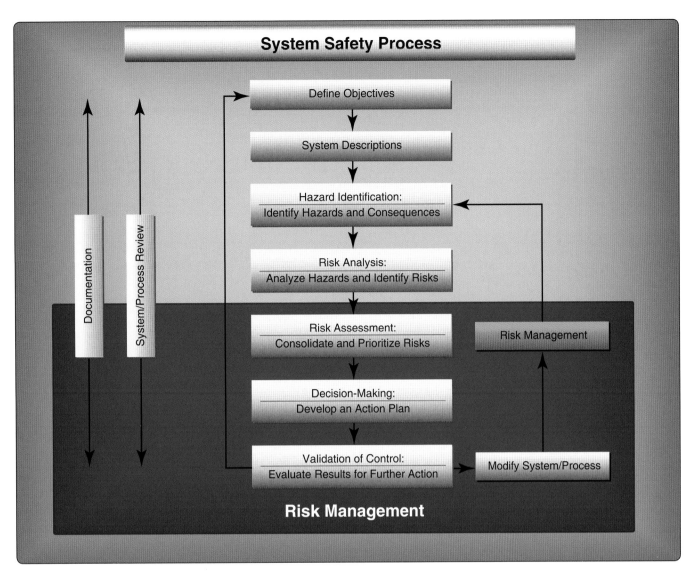

**Figure 1-10.** *An example of a system safety process an instructor could use in flight training.*

## Chapter 2

# Introduction to the Helicopter

## Introduction

The objective of the first flight lesson is to determine the student's motivation and goals and introduce the student to:

- Training procedures
- The helicopter
- Local flying area and prominent landmarks
- The relationship between control inputs and aircraft attitude

## Introduction to the Helicopter

### Objective

The purpose of this lesson is to introduce the student to rotary-wing flight. The student demonstrates a basic knowledge of the main components, safe helicopter entry and exit, and use of flight controls in cruise flight.

### Content

1. Preflight Discussion
   a. Discuss lesson objective and completion standards
   b. Normal checklist procedures coupled with introductory material
   c. Weather analysis

2. Review

3. Instructor Actions
   a. Preflight used as introductory tool
   b. Short, familiarization flight

   ...ent Actions
   ...d exits helicopter safely
   ...ls in cruising flight

...ssign Helicopter Flying...

## Training Procedures

The introduction to training procedures offers the certificated flight instructor (CFI) an opportunity to better ascertain the student's experience and background, which influence training. At this time, the CFI explains the general safety procedures, lay out of the school, the course syllabus and how it is used. This includes how, when, and where instruction will take place. The CFI also discusses the role of preflight and postflight briefings in the training program, as well as how he or she monitors the student's progress. Additionally, the introduction to training should be used to determine the student's motivation for flight training and learn what their goals are. By understanding what the student would like to gain through flight training, the CFI will be better prepared to tailor their training plan to the needs of the student.

## Introduction to the Helicopter

Walking the student through a preflight provides an excellent opportunity to introduce or review the main components of the helicopter. *[Figure 2-1]* Refer the student to the Helicopter Flying Handbook for in-depth information on the rotor systems, landing gear, and flight controls. During the discussion, the CFI should demonstrate how to enter and exit the helicopter properly while the rotors are turning. This is also a good time to explain or review:

- General helicopter hazards, such as main and tail rotor blades. A simple demonstration of how low main rotor blades can droop is possible by manually pulling down on the tip of a static blade. In aircraft equipped with retractable droop stops, the CFI must explain that actual droop can be greater once the stops retract with greater rotor revolutions per minute (rpm). Ensure that all demonstrations comply with restrictions found in the appropriate rotorcraft flight manual.

**Figure 2-1.** *A CFI provides an overview of the helicopter to introduce the main components and discuss how to enter and exit a helicopter properly.*

- Emergency egress.

- Foreign object damage (FOD) hazards associated with items, such as hats, jackets, and loose paperwork.

- Seat belt use at all times during flight.

- Proper wear and use of the headset.

- Proper sitting posture and position of the hands and feet.

- Positive exchange of controls procedures and acknowledgments.

- The see-and-avoid concept.

- The clock method of reporting aircraft and other hazards to flight to the other crewmember.

- The need for clothing suitable for the location and weather. It is always good practice to have sufficient clothing for walking back to the starting point. Helicopters can readily take a pilot far beyond populated areas. The pilot should always have enough resources to survive or to wait for a repair crew to arrive, in case of emergency. (Please refer to Chapter 12, Helicopter Emergencies.)

- Suitable eye protection, such as good sunglasses to protect the eyes from harmful rays that produce cataracts in later years. Helicopters admit much more sunlight than almost any other aircraft, due to the bigger bubble or cockpit plexiglas area and chin window areas. Additionally, many helicopters fly with the doors off in warmer climates, thereby exposing the student's eyes to much more radiation.

- Seat and pedal adjustment in the helicopter to achieve full control travel.

- Headset and commonly used noise-canceling microphone function, so that the headset and microphone can be properly fitted and adjusted. The student should know how to adjust the volume of the headset and be able to understand the instructor and radios through the headset. If a voice-activated intercom is installed, the student should be taught what the squelch control function does and how to adjust it when necessary. Headsets not utilized should be disconnected and stowed away to prevent unwanted noise and reduce the risk of FOD. Also, loose items such as seatbelts, bags, jackets, hats, and flight publications should be stowed.

- Controls and buttons located on the cyclic and collective. Most of this preflight instruction should be done in as quiet a location as possible before engine start. After engine start, student perceptions will probably be overloaded quickly with new experiences and sensations from their first helicopter flight. Effective instruction would have the ground instructor bringing the class out to the helicopter after every lesson to have them locate, examine, and describe the function of each part described in that lesson. The students should be able to explain the relationship between a component of the helicopter and the aerodynamics requiring that component.

The importance of good prebriefings can never be overstated. In almost every case, if the student does not learn from the briefing what is expected and the contents of the flight lesson for that day before going to the helicopter, then that student will not learn after getting into and starting the helicopter. Instructors forget that the new student pilot is constantly barraged by new information. Newly experiencing the sights, sounds, vibrations, and other sensory inputs of helicopter flight, the beginning student has great difficulty understanding and remembering what the instructor says. If the instructor merely reinforces what the student learned in the classroom, the student is more likely to recall the instructions and procedures for the maneuvers amid the new experiences.

Likewise, during the prebriefing, the student should be introduced to the flying area. The time required for a review of the chart to be used depends on the experience level of the student and when charts and maps were taught during the training. The instructor should also remember that the student may not remember as well if the student is always on the flight controls. The instructor may need to relieve the student of the flight controls for a few moments near each boundary marker or checkpoint for the student to have time to fully absorb the view and relate the sight to the chart or map being used.

If the student has airplane experience, the instructor should be aware of negative transfer of airplane skills to helicopter flying. The first flight should set the stage for the remainder of the flight course. A shorter flight is always better than a long flight. If the student becomes warm or hot, the likelihood of airsickness is greater. Some students have an aversion to heights, which can be overcome by determination and gradual exposure.

The instructor has the duty always to give the student just enough—just enough encouragement, or just enough challenge for that stage of training, or just enough critique—for the student to learn but not to discourage. The instructor should always have enough understanding of the student's progress to discuss the student's problems and explain how or why the error is occurring and what corrective or different action to take to have a better outcome. Especially on the ground, the instructor should always strive for the student

to comprehend, not just remember and perform by rote memorization.

## Introduction to Flying

For the first flight, the instructor should give the student just enough flight experience to make the student want to come back for more. During the first flight, the CFI should allow the student to fly the aircraft and have fun doing it. An enjoyable introductory flight builds student motivation and the student will be more ready to learn. Solo flight comes after more flight experience, so there is plenty of time for the student to learn local landmarks. *[Figure 2-2]* This flight should be an introduction to flying itself and should follow the pattern of learning simple tasks before more complex tasks. The student should learn to fly first, and then learn where to fly.

**Figure 2-2.** *During an introductory flight, a student pilot receives the first helicopter flight experience while being shown the general functions of the controls and instruments and is also introduced to the local landmarks and the layout of the airfield.*

The CFI must show the student that flying can be fun, and then introduce the student to the local flying area. During this flight, seat the student at the pilot's seat. (Seat the new student at the copilot's seat for the first few flights if access to engine starting or flight control friction is not easily accessible from the copilot's seat.) Explain the general function of the controls and instruments. Demonstrate adjustment of the controls for comfort and safety, as applicable to the make and model of helicopter being flown. Relate this flight to the student's flying background and level of experience. For example, a new student's introductory flight can also be used to discuss basic air traffic control (ATC) functions and procedures.

A brief "hands on" for the student during cruising flight helps the CFI further evaluate the student's level of ability. If at all possible, the first flight or at least the first portion of the first flight should be conducted in a calmer environment, such as in the morning or at a higher altitude, so the student has a chance to experience the helicopter flight without the turbulence that is often confusing to the student.

This flight also provides the CFI with an opportunity to evaluate the student's attitude, tolerance, and temperament. The student should enjoy this first trip, creating a positive foundation for the rest of the course. Explain that procedures that seem complicated at this time become easier with more exposure and training.

Try to avoid confusing the student by presenting too much detailed information at this early stage in training. As discussed in the Aviation Instructor's Handbook, students tend to acquire and memorize facts when exposed to a new topic. As learning progresses, they begin to organize their knowledge to formulate an understanding of the things they have memorized. Progressing further still, students learn to use the knowledge they have compiled to solve problems and make decisions. Encourage and praise such behavior whenever students exhibit the pilot in command (PIC) input. Keep in mind that student performance should not be criticized or corrected at this stage; explain in general terms what occurs during flight to clarify student's understanding.

In the early stages of flight training, the traditional lesson plan (see the Aviation Instructor's Handbook) provides the CFI with a teaching delivery method more in tune with the student's level of knowledge. Scenario-based training (SBT) works better with learners who have mastered the basic knowledge needed to make more advanced decisions. The samples used in the early chapters utilize the traditional lesson plan.

The most important lesson for helicopter pilots to learn is to be wary. As training progresses, the instructor can incorporate discussions of documented helicopter accidents related to the lesson of the day. The instructor can offer techniques and procedures that would prevent that type of incident from happening. While it is important to relate some of these stories to student pilots, the instructor should avoid too many accident discussions early in the training as they may instill fear in students that may not understand the details. Respect for the dangers in aviation can aid a student's progression, but fear acts as a barrier to learning.

Instructors in the debriefing after the flight should always discuss what was satisfactory and then discuss what improvements the student could make and, even more important, how to make improvements. It does not help the student to say the flight was unsatisfactory that day if the instructor cannot describe in detail how the student could correct any responses or maneuvers.

After the debriefing, most successful instructors begin to brief the student on the contents of the next day's training flight. This allows:

1. The student time between flights to study and think about the next maneuver to learn at their own pace.

2. The student to recall questions from the current flight about a specific point during the flight.

3. The student to formulate questions concerning practices or procedures for the instructor to address before the next flight.

4. The instructor to relate the current flight to the upcoming flight's goals and maneuvers to further the student's understanding of the relationship of the procedures.

## Instructor Tips

- For the airplane pilot transitioning to helicopters, remind the student that a helicopter is very different from an airplane and much negative transfer is possible if they do not continually remind themselves of which aircraft that they are flying at the time. Helicopters are designed and built to be controllable. Airplanes are designed and built to be stable. Helicopter flight controls are considerably more sensitive than those in an airplane, which can be difficult for a former airplane pilot to adjust to. CFI's must also explain the aerodynamic effects that must be controlled by the helicopter pilot due to the main rotors' blade tip speed. [Figure 2-3] The Helicopter Flying Handbook is a good reference for detailed explanations on the calculations of the main rotors' blade tip speeds and the magnitude of the aerodynamic effects that must be controlled by the helicopter pilot.

- Avoid sudden or violent maneuvers that might make a newcomer to flying nervous. Emphasize how little movement is required on the cyclic and collective controls. This is critical for prior airplane pilots and can be demonstrated by calculating and explaining the blade tip speed to emphasize the magnitude of the aerodynamic effects controlled by a helicopter pilot. Introduce the pedal requirement immediately with short quick inputs rather than slow and long inputs. Demonstrate and point out to the student which part of the body should be used and are necessary in order to achieve the proper input.

- Monitor the student pilot's hand grip pressure on the flight controls, foot position on the pedals, body posture, and eyes regularly for clues of nervousness, lack of progress, improper reaction to the situation, and situational unawareness.

## Introduction to the Helicopter

### Objective

The purpose of this lesson is to introduce the student to rotary-wing flight. The student demonstrates a basic knowledge of the main components, safe helicopter entry and exit, and use of flight controls in cruise flight.

### Content

1. Preflight Discussion
   a. Discuss lesson objective and completion standards
   b. Normal checklist procedures coupled with introductory material
   c. Weather analysis

2. Review

3. Instructor Actions
   a. Preflight used as introductory tool
   b. Short, familiarization flight

4. Student Actions
   a. Enters and exits helicopter safely
   b. Handles controls in cruising flight

### Postflight Discussion

Preview and assign the next lesson. Assign *Helicopter Flying Handbook,* Chapter 1, Introduction to the Helicopter, and Chapter 2, Aerodynamics of Flight.

**Figure 2-3.** *Example of a traditional lesson plan.*

- Always practice positive transfer of control procedures and acknowledgments. This is particularly important in the early stages of training to instill good habits and safety when either the student or the CFI is on the controls for a long period of time.

- Helicopters are not acrobatic in the general sense. Therefore, positive "G" loads are the normal condition. All good helicopter pilots are smooth flyers because they know smooth flight is good for the machine and passengers/cargo.

- Sudden or violent maneuvering is usually the precursor for main rotor or tail rotor strikes. The helicopter pilot should always be planning the flight path to avoid close, tight situations requiring rapid maneuvering.

The helicopter instructor should be relating the ongoing training to student plans for after they earn their certificates. This encourages learning and help students relate the training to a positive personal goal.

The goals of the first flight should be for the student to recognize flight attitude of the helicopter relative to the horizon and to relate the control inputs necessary to achieve changes in the aircraft's attitude. On the first flight, acquaint the student with the basic flight instruments, such as the rotor tachometer, engine tachometer, compass, airspeed indicator, altimeter, and power gauge (manifold pressure or torque). Additionally, show the student how the helicopter responds to pedal inputs at a hover versus in forward flight, and how the power changes depending on tail rotor power demands.

The first helicopter flight should be rewarding, and not overwhelming or boring. If possible, one day should be the detailed preflight and prebriefing and the next the regular preflight and actual first flight.

Another item to include on the first flight is the engine cooling time, including the reasons the student sees airplanes come to a full stop and kill the engines immediately, while the helicopter pilot must sit for some minutes before the engine can be shut down. Explain that a helicopter requires relatively more power to hover taxi than an airplane requires to ground taxi. Helicopters require more of their available power to hover so the powerplant is relatively hotter and requires a longer cool-down period. Generally, airplane engines begin to cool during descent to landing and require little or no time to cool down after landing.

## Chapter Summary

This chapter provided the CFI with the objectives of an introductory flight. It also set the stage for the future training sessions, leading to the development of a competent, safety conscious, and cordial pilot.

# Chapter 3
# Aerodynamics of Flight

## Introduction

All helicopter pilots must have a basic knowledge of the aerodynamic principles that enable helicopter flight. While the principles that apply to a helicopter are the same as those that apply to other aircraft, the application of these principles is more complex due to the rotating airfoils. Chapters 2 and 3 of the Helicopter Flying Handbook (FAA-H-8083-21) and Aircraft Weight and Balance Handbook (FAA-H-8083-1), form the foundation for this chapter.

As with any training, begin the presentation of new material at the student's level of understanding. This can be determined throughout the introductory meeting with the student simply by engaging conversation about helicopters and general flight. Any previous flight experience will be apparent during preflight and while flying. Written or oral testing on the first day of flight school could deter a student from further flight training. A proficient, certificated flight instructor (CFI) should be able to determine the background and expertise of a student by careful use of the initial introductory meeting.

The student's aviation background determines when to introduce different aspects of aerodynamics. The student must have the appropriate background knowledge to comprehend the subject matter. Periodic reviews during the course of instruction help the instructor tailor the lesson to the student's comprehension and arrange the material to fit the student's needs. Define new terms when first introduced.

The overall objective of this chapter is to help the instructor review the aerodynamics found in the Helicopter Flying Handbook (FAA- 8083-21, as revised) and help the student understand how those effects practically affect their helicopter flight. In order to control a helicopter in flight, the student must have the consistent ability to identify and compensate for varying aerodynamic forces in flight.

## Forces Acting on the Aircraft

Define and discuss the four forces acting on an aircraft in straight-and-level and unaccelerated flight. Give examples of how the combinations of these forces act on the airframe.

1. Thrust—the forward force produced by a powerplant/ propeller or rotor. It opposes or overcomes the force of drag.

2. Drag—a rearward, retarding force caused by disruption of airflow by the wing, rotor, fuselage, and other protruding objects. Drag opposes thrust and acts rearward parallel to the relative wind.

3. Weight—the combined load of the aircraft itself, the crew, the fuel, and the cargo or baggage. The earth's gravitational force, which creates the weight, pulls the aircraft downward.

4. Lift—overcomes the downward force of weight to allow flight to occur and is produced by the dynamic effect of the air acting on the airfoil and acts vertically through the center of gravity.

### Lift

A very easy way to confuse new flight students is to throw a lot of obscure information at them with no concrete references or examples. Aerodynamics can be very difficult for the new student to understand because it is difficult to visualize what is happening to the rotor blades or tail rotor in flight. When teaching the student about lift and how the helicopter is able to obtain lift, the instructor must be creative and find ways to explain the theories, such as Bernoulli's Principle and Newton's Laws of Motion, in direct relation to the helicopter and how every flight control movement affects lift.

### *Bernoulli's Principle*

Instructors should introduce Bernoulli's Principle to the student in simple terms and attempt to relate the theory directly to the production of lift that is created from the main and tail rotor blades. The discussion should begin with Bernoulli's initial discovery that air moving over a surface decreases air pressure on the surface, and show the student an example of the differences in air pressure when an object moves through the air. Further discussion should include the following points and examples:

1. Show the student a picture of an airfoil and how the air pressure changes when the air is disrupted. A picture of an airfoil is usually a small cutout or slice of the entire wing or rotor blade. The instructor should explain that the entire rotor blade(s) are essentially one large airfoil.

2. As airspeed increases, surface air pressure decreases accordingly and this difference in pressure around the airfoil is directly related to the flight of an aircraft.

3. As an airfoil starts moving through the air, it divides the mass of air molecules at its leading edge. The distance over the top of the blade with the angle of attack is greater than the distance along the bottom surface of the rotor blade. Air molecules that pass over the top must move faster than those passing under the bottom to meet at the same time along the trailing edge. The faster airflow across the top surface creates a low-pressure area above the airfoil.

4. Air pressure below the airfoil is greater than the pressure above it and tends to push the airfoil up into the area of lower pressure. As long as air passes over the airfoil, this condition exists. It is the difference in pressure that causes lift. When air movement is fast enough over a wing or rotor blade, the lift produced matches the weight of the airfoil and its attached parts. This lift is able to support the entire aircraft. As airspeed across the wing or rotor increases further, the lift exceeds the weight of the aircraft and the aircraft rises.

5. Not all of the air met by an airfoil is used in lift. Some of it creates resistance, or drag, which hinders forward motion. Lift and drag increase and decrease together. They are affected by the airfoil's angle of attack in the air, the speed of airflow, the air density, and the shape of the airfoil or wing.

### *Newton's Laws of Motion*

Newton's laws of motion provide the foundation for the student's understanding of basic aerodynamic principles. The instructor should develop multiple ways of explaining these laws to ensure that if the student does not comprehend one explanation, the instructor has an alternate explanation that relates to something that the student will understand. Begin with relating the laws to helicopter flight, such as the requirements for lift, thrust, and power to overcome the effects of the three laws and the energy state of the helicopter. If the student has a difficult time understanding flight examples, try using an example that is more familiar, such as a car or motorcycle. This helps the student better understand the laws when the instructor applies it to flight.

## First Law—the Law of Inertia

A body at rest remains at rest, and a body in motion remains in motion at the same speed and in the same direction unless acted upon by some external force. The key point to explain is that if there is no net force resulting from unbalanced forces acting on an object (if all the external forces cancel each other out), then the object maintains a constant velocity. If that velocity is zero, then the object remains at rest. And, if an additional external force is applied, the velocity changes because of the force.

A helicopter in flight is a particularly good example of the first law of motion. There are four major forces acting on an aircraft: lift, weight, thrust, and drag. If we consider the motion of an aircraft at a constant altitude, we can neglect the lift and weight. A cruising aircraft flies at a constant airspeed and the thrust exactly balances the drag of the aircraft. This is the first part sited in Newton's first law; there is no net force on the helicopter and it travels at a constant velocity in a straight line.

Now, if the pilot changes the thrust of the engine, the thrust and drag are no longer in balance. If the thrust is increased, the helicopter accelerates and the velocity increases. This is the second part sited in Newton's first law; a net external force changes the velocity of the object. The drag of the helicopter depends on the square of the velocity. So, the drag increases with increased velocity. Eventually, the new drag equals the new thrust level and at that point, the forces again balance out, and the acceleration stops. The helicopter continues to fly at a new constant velocity that is higher than the initial velocity. We are again back to the first part of the law with the helicopter traveling at a constant velocity.

In this example, only the motion of the helicopter in a horizontal direction is explained and as the student becomes comfortable with aerodynamics, further discussions should include the effects of the thrust on weight and on lift. For example, increasing the throttle setting increases the fuel usage and decreases the weight, and the increase in velocity increases the lift as well as the drag. Each of these changes effect the vertical motion of the helicopter.

It is important to point out the role of engine power when explaining the law of inertia. Power is used to accelerate the helicopter, to change its velocity, and thrust is used to balance the drag when the helicopter is cruising at a constant velocity. When a helicopter is on a normal approach, the power demand is generally in the middle range and the total drag is at the lowest. As the aircraft decelerates to effective translational lift airspeed and terminates to a hover, the power demand is quite significant, generally the highest of all maneuvers. An airplane makes minimal power demands at the termination of its approach through the flare and landing.

## Second Law—The Law of Acceleration

A change in velocity with respect to time. The force required to produce a change in motion of a body is directly proportional to its mass and rate of change in its velocity.

For example, for a given helicopter, acceleration would be slower when loaded to maximum gross weight than when loaded to a lesser gross weight. During a normal takeoff, the power margin available between maximum torque available and hover power can be quite small based on helicopter weight and environmental factors. During the transition to forward flight and through effective translational lift airspeed, acceleration is limited until the aircraft is in smooth undisturbed air and the influence of induced drag begins to subside. Once the aircraft reaches its maximum endurance/rate of climb airspeed, acceleration potential is increased as total drag is at its lowest point. *[Figure 3-1, Point E]*

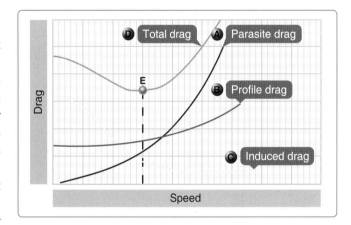

**Figure 3-1.** *Drag graph.*

Total drag is the sum of parasite drag and induced drag as shown in *Figure 3-1, Point A and C*. The total drag curve can also be referred to as the thrust required curve because thrust is the force acting opposite drag. At the point where total drag *[Figure 3-1, Point D]* and thrust required are at a minimum, the lift-to-drag ratio is maximum and is referred to as L/D$_{MAX}$. At L/D$_{MAX}$, the entire airframe is at its most efficient, producing the most lift for the least drag. Maximum endurance is found at L/D$_{MAX}$, because thrust required and thus fuel flow (fuel required) are at a minimum, giving maximum time airborne.

## Third Law—Action and Reaction

For every action, there is an equal and opposite reaction. The instructor should relate the third law to the amount of power applied to the rotor system and the need for the antitorque or tail rotor to supply the equal and opposite reaction to the torque of the engine(s) applied to the main rotor. The rotor system of a helicopter accelerates air downward, resulting in an upward thrust. A single-rotor helicopter demonstrates this law perfectly. Consider a helicopter on floats that is not moored to a dock. As the main rotor begins to turn counterclockwise during aircraft start, the fuselage reacts by turning in a clockwise direction until the point at which the tail rotor has reached sufficient rpm to provide the thrust necessary to counteract that force.

Torque effect is a result of Newton's laws and an aspect of helicopter flight that a student must thoroughly understand. The turning of the helicopter's main rotor blades in one direction causes the helicopter to turn in the opposite direction. In most helicopters, this is counteracted by the use of a second rotor (tail rotor) to provide the thrust to limit the rotation. Some helicopters use vectored air, while others use a counterrotating main rotor system. All have one thing in common—a method of counteracting the torque of the main rotor system. *[Figure 3-2]*

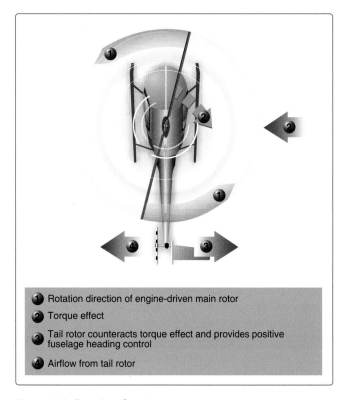

1. Rotation direction of engine-driven main rotor
2. Torque effect
3. Tail rotor counteracts torque effect and provides positive fuselage heading control
4. Airflow from tail rotor

**Figure 3-2.** *Rotation direction.*

At some point in training, the instructor should have the student bring the helicopter to a high hover and explain that work load is greater and an increased left pedal requirement exists to hold a constant heading. The opposite can be shown at a lower hover with a decrease in left pedal requirement to hold the same heading.

## Weight

As weight increases, the power required to produce lift needed to compensate for the added weight must also increase. This is accomplished through the use of the collective. Most

performance charts include weight as one of the variables and students must be aware of the importance of managing aircraft weight to obtain optimum performance. By reducing weight, the helicopter is able to safely take off or land at locations that would otherwise be impossible.

Explain to students how maneuvers that increase the G loading such as steep turns, rapid flares, or pulling out of a dive create greater load factors and act as a multiplier of weight. The load factor is the actual load on the rotor blades at any time, divided by the normal load or gross weight. *[Figure 3-3]* At 30° of bank, the load factor is 1G, but at 60°, it is 1.8G, an increase of 80 percent. If the weight of the helicopter is 1,600 pounds, the weight supported by the rotor in a 30° bank at a constant altitude would be approximately 1,600 pounds. In a 60° bank, it would be 2,880 pounds and in an 80° bank, it would be 8,000 pounds. Emphasize to students that an additional cause of large load factors is rough or turbulent air. The severe vertical gusts produced by turbulence can cause a sudden increase in angle of attack (AOA), resulting in increased rotor blade loads that are resisted by the inertia of the helicopter.

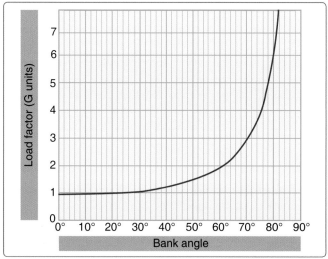

**Figure 3-3.** *Load factor.*

## Thrust

Thrust, like lift, is generated by the rotation of the main rotor system. Point out to the student that in a helicopter thrust can be forward, rearward, sideward, or vertical. The direction of the thrust is controlled with the cyclic. If cyclic control to produce thrust is too great, lift is lost and the aircraft descends. Conversely, if too little cyclic control is made, the aircraft begins a climb. Using visual aids, demonstrate how the resultant lift and thrust determines the direction of movement of the helicopter. *[Figure 3-4]*

Explain to the student that the tail rotor also produces thrust. The amount of thrust is variable through the application of

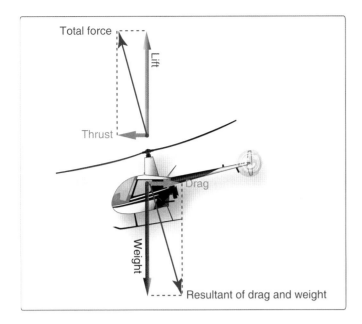

**Figure 3-4.** *Thrust.*

the antitorque pedals and is used to control the helicopter's heading during hovering flight and trim during cruise flight.

## Drag

No discussion of aerodynamics is complete without its defining the three types of drag, how drag is created, and its effect on the aircraft. A certificated flight instructor (CFI) must become intimately familiar with the drag chart and how it relates to airspeed and power demands. Demonstrate this during the performance planning phase as the student has actual torque values to compare. Then, when the student is flying the helicopter, apply the values that were computed and show the effect on the helicopter. A technique is to show how each flight control is affected by simple hover flight maneuvers. Demonstrate the change in torque that occurs between left and right pedal turns and explain why. Discuss how the cyclic is utilized to hold position over the ground, while the pedals rotate the fuselage and control heading. When excess power is available, demonstrate how the collective pitch can be applied to vary the hover height, or to accelerate the helicopter. It would be prudent to discuss here that if no excess power is available, application of the collective then may be used to control the rotor rpm. This is done by changing the pitch in the blades. Over application of the collective in a low power margin setting results in rotor rpm decay and a loss of lift. Rotor rpm is the key to sustaining the aircraft in a steady state profile and should never be allowed to decay below minimum operating levels. It is the key to life for a helicopter pilot!

The types of drag are:

1.  Parasite drag—drag created by the fuselage or any nonlifting components (e.g., strut, skin friction, interference).

2.  Profile drag—caused by the frictional resistance of the rotor blades passing through the air.

3.  Induced drag—results from producing lift.

    a.  Blade tip vortices—pressure differential at tips of blades trying to equalize and produce a stream of vortices (turbulence).

    b.  Induced flow—causes lift and total aerodynamic force to tilt further rearward on the airfoil.

    c.  Total aerodynamic force tilted further backward at higher angles of attack.

4.  Total drag—sum of induced, profile, and parasite.

Use a graph that depicts drag/power relationship, and have the student identify the power requirements to overcome drag at various airspeeds. *[Figure 3-1]*

The following describes the relationship of each of the different types of drag to the airspeed of the aircraft.

1.  Parasite drag—lowest point at a hover, but increases with airspeed. The major source of drag at higher airspeeds.

2.  Profile drag—remains relatively constant at low airspeed, but increases slightly at higher airspeed ranges.

3.  Induced drag—major source of drag at a hover, but decreases with forward airspeed.

4.  Total drag—the sum total of induced, profile, and parasite drag.

    a.  Total drag decreases with forward airspeed until best rate of climb speed is reached. *[Figure 3-1, Point E]*

    b.  Speeds greater than best rate of climb causes a decrease in overall efficiency due to increasing parasite drag.

Once the student understands the forces acting on the helicopter, provide examples of balanced and unbalanced flight forces. For example, when hovering stationary in calm wind at a constant altitude, thrust is equal to drag and lift is equal to weight. The aircraft is not moving vertically or horizontally. The aerodynamic forces are balanced. *[Figure 3-5]*

The student will also notice during hovering flight in a calm wind condition that with smaller American made helicopters like the Robinson R-22, Bell 206, and Schweizer 300, the left side of the aircraft will probably hang lower than the right. This is due to the direction of the tail rotor thrust and the engineered mast tilt to compensate for translating tendency.

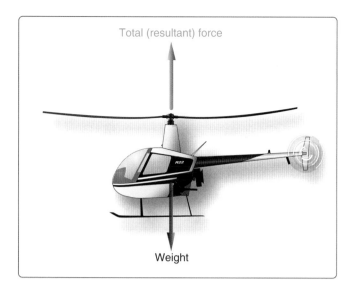

**Figure 3-5.** *Balanced forces in hover.*

On much larger helicopters such as the BH-205, S-76, and BK-117, in which an additional gearbox is used to raise the tail rotor up to the main rotor plane, the tilting of the fuselage is not as prevalent.

The pitch attitude will vary depending on the loading of the helicopter. Many helicopters when flown single pilot will be nose high at a hover. Conversely, they may be nose load when fully loaded. The center of gravity (CG) of the helicopter determines which portion of the landing gear will come off the ground first. The CFI must pay particular attention to the attitude of the helicopter as the student lifts it off the ground. If excess power is applied in other than a level attitude, the helicopter may proceed to roll beyond its dynamic rollover limits. When lifted off the surface correctly and safely, the pilot has the opportunity to lower the collective if a portion of the landing gear is attached or hung on the surface, thus preventing a rollover incident from occurring. It is imperative that the CFI closely monitor the attitude of the helicopter and not the actions of the student. This simple action may determine whether or not the helicopter is allowed to stray beyond the comfort level of the instructor to recover from a particular action by the student. Never allow a student to go beyond your comfort level.

Several inputs are required simultaneously as the aircraft is brought to a hover. Stress to the student that these actions must occur without delay or coordinated flight will not occur. For example, as the collective is increased to lift the helicopter off the surface, the throttle must also be increased. Even if a governor accomplishes that action, the pilot still must monitor the power instruments to ensure that no limits are exceeded. With the increase in power, there is also an increase in torque and the tendency for the nose to turn to the right. The pilot must apply sufficient left pedal to maintain the helicopter heading. While this is occurring and the lift in the rotor system is changing, the pilot must apply cyclic to maintain position over the ground and not allow the helicopter to drift in any one direction. The helicopter bank attitude might not be level due to crosswinds and translating tendency. The pitch attitude might not be level due to tailwinds or CG. The pilot must ensure that the tail rotor is clear of all obstacles and is not allowed to hang so low that it impacts the ground or other objects.

For example, in steady state flight, the aircraft is maintaining a constant airspeed and constant altitude. The aerodynamic forces are balanced. Although the helicopter is moving, it is not accelerating or climbing. *[Figure 3-6]*

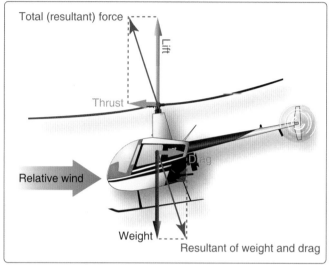

**Figure 3-6.** *Steady state—balanced forces.*

Any time opposing forces become unequal (unbalanced), acceleration results in direction of the greater force. If lift is greater than weight the helicopter climbs. If thrust is greater than drag, the helicopter moves horizontally. Point out that thrust can occur in any or all directions. For example, if the helicopter is moving sideways or backwards, thrust is in the direction that it is moving. *[Figure 3-7]*

## Airfoil

Define and discuss the different types of airfoils with the student and stress the importance of using standardized terminology. An airfoil is a curved surface body or structure designed to produce a lift or thrust force when subjected to an airflow. An instructor can check the student's understanding of airfoils and the terminology used to describe them by having the student draw and label the parts of an airfoil. *[Figure 3-8]* Refer to the Pilot's Handbook of Aeronautical Knowledge and the Helicopter Flying Handbook (FAA-8083-21) for definitions and illustrations of airfoil design.

### Blade Twist

Explain to the student that the rotor blade of a helicopter is designed with a twist to relieve the stresses on the blade

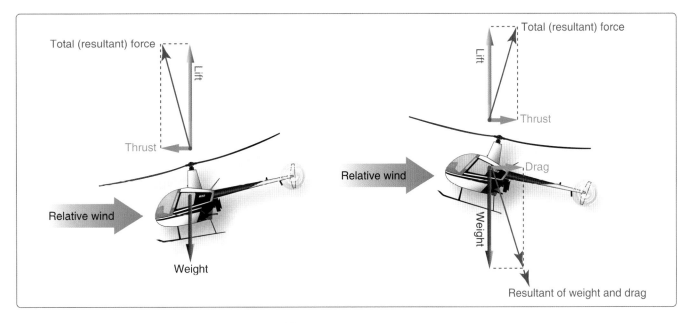

**Figure 3-7.** *Acceleration or deceleration—unbalanced forces.*

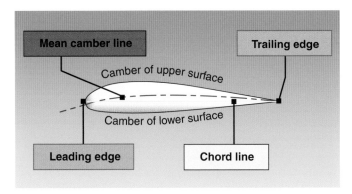

**Figure 3-8.** *Elements of an airfoil.*

and distribute lifting force more evenly along the blade due to the lift differential along the blade. Blade twist provides greater pitch angles at the blade root where velocity is low and smaller angles at the tip where blade velocity is higher. This increases the induced air velocity and blade loading near the inboard section of the blade.

## Rotor Blade and Hub Definitions

The CFI must be familiar with the following basic terms and be able to explain them to the student.

1.  Hub—the attachment point of the rotor blades.

2.  Tip of the blade—the farthest outboard section of the rotor blade.

3.  Root of the blade—the section of the blade closest to the hub and where the attachment point is located.

4.  Twist—the change in blade angle with respect to the angle at the hub outward to the tip.

5.  Taper—the change (decrease) in blade chord with radial distance.

These terms related to the rotor hub and blades are best discussed in the classroom and identified on the aircraft during a preflight.

## Airflow and Reactions in the Rotor System

When introducing and describing the airflow in a rotor system, the instructor must first identify the types of relative wind. By defining and explaining the various air movements in a rotor system *[Figure 3-9]* and the relationship of air movement to an airfoil, the instructor establishes a foundation for more detailed discussions of aerodynamic principles.

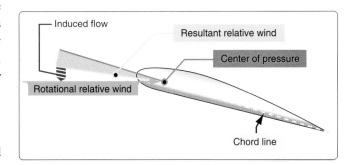

**Figure 3-9.** *Air movements in a rotor blade system.*

*   The movement of a rotor blade through the air creates relative wind. Relative wind moves in a parallel but opposite direction to the movement of the rotor blade.

*   The flow of air parallel to and opposite the flightpath of an airfoil is rotational relative wind. It always meets the airfoil at a 90° angle.

*   The component of the total relative wind velocity created by forward flight velocity/airspeed is airspeed relative wind.

- Induced flow (downwash) is a downward component of air that is added to the rotational relative wind.

- Resultant relative wind is the airflow from rotation (rotational relative wind) that is modified by induced flow.

- Up flow (inflow) is airflow approaching the rotor disk from below as the result of some rate of descent. Up flow also occurs as result of blades flapping down or an updraft.

A demonstration of the airflow in the following instances helps the student understand the concept of relative wind:

- Airfoil moving in one direction

- Rotating rotor blades

- Advancing blade

- Retreating blade

- Relative winds are the same for tail rotor

## Rotor Blade Angles

Angle of incidence is the acute angle between the chord line of the airfoil and the plane of rotation (tip path plane) or the angle between the chord line of a blade and the relative wind. Sometimes, this is referred to as the blade pitch angle. The angle is changed through rotation of the rotor blade around its spanwise axis, which is known as feathering. *[Figure 3-10]* An instructor can use training aids to discuss the angle of incidence in the classroom, but it is best demonstrated at the aircraft. Show the student how the angle of incidence

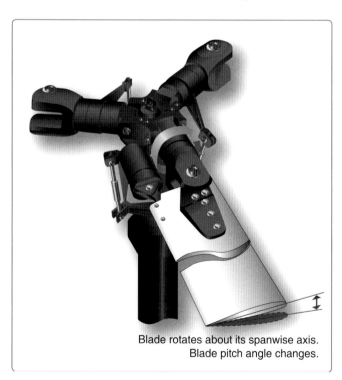
Blade rotates about its spanwise axis.
Blade pitch angle changes.

**Figure 3-10.** *Rotor blade feathering.*

is changed on all blades (except tail rotor) simultaneously by using the collective pitch control. Define this action as collective feathering and explain how it affects the overall lift of the rotor system.

Demonstrate how the cyclic pitch control causes a differential change in the angle of individual blades (except tail rotor) and define it as cyclic feathering. Stress to the student that cyclic feathering changes the attitude of the rotor system but does not change the amount of lift. *[Figure 3-11]*

Point out how the angle of incidence for the tail rotor is changed on all tail rotor blades simultaneously by using the antitorque pedals.

- Stress that angle of incidence is a mechanical angle. *[Figure 3-12]*

- Remind the student that the AOA is the acute angle between the chord line of an airfoil and the resultant relative wind. It can change with no change in the angle of incidence due to blade flapping and up/down drafts.

- Stress that angle of attack is an aerodynamic angle. *[Figure 3-12]*

Discuss lift at different AOAs. With the use of diagrams, an instructor can explain how the AOA affects the amount of lift. An easy demonstration of how the AOA affects lift is to remind the student of what happens if an arm is extended out of the window of a moving vehicle. Using guided discussion and demonstration, ask the student what happens when the palm of the hand is parallel to the ground and when it is rotated forward. Show the student how the hand rises until reaching the point at which it stalls and is just pulled rearward. Emphasize the following principles:

1. Larger angles of attack create more lift on an airfoil.

2. Smaller angles of attack result in a reduction of lift on the airfoil.

3. Exceeding the maximum (critical) angle can produce a stall. Maximum angle of attack is 15° to 20° on most airfoils.

## Hovering Flight

It is essential for the student to understand the aerodynamics of hovering. Explain that for a helicopter to hover, lift produced by the rotor system must equal the total weight of the helicopter. An increase of blade pitch through application of collective increases the angle of incidence and generates the additional lift necessary to hover. As forces of lift and weight are in balance during stationary hover, those forces must be altered through application of collective either to climb or to descend.

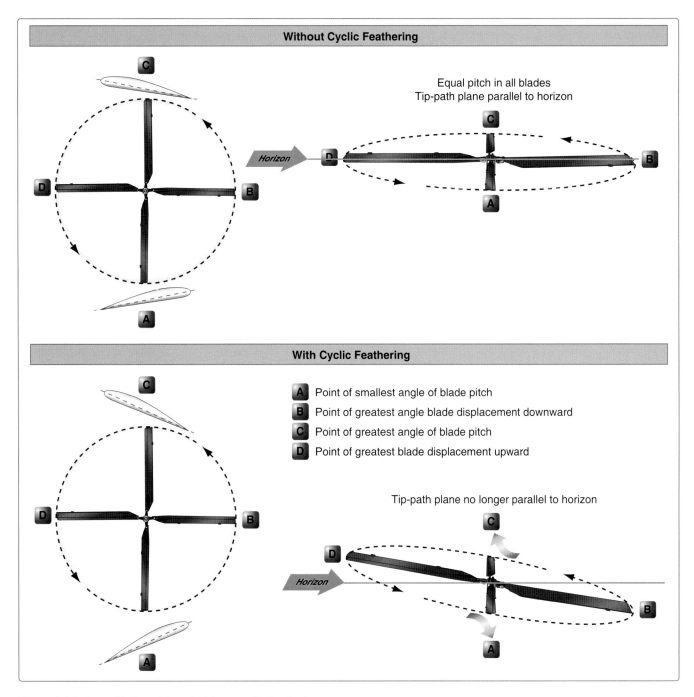

**Figure 3-11.** *Rotor blades with and without cyclic feathering.*

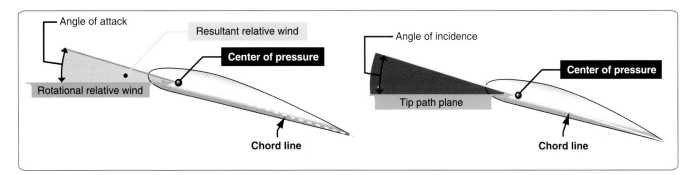

**Figure 3-12.** *Angle of attack and angle of incidence.*

Describe to the student how, at a hover, the rotor-tip vortex reduces effectiveness of the outer blade portions. *[Figure 3-13]* Vortices of the preceding blade affect the lift of the other blades in the rotor system. When maintaining a stationary hover, this continuous creation of vortices combined with the ingestion of existing vortices is the primary cause of high power requirements for hovering. Rotor-tip vortices are part of the induced flow and increase induced drag. Ensure that the student understands that, during hover, rotor blades move large amounts of air through the rotor system in a downward direction. This movement of air also introduces induced flow into relative wind, which alters the AOA of the airfoil. If there is no induced flow, relative wind is opposite and parallel to the flightpath of the airfoil. With a downward airflow altering the relative wind, the AOA is decreased so that less aerodynamic force is produced. This change requires an increase in collective pitch to produce enough aerodynamic force to hover.

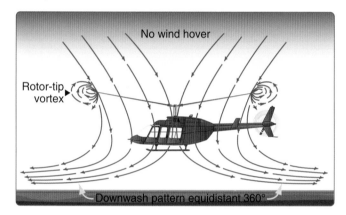

**Figure 3-13.** *Air movements in a rotor blade system.*

## Translating Tendency or Drift

Explain to the student that the thrusting characteristics of a tail rotor during hovering flight create a tendency for the helicopter to drift laterally, which is called translating tendency. A single-rotor helicopter with a counterclockwise rotating main rotor tends to drift laterally to the right. Stress the cause: thrust exerted by the tail rotor compensates for main rotor torque. Translating tendency is to the left in a helicopter with a clockwise rotation of the main rotor.

Explain to the student that the helicopter fuselage will remain relatively level to slightly left side low. The amount of fuselage tilt varies between types and design of helicopters. The tip path plane of the main rotor will not be level and will have to be adjusted accordingly with cyclic to counteract translating tendency and adverse wind conditions. The ability to tilt or adjust the wings of the helicopter allows the helicopter to maintain its position over the ground.

Describe the methods used to correct for translating tendency:

1. Flight control rigging may be designed by the manufacturer so the rotor disk is tilted slighted when the cyclic control is centered to compensate for drift.

2. Transmission may be mounted so the mast is tilted slightly when the helicopter fuselage is laterally level.

3. Pilot applies cyclic in the opposite direction to arrest the drift.

## Pendular Action

Pendular action is the result of the CG being below the supporting structure (rotor system). Tilting the rotor in one direction results in the fuselage swinging in the opposite direction. Stress to the student that this swinging is normal for helicopter operation since the helicopter fuselage is below the rotor system and overcontrolling can result in exaggerated pendular action and should be avoided. The cyclic should always be moved at a rate that allows the main rotor and fuselage to move as a unit. Emphasize that the student should use slow, smooth, cyclic inputs while hovering. The student must understand that it is the relationship of the tip path plane to the horizon, and not the position of the fuselage, that determines the helicopter's direction of travel.

## Coning

Coning is the upward flexing of the rotor blades. Point out to the student that coning is a normal phenomenon in all rotors producing lift. The amount a blade cones is a resultant between lift and centrifugal force. When lift is stronger than centrifugal force, the blade cones upward. When centrifugal force is stronger than lift, the blade moves downward, reducing the coning angle. *[Figure 3-14]*

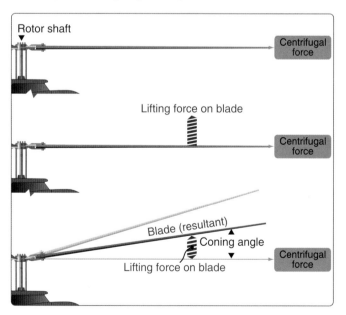

**Figure 3-14.** *Effect of centrifugal force and lift.*

Explain the relationship between lift and excessive coning and describe the causes of excessive coning to the student:

- Low revolutions per minute (rpm)—less centrifugal force

- High gross weight—more lift needed

- High G maneuvers—more lift needed

- Turbulent air—point out to the student that any maneuvers requiring additional lift could lead to excessive coning.

Give examples of excessive coning. Ensure the student understands:

- Flight conditions that require large amounts of lift may lead to an excessive coning condition in the rotor.

- As lift forces increase in the rotor, they overcome the rigidity produced by centrifugal force. The rotor blades begin flexing upward, which could lead to an excessive coning angle.

Guide the student in identifying the adverse effects of excessive coning in the rotor system. *[Figure 3-15]*

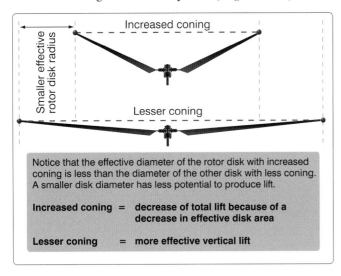

**Figure 3-15.** *Effects of coning.*

1. Loss of disk area.

2. Loss of total lift available.

3. Stress on blades.

4. Excessive stress forces in the rotor could lead to blade cracking or blade separation from the rotor system.

5. Excessive coning combined with low rotor rpm may cause the blades to droop much lower than normal. This condition is likely to occur at the end of an autorotation and may allow the rotor blades to damage or remove the tail boom.

6. Excessive coning may become unrecoverable in flight.

## Coriolis Effect (Law of Conservation of Angular Momentum)

The law of conservation of angular momentum states that the value of angular momentum of a rotating body will not change unless external torques are applied. Explain to the student that, in other words, a rotating body continues to rotate with the same rotational velocity until some external force is applied to change the speed of rotation. Angular momentum can be expressed by the formula:

$$\text{Mass Angular} \times \text{Velocity} \times \text{Radius Squared}$$

Discuss how changes in angular velocity, known as angular acceleration or deceleration, take place if the mass of a rotating body is moved closer to or further from the axis of rotation. The speed of the rotating mass increases or decreases in proportion to the square of the radius. These forces cause acceleration and deceleration.

Tell the student that the coriolis effect may be stated in the following terms.

A mass moving radically—

- Outward on a rotating disk exerts a force on its surroundings in the direction opposite to rotation.

- Inward on a rotating disk exerts a force on its surroundings in the direction of rotation.

The major rotating elements in the system are the rotor blades. As the rotor begins to cone due to G-loading maneuvers, the diameter of the disk shrinks. Due to conservation of angular momentum, the blades continue to travel the same speed even though the blade tips have a shorter distance to travel due to reduced disk diameter. This action results in an increase in rotor rpm. Most pilots arrest this increase with an increase in collective pitch.

Conversely, as G-loading subsides and the rotor disk flattens out from the loss of G-load induced coning, the blade tips now have a longer distance to travel at the same tip speed. This action results in a reduction of rotor rpm, and is corrected by reducing collective pitch.

## Ground Effect

Define ground effect for the student as the increased efficiency of the rotor system caused by interference of the airflow when near the ground. Discuss how ground effect permits relative wind to be more horizontal, the lift vector to be more vertical, and induced drag to be reduced, all allowing the rotor system to be more efficient. Maximum ground effect is achieved when hovering over smooth hard surfaces. When hovering over such terrain as tall grass, trees, bushes, rough terrain, and water, ground effect is reduced. Explain the two reasons for this phenomenon: induced flow and vortex generation. *[Figure 3-16]*

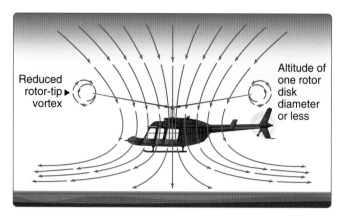

**Figure 3-16.** *Airflow at altitude of one rotor disk diameter or less.*

## Gyroscopic Precession

Explain to the student that precession occurs in rotating bodies that manifest an applied force 90° after application in the direction of rotation. Point out that although precession is not a dominant force in helicopter aerodynamics, pilots and designers must consider it since turning rotor systems exhibit some of the characteristics of a spinning gyro. *Figure 3-17* illustrates effects of precession on a typical rotor disk when force is applied at a given point. A downward force applied to the disk at point A results in a downward movement of the disk at point B. Aircraft designers take gyroscopic precession into consideration and rig the cyclic pitch control system to create an input 90° ahead of the desired action.

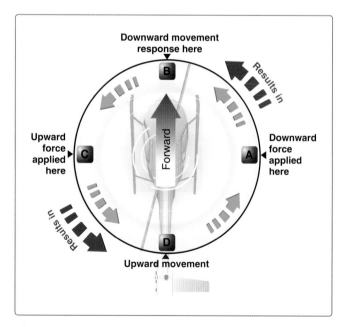

**Figure 3-17.** *Gyroscopic precession.*

*Figure 3-18* shows reactions to forces applied to a spinning rotor disk by control input or wind gusts.

| Force Applied to Rotor Disk | Aircraft Reaction |
|---|---|
| Up at nose | Roll right |
| Up at tail | Roll left |
| Up on right side | Nose up |
| Up on left side | Nose down |

**Figure 3-18.** *Reactions to forces on a rotor disk.*

## Vertical Flight

A student must understand that for climbing flight to occur, lift must be greater than weight. This is true whether at a hover or in steady state flight. Refer back to the forces acting on an aircraft in flight when explaining this concept.

## Forward Flight

When explaining forward flight to the student, refer to the section on forces acting on an aircraft in flight. Remind the student that flight is the result of all forces, and that lift and thrust must be equal to the result of weight and drag for steady state flight. Point out that acceleration in forward flight is the result of thrust being greater than drag.

## Translational Lift

Translational lift is the additional lift obtained from increased efficiency of the rotor system with airspeed obtained either by horizontal flight or by hovering into a wind. Describe the airflow patterns during directional flight and explain the causes of transitional lift.

The relative wind entering the rotor system becomes more horizontal and results in the following:

1. A more vertical lift component

2. Less induced drag

3. An increased AOA

4. Less turbulent air entering the rotor system

The airspeed range at which effective translational lift occurs is approximately 16–24 knots. As rotor efficiency increases and additional lift is produced due to more beneficial AOA, the rotor disk flaps upward causing the nose to pitch up; additional forward cyclic pressure is necessary at this point. As the airspeed increases and more lift is produced in the aft portion of the rotor disk, the nose tends to lower, requiring some aft cyclic to maintain an accelerative attitude and safe climb angle.

Provide the student with a graph depicting drag at different airspeeds. Using a graph like *Figure 3-1* and guided discussion, ensure the student understands:

1. Each knot of forward airspeed increases the efficiency of the helicopter rotor system up to a point where retreating blade stall aerodynamics negate any further rotor system gains.

2. At effective translational lift (ETL), the rotor system completely outruns the recirculation of old vortices and begins to operate in smooth, undisturbed air.

3. Induced drag and total drag are reduced and overall rotor efficiency increases.

4. Increased efficiency continues with increased airspeed until best climb speed is reached. *[Figure 3-1, Point E]*

5. Airspeeds greater than best rate of climb speed result in lower efficiency of the helicopter due to increased parasite drag.

## Translational Thrust

Translational thrust occurs as the helicopter transitions to forward flight and the tail rotor begins to operate in smooth undisturbed air. As the takeoff proceeds, the pilot notices the nose yaw (to the left in a counterclockwise turning system). This is the result of the increased translational thrust. To regain trimmed flight, a little right pedal is normally required. At about this same aerodynamic point, the airflow begins to smooth over the vertical stabilizer which carries some of the antitorque load in forward flight. This allows for slightly more reduction in tail rotor thrust, requiring further reduction in left pedal application. If there is no governor, a throttle change may be required to reduce the rpm slightly since the power demand was reduced. Depending on the helicopters position and airspeed, the rotor resultant rpm increase can be controlled by a slight increase in collective to maintain the rpm setting.

## Induced Flow

Explain to the student that, at flat pitch, air leaves the trailing edge of the rotor blade in the same direction it moved across the leading edge; thus, no lift or induced flow is being produced. Demonstrate how, as blade pitch angle is increased, the rotor system induces a downward flow of air through the rotor blades, creating a downward component of air that is added to the rotational relative wind. Point out that because the blades are moving horizontally:

- Some of the air is displaced downward.
- The blades travel along the same path and pass a given point in rapid succession.

- Rotor blade action changes still air to a column of descending air.

This downward flow of air is called induced flow (downwash). Emphasize that it is most pronounced at a hover under no-wind conditions.

## Transverse Flow Effect

Advise the student that in forward flight, air passing through the rear portion of the rotor disk has a greater downwash angle than air passing through the forward portion. Explain that this difference in downwash angle is due to the fact that the greater the distance air flows over the rotor disk, the longer the disk has to work on it and the greater the deflection is on the aft portion.

Ensure the student understands:

- Downward flow at the rear of the rotor disk causes a reduced AOA, resulting in less lift.
- The front portion of the disk produces an increased AOA and more lift because airflow is more horizontal.
- These differences in lift between the fore and aft portions of the rotor disk are called transverse flow effect.
- Transverse flow effect causes unequal drag in the fore and aft portions of the rotor disk and results in vibration easily recognizable by the pilot.
- Transverse flow occurs between 10 and 20 knots.

Stress to the student that transverse flow effect is most noticeable during takeoff and, to a lesser degree, during deceleration for landing. Demonstrate how gyroscopic precession causes the effects to be manifested 90° in the direction of rotation, resulting in a right rolling motion requiring left cyclic input to maintain a more level fuselage attitude and proper ground track.

## Dissymmetry of Lift

Dissymmetry of lift is the difference in lift that exists between the advancing half of the rotor disk and the retreating half. Explain to the student how to determine the total relative wind velocity on the advancing and retreating blades.

Discuss the relative wind velocity of blades at a hover and during translational flight.

### Hover

At a hover, relative wind velocity is:

- Approximately 400 knots at the tips.
- Approximately 300 knots one-fourth of the way in from the tips.

- Approximately 200 knots one-half of the way in from the tips.

- Approximately 100 knots three-fourths of the way in from the tips.

- 0 knots at the center of the hub.

## Translational Flight

In translational flight, relative wind velocity:

- Is a combination of blade speed and airspeed.

- Of the advancing blade is blade speed plus airspeed.

- Of the retreating blade is blade speed minus airspeed.

Develop the relative wind velocity for the advancing and retreating blades in the 090° to 270° position. [Figure 3-19] Show area of reverse flow. Emphasize that equal lift is created by advancing and retreating blades.

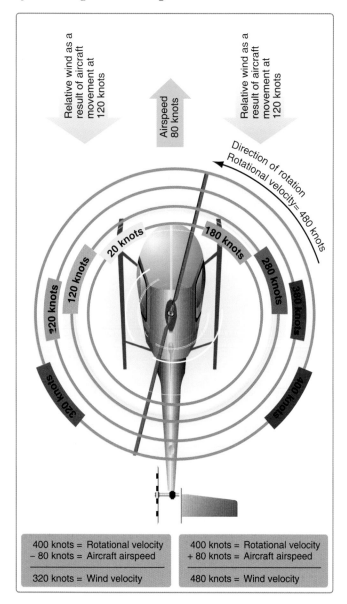

Figure 3-19. Relative wind velocity in forward flight.

1. Advancing blade—greater lift.

2. Retreating blade—less lift.

Discuss roll and explain that American-designed helicopters (counterclockwise rotation) would roll to the left and pitch up if transverse flow and dissymmetry of lift were not overcome. Explain main rotor method of overcoming dissymmetry of lift (flapping).

1. Advancing blade produces more lift; when flapping up, AOA decreases due to an increase in induced flow—loses lift.

2. Retreating blades produce less lift; when flapping down, AOA increases due to a decrease in induced flow—gains lift.

3. When blade flapping has compensated for dissymmetry of lift, the rotor disk is tilted to the rear.

4. Cyclic feathering also compensates for dissymmetry of lift (changes AOA) in the following ways:

    a. Cyclic feathering changes the angle of incidence differently around the rotor system.

    b. Forward cyclic decreases angle of incidence on advancing blade, resulting in reduced AOA, and increases angle of incidence on retreating blade resulting in increased AOA.

5. Tail rotor compensates for dissymmetry of lift by both flapping and feathering at the same time, accomplished by rotor design and mounting. A delta hinge allows for flapping, which automatically introduces feathering of the tail rotor.

Exercise caution during a low-altitude, high-speed takeoff as pitch attitude is very low. If an engine failure or partial power condition were experienced, the pilot would not be able to safely place the aircraft in an autorotative profile. A quick review of the height velocity diagram would be very useful here.

## Sideward, Rearward, and Turning Flight

Explain to the student that to accomplish these different modes of flight, the rotor disk is tilted in the desired direction. The forces acting on the helicopter remain the same, only the resultant vectors are different. Sideward hovering flight requires more pedal control to maintain heading. Depending on the lateral speed of travel, some fuselage tilting can be expected. Rearward flight must be accomplished slowly and cautiously due to wind effects on the horizontal stabilizer and the lowering of the tail rotor making surface contact easier to occur.

# Autorotation

To help students better understand autorotation, divide it into four distinct phases: entry, steady-state descent, deceleration, and touchdown. Guide the student through each phase, stressing how it is aerodynamically different from the others.

## Entry

Guide the student through the entry or first stage of autorotation and explain that this phase is entered after loss of engine power. The loss of engine power and rotor rpm are more pronounced when the helicopter is at high gross weight, high forward speed, or in high density altitude conditions. Any of these conditions demand increased power (high collective position) and a more abrupt reaction to loss of that power. In most helicopters, it takes only seconds for rpm decay to bring rpm to a minimum safe range, requiring a quick collective response from the pilot. The entry into autorotation must be immediate and smooth by lowering the collective, adjusting the pedals for the loss of torque, and adjusting the airspeed for the proper glide angle. The instructor should never initiate an autorotation, or simulated forced landing, unless there is suitable landing within glide distance in the event of a powerplant or drive line failure.

Discuss with the student the airflow and force vectors for a blade in this configuration. Remind the student that lift and drag vectors are large and the total aerodynamic force (TAF) is inclined well to the rear of the axis of rotation. An engine failure in this mode causes rapid rotor rpm decay. Inform the student that to prevent this, a pilot must lower the collective quickly.

Explain to the student that as the helicopter begins to descend, the airflow begins to flow upward and under the rotor system.

## Steady-State Descent

Airflow is now upward through the rotor disk because of the descent. Once equilibrium is established, rate of descent and rotor rpm are stabilized, and the helicopter is descending at a constant angle. Angle of descent is normally 17° to 20°, depending on airspeed, density altitude, wind, and type of helicopter. The instructor should guide the student through any RFM procedures or charted values for minimum rates of descent versus maximum glide distance if provided for that helicopter.

During this phase of the autorotation, the aircraft is maneuvered to reach a safe landing area by adjusting airspeed and making turns as appropriate while maintaining rotor rpm at the proper range for the type of helicopter. Checklist items are also completed as time permits and a Mayday call made.

Explain to the student how the loss of engine power during the autorotation requires the pilot to use the pedal controls to keep the helicopter in trim throughout the descent until the deceleration and touchdown point is reached, otherwise the increased drag would greatly increase the rate of descent. Also explain how the fuselage tends to weathervane into the wind due to the vertical fin.

## Deceleration

Explain to the student that to make an autorotative landing, the pilot reduces airspeed and rate of descent just before touchdown. Both actions can be partially accomplished by applying aft cyclic, which changes the attitude of the rotor disk in relation to the relative wind. During this maneuver, the goal of the pilot shifts from maintaining an airspeed to attaining a minimum ground speed for touchdown while decreasing the rate of descent. Ensure the student understands that this attitude change:

- Inclines the lift vector of the rotor system to the rear, slowing forward speed.

- Increased airflow results in increasing rpm, which must be controlled with the collective.

- The lifting force of the rotor system is increased and rate of descent is reduced.

- During this stage of the autorotation, the lack of torque is noticeable and the aircraft fuselage may rotate counterclockwise with application of the collective due to frictional drag in the transmission, drive train, associated pumps, and generators (depending on type of helicopter). Pedal application will be required to maintain a heading aligned with the touchdown area. Any crosswind also causes the nose to weathervane into the wind due to lift produced by the vertical fin.

## Touchdown

During this final phase of the autorotation with the airspeed at a minimum as required for the conditions of the landing area, the cyclic stick is moved forward to place the aircraft in a landing attitude while applying collective pitch to cushion the landing. The height at which this phase is entered depends on the size of the helicopter and the length of the tailboom. The landing attitude varies between helicopter designs from touching the aft portion of the landing gear first as in an airplane, to a level attitude with all surfaces touching down at once. Each manufacturer has a preferred landing attitude that must be used to. Heading control must be maintained with the pedals to preclude the aircraft from rolling over once ground contact is made.

The instructor must ensure several conditions are met to allow the helicopter to arrive at this point:

1. The rate of descent, rotor rpm, and airspeed are all within established parameters, as well as landing area alignment and positioning. If any of these conditions are not within limits, re-engage the engine and make a power recovery or go-around.

2. The landing gear is and stays aligned with the ground track of the helicopter.

3. The decelerating flare did not result in an increase in altitude (ballooning) or was not begun at too high of an altitude.

4. The student cannot be allowed to increase the collective too soon, and the student must be prompted to use available collective to cushion the landing soon enough.

5. The collective must be used to cushion the touchdown, but the student should not be allowed to hold the helicopter off the surface.

6. The student must not be allowed to retain an excessive decelerating attitude at too low an altitude allowing a tailboom or tail rotor strike. The student must be taught how to begin the flare and then decrease the nose high attitude to a landing attitude.

7. The student cannot be allowed to move the cyclic aft after touchdown. This generally allows the rotor blades to dip aft over the tailboom and when occurring at the same time as the actual touchdown, results in a tailboom strike.

8. The student cannot be initially allowed to lower the collective after touchdown. Once the helicopter is completely down and no longer subject to bouncing and flexing, some helicopter RFMs allow a slight decrease in collective to aid stopping and to decrease low rpm blade flexing.

For a detailed description and illustration of autorotation, refer to chapter 11 of the Helicopter Flying Handbook (FAA-8083-21, as revised).

## Instructor Tips

- Start the presentation of new material at the student's level of understanding. *[Figure 3-20]*

- Check out Internet sites such as the National Aeronautics and Space Administration (NASA) Beginner's Guide to Aeronautics (www.lerc.nasa.gov/WWW/K-12/airplane/index.html) for graphics and simulations for use in explaining aeronautics.

## Chapter Summary

This chapter reviewed essential points to be taught during aerodynamics instruction. It provided the instructor with additional material that can be used in explaining aerodynamic principles, as well as examples to enhance the learning process.

### Aerodynamics of Flight

**Objective**

Identify characteristics of translating tendency and methods of compensation in a single-rotor helicopter. The student demonstrates the consistent ability to identify and compensate for characteristics of translating tendency in a single-rotor helicopter.

**Content**

Classroom discussion:
- Define translating thrust.
- Provide instruction on the thrusting characteristics of a tail rotor during hovering flight.
- Identify the methods used to overcome translating tendency.

**Postflight Discussion**

Critique student performance. Preview the next lesson. Assign *Helicopter Flying Handbook* Chapter 3, Helicopter Flight Controls.

**Figure 3-20.** *Sample lesson plan.*

# Chapter 4
# Helicopter Flight Controls

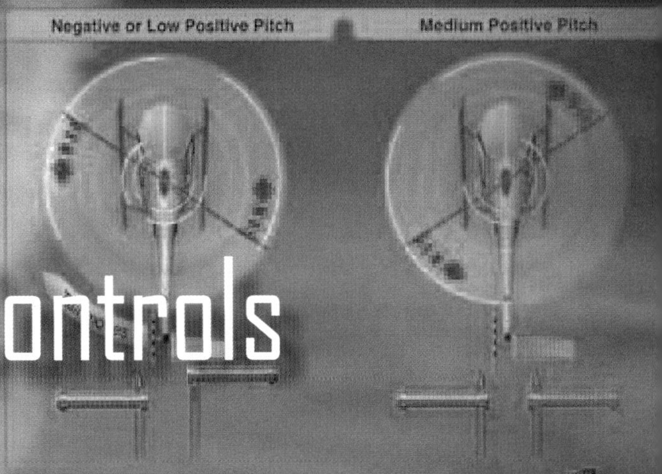

## Introduction

When introducing a student to the flight controls of a helicopter, the instructor must ensure the student understands how each control affects the flight of the aircraft. *[Figure 4-1]* The student may not be comfortable with the helicopter controls for some time, but must understand the function of each control and the reactions of the other controls when control movements are made. For example, if increasing the collective pitch increases power. If the engine is manually controlled, the throttle must be adjusted to maintain

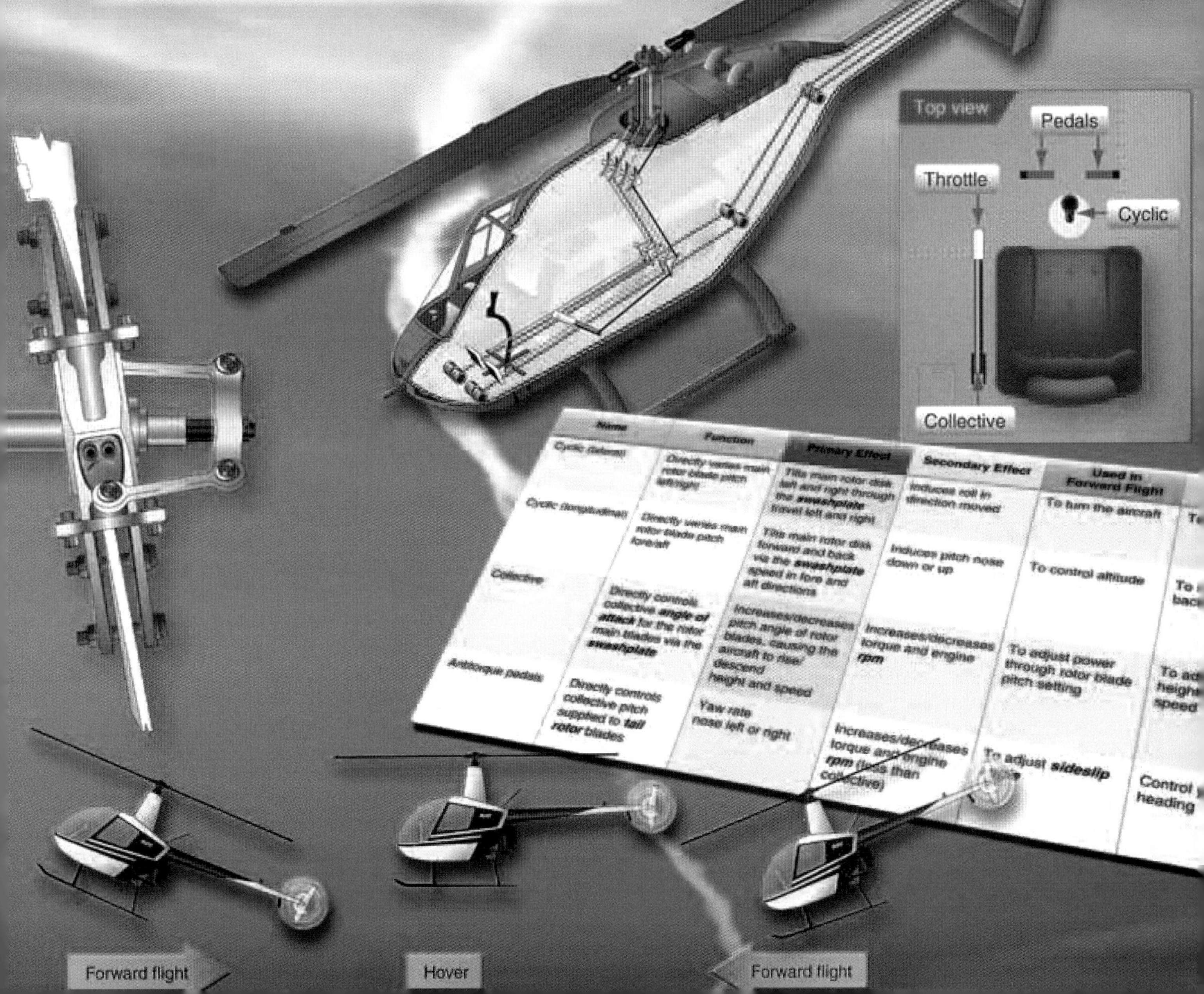

Top view
Pedals
Throttle
Cyclic
Collective

| Name | Function | Primary Effect | Secondary Effect | Used in Forward Flight | |
|------|----------|----------------|------------------|------------------------|---|
| Cyclic (lateral) | Directly varies main rotor blade pitch laterally | Tilts main rotor disk left and right through the *swashplate* travel left and right | Induces roll in direction moved | To turn the aircraft | Te... |
| Cyclic (longitudinal) | Directly varies main rotor blade pitch fore-aft | Tilts main rotor disk forward and back via the *swashplate* speed in fore and aft directions | Induces pitch nose down or up | To control altitude | To... back |
| Collective | Directly controls collective *angle of attack* for the rotor main blades via the *swashplate* | Increases/decreases pitch angle of rotor blades, causing the aircraft to rise/descend — height and speed | Increases/decreases torque and engine *rpm* | To adjust power through rotor blade pitch setting | To a... heigh... speed |
| Antitorque pedals | Directly controls collective pitch supplied to tail rotor blades | Yaw rate nose left or right | Increases/decreases torque and engine *rpm* (less than collective) | To adjust *sideslip* | Control heading |

| Name | Function | Primary Effect | Secondary Effect | In Forward Flight | In Hover Flight |
|------|----------|----------------|------------------|-------------------|-----------------|
| Cyclic (lateral) | Directly varies main rotor blade pitch (left versus right) | Creates left/right directional thrust and tilted rotor system | Induces rolling/tilting moment | Turns the helicopter/ holds ground track | Moves helicopter sideways |
| Cyclic (longitudinal) | Directly varies main rotor pitch forward versus aft | Creates forward/aft directional thrust and tilted rotor system | Induces nose-down or nose-up pitching action often neutralized in stable forward flight by horizontal stabilizer down force | Controls altitude or airspeed | Moves helicopter forward or backward |
| Collective | Directly controls main rotor pitch angle/angle of incidence | Increases/decreases total main rotor thrust | Increases/decreases total thrust for altitude and/or airspeed, power, and antitorque requirements | Adjusts/maintains altitude and/or airspeed settings | Adjusts/maintains hover altitude |
| Antitorque pedals | Directly controls antitorque thrust | Varies antitorque thrust | Controls yaw | Maintains trim for coordinated flight | Maintains heading |
| Throttle controls/ power levers | Maintains rotor rpm | Affects powerplant output range | Determines performance power limits of helicopter | Sets cruise rpm | Sets hover rpm |

**Figure 4-1.** *Helicopter controls and effects.*

revolutions per minute (rpm). If the helicopter powerplant has a governor, then the pilot must ensure that power stays within limitations. If the cyclic is moved, then the collective must be moved to maintain altitude because lift has now been redirected into thrust for travel. Anytime the collective is moved, the pedals must be adjusted for heading or trim. Training for this control coordination can be accomplished by using a simulator or a helicopter. Use of a simulator for this instruction reduces student stress levels and may enhance learning. If a simulator is not available and instruction takes place in a helicopter, the instructor should ensure the student understands the location and function of each control. It is also imperative that the instructor stay close to the flight controls during all phases of flight.

Flying a helicopter is inherently demanding due to all of the moving parts and the controls available to the student and the instructor alike. It is paramount that the instructor be able to manipulate the controls to keep the aircraft in a safe flight profile at altitude and as the student is moved to flight modes requiring increasingly more vigilance, such as a hover in proximity to the ground, other aircraft, and personnel. As the instructor, develop a safety-focused teaching style while being inconspicuous to the student. This is called the "instructor pilot ready position." It is recommended that the instructor be very close to the controls so the student cannot move the controls too far or the controls will hit the instructors waiting hand or foot. A good instructor

forms a boundary area around the controls in which the student can operate the controls without interference from the instructor's fingers and feet. This boundary formation should ensure the helicopter stays within the instructor's personal limits, yet allow the student to develop a control touch without interference. The instructor should always judge the situation by the flight status and condition of the helicopter, not by what the student is doing. It is what the helicopter is doing that is important.

Whether using a simulator or helicopter, beginning the flight instruction at altitude is a good way to allow the student to manipulate all of the controls at one time and with a larger margin of error than beginning the flight instruction at a hover. As the student's proficiency increases and the flight control inputs become smaller, the student can then be allowed to fly lower and slower, ultimately terminating an approach to a hover. A less preferred but widely used technique is to let the student operate one control at a time while the instructor operates the others so the student can get the feel of a control and its function in flight. Always emphasize making smooth, coordinated control inputs.

## Collective Pitch Control

Explain to the student that the collective changes the pitch of the main rotor blades (angle of incidence) and, as a result of that pitch angle change, is used to increase or decrease the blade angle of attack (AOA). This is accomplished through a series of mechanical linkages that changes the angle of incidence of all blades simultaneously, or collectively. *[Figure 4-2]* Demonstrate on a static helicopter how pulling up on the collective increases the pitch of the rotor blades while lowering the collective decreases the pitch. Explain how the collective is used to increase both lift and thrust by changing the lift vector.

Stress to the student that the collective must be kept free of obstructions at all times. The instructor must ensure the student understands the importance of ensuring the collective is free to move through its full range of travel and is kept clear of anything that could limit movement, such as a thigh, map, cell phone, camera, or even an article of clothing.

An instructor may demonstrate how to use the collective to initiate takeoff, climb, and descent. One technique for practicing the application of collective pitch occurs during flight. Climb to a safe altitude and allow the student to operate the collective to climb, descend, and maintain altitude during a turn. Explain the proper application and use of collective friction. Demonstrate how the collective is used to maintain a constant altitude during accelerations and decelerations. During this demonstration, the instructor initially maintains level flight with the other controls and gradually allows the student to have the others controls as proficiency is gained.

## Throttle Control

A student must thoroughly understand the intricacies of the helicopter being flown. While some helicopters have a governor to control the engine revolutions per minute (rpm), or a correlator to increase/decrease throttle inputs automatically to an acceptable range that generally requires some pilot input, other models rely solely on the pilot's manual input of twisting the throttle. *[Figure 4-3]* Even when rpm is controlled by a governor or fuel control system, emergency procedures require manual operation of the throttle to control engine and, ultimately, rotor rpm.

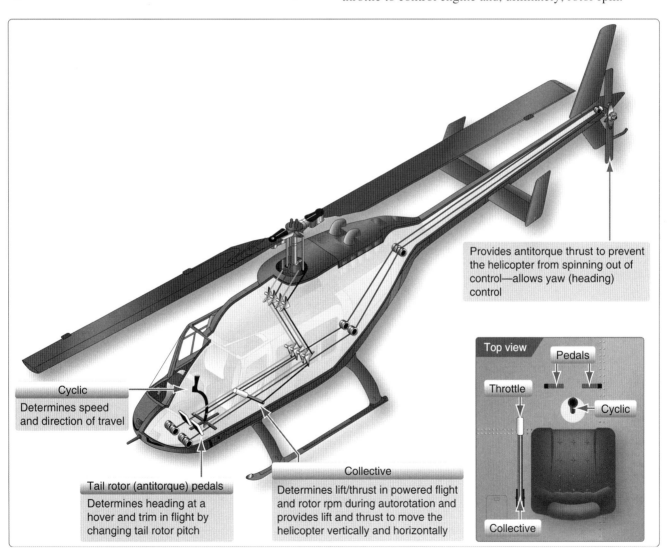

Provides antitorque thrust to prevent the helicopter from spinning out of control—allows yaw (heading) control

Cyclic
Determines speed and direction of travel

Tail rotor (antitorque) pedals
Determines heading at a hover and trim in flight by changing tail rotor pitch

Collective
Determines lift/thrust in powered flight and rotor rpm during autorotation and provides lift and thrust to move the helicopter vertically and horizontally

Top view
Pedals
Throttle
Cyclic
Collective

**Figure 4-2.** *There are four major controls in the helicopter that the pilot must use during flight: collective pitch, throttle, cyclic pitch, and antitorque.*

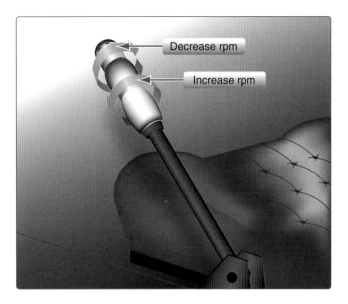

**Figure 4-3.** *A typical throttle that requires manual twisting.*

Manual operation of a nongoverned throttle can be explained and demonstrated during instruction on the collective. A simple explanation that students may be able to relate to is comparing the manually controlled engine to a manual car transmission and a governed engine to an automatic transmission. Proper use of the throttle is an integral part of maintaining both engine and rotor rpm during flight. Explain the use of throttle friction in reducing the sensitivity of the throttle. While students who have experience riding motorcycles or other powered recreational type vehicles are familiar with the concept of a twist-grip throttle, instructors must guard against twisting the throttle in the wrong direction for a given application. Training on governor override

or manual throttle operation should be explained, but a demonstration and practice should occur only after the student has mastered all the control inputs required to fly. Explain that revision to the manual mode of operation is an abnormal, or emergency, procedure and is almost exclusive to reciprocating powered helicopters. Few large turbine powered helicopters have a manual override function suitable for training.

Stress to the student the importance of checking the throttle during the preflight. The throttle, whether governed or not, must have freedom of movement from stop to stop. There should be no binding or excessive stiffness in the operation of the throttle. Point out that throttle control friction must be decreased before checking the throttle.

Special attention should be given to ensure that the throttle/power lever is set in the "start" position prior to starting. This position varies between aircraft design and is explained in the Rotorcraft Flight Manual for the particular helicopter being flown. Improper throttle or power lever settings can lead to overspeed of a reciprocating engine due to the clutch, or engine temperature exceeding limits with turbine powered helicopters.

## Cyclic Pitch Control

A student should understand that moving the cyclic control tilts the rotor system in the direction the cyclic is displaced whether it is fore or aft, or in a side to side motion thereby providing thrust in the direction the rotor is tilted. *[Figure 4-4]* This cyclic movement from the pilot's right

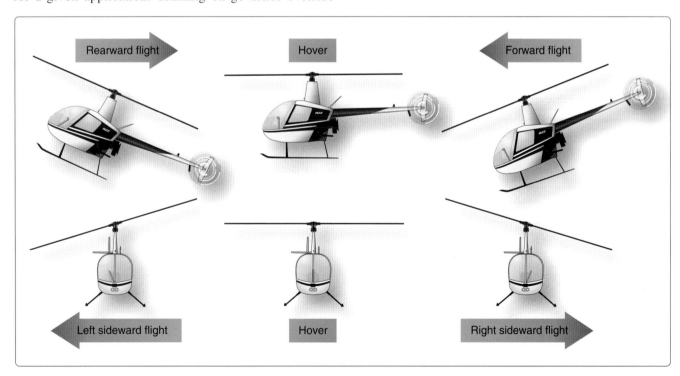

**Figure 4-4.** *Cyclic control stick position and the main rotor disk position relative to pilot in helicopter.*

hand results in the aircraft moving in the direction the pilot desires as the cyclic is displaced. Any movement from the cyclic has a corresponding effect on the pitch of each blade either increased or decreased pitch as they move or cycle thru every rotation 360°. This enables the aircraft to move in the direction the pilot desires.

Emphasis must be placed on keeping the cyclic free of obstructions. Students must understand the importance of ensuring the cyclic is free to move its full travel and to keep it clear of anything that could interfere with or limit its movement such as knee boards or passenger legs.

Demonstrate the cyclic input required to hover. Initiate a takeoff and climb to a safe altitude. Demonstrate how the cyclic is used to maintain the pitch and bank attitude of the helicopter and maintain a constant airspeed during climbs, descents, and turns. Explain the use of the cyclic trim system to relieve cyclic pressures and reduce pilot fatigue.

During the preflight inspection, demonstrate movement of the cyclic in all quadrants and allow the student to observe the inputs made to the swashplate and main rotor system.

## Antitorque Control

Discuss the primary purpose of the antitorque system: to counteract the torque effect created by the rotation of the main rotor system. The antitorque system could consist of vectored thrust from the engine or be provided by a tail rotor. Explain that in either case, the function is the same. Most training helicopters will utilize a tail rotor for this purpose. *[Figure 4-5]* Antitorque pedals change the pitch of the tail rotor and provide the thrust required to counteract the torque effect. Discuss how the pedals are used to maintain coordinated flight during cruise flight, but are used for heading control during hovering flight. Operation of the antitorque pedals through the full range of travel allows the student to observe the pitch change in the tail rotor. Always remind the student of the safety hazards of pinching, moving parts and to keep well clear while the controls are being moved.

Explain to the student the importance of keeping the antitorque pedals free of obstructions and having full range of movement. Emphasize that loose objects that fall during flight and are not retrieved could jam the pedals and reduce aircraft controllability.

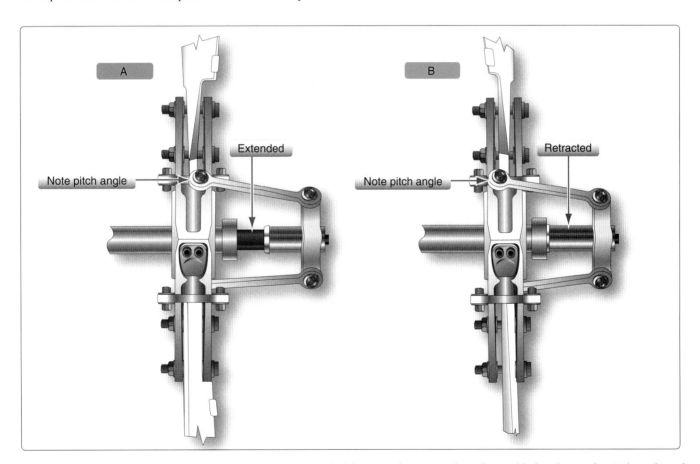

**Figure 4-5.** *When the right pedal is pressed or moved forward of the neutral position, the tail rotor blades change the pitch angle and the nose of helicopter yaws to the right. With the left pedal pressed or moved forward of the neutral position, the tail rotor blades change the pitch angle opposite to the right pedal and the nose of helicopter yaws to the left.*

Demonstrate pedal inputs during a hover. *[Figure 4-6]* Climb to a safe altitude and allow the student to operate the pedals to maintain coordinated cruise flight. Demonstrate how the pedals are used during climbs, descents, and coordinated turns in cruise flight. Explain that when increasing collective pitch, antitorque requirements are greater; when reducing collective pitch, antitorque requirements are less. During this demonstration, the instructor maintains coordinated flight with the other controls.

## Practice

Once the student has practiced with each of the controls individually, while still at a safe altitude, gradually turn over control of the aircraft one control at a time. Remember, STAY CLOSE to the flight controls. When the student has a basic understanding and demonstrates the ability to control the aircraft at altitude in cruise flight, use the same procedures to introduce aircraft control during hovering flight.

Allowing a beginning student to fly down a runway or taxiway, or other set ground track. Making slower and lower approaches will almost naturally lead the student into the hovering mode, allowing a better understanding of control response while avoiding the overly early and frightening attempts at hovering flight. This technique allows them to learn the changes in the helicopter's response to lower airspeeds and ground effects at their own pace.

## Instructor Tips

- Always practice positive exchange of controls procedures and acknowledgments by using a three way positive transfer of controls. *[Figure 4-7]* This is particularly important in the early stages of training when either the student or the flight instructor is on the controls for a long period of time. If the instructor or the student is following along on the controls, ensure that both understand who has ultimate control over the flight controls.

- Always emphasize making smooth, coordinated control inputs.

- Always stay close to the controls. Be ready to take control of the aircraft and never underestimate the student's ability to make a mistake.

- Always practice initial hovering over smooth surfaces, free of any protrusions that might catch the landing gear. A lack of protrusion may allow the landing gear to slide freely in case of accidental ground contact.

## Chapter Summary

This chapter provided the instructor with techniques of introducing flight controls and their application to a student. It also offered the instructor safety related points of emphasis regarding helicopter flight controls.

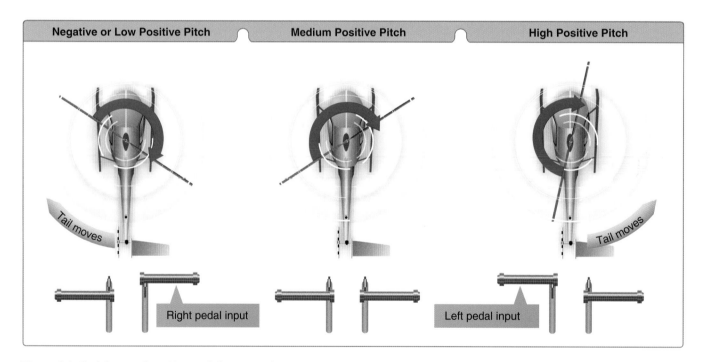

**Figure 4-6.** *Pedal control position and thrust at tail rotor.*

## Helicopter Flight Controls

### Objective

The purpose of this lesson is to introduce the student to helicopter flight controls. The student will demonstrate a basic knowledge of the aerodynamics effects of and helicopter reaction to the movement of each flight control.

### Content

1. Preflight discussion:
   a. Discuss lesson objective and completion standards
   b. Normal checklist procedures coupled with introductory material
   c. Weather analysis

2. Review basic helicopter aerodynamics.

### Schedule

- Preflight Discussion: 10
- Instructor Demonstrations: 25
- Student Practice: 45
- Postflight Critique: 10

### Equipment

Chalkboard or notebook for preflight discussion.

### Instructor actions

   a. Preflight used as introductory tool
   b. Establish a straight-and-level cruise
   c. Demonstrate the effects of each flight control, pointing out the visual and flight instrument indications during cruise flight.

### Student actions

- Practices with each flight control individually and various combinations of all flight controls during cruise flight.

### Postflight Discussion

Review the flight; preview and assign the next lesson. Assign the *Helicopter Flying Handbook*, Chapter 4, Helicopter Components, Sections, and Systems.

**Figure 4-7.** *Sample lesson plan.*

# Chapter 5
# Helicopter Components, Sections, and Systems

## Introduction

When introducing the student to helicopter components, sections, and systems, the instructor must ensure that the student is familiar with and understands the basic functions and use of each system. Ideally, this is accomplished by use of a static aircraft or a mock-up of the aircraft depicting most of the parts. Allowing the student to touch and explore these components, sections, and systems enhances learning. Knowing the basics of each major component, section, and system gives the student a better ability to recognize malfunctions and possible emergency situations. Understanding the interactions of these systems allows the student to make an informed decision and take appropriate corrective action should a problem arise. Point out to the student what should be checked during preflight.

In addition to introducing the student to the various helicopter components and systems, the instructor should also educate the student on the different types of materials that are used to make the components and the positive and negative factors of each. If the student understands the various factors of the components, the information can begin to form the basis of component condition knowledge. This should enable the student to better determine the flight status of the components and what failure modes can appear as well as what the record of the material is. Instructors should be creative and attempt to explore comparisons for all helicopter components and systems. There are no two helicopters alike from one manufacturer to another. Flying a helicopter new to the pilot should always include some ground school instruction on the systems and a flight checkout on the specific characteristics of that helicopter. Students should never assume that knowledge of one helicopter's systems should transfer to another.

### Airframe Design

Airframe design is a field of engineering that combines aerodynamics, materials technology, and manufacturing methods to achieve balances of performance, reliability and cost, which can affect both maintenance and flight. Composites, for example, are very sensitive to ultraviolet (UV) radiation from the sun and must be painted to protect them, whereas aluminum has minimum UV degradation from the sunlight but will corrode over time if flown around salt water. Aluminum is light and reasonable easy to fabricate into parts, but composites can be much stronger although more difficult to manufacture. In addition, composite materials seem to suffer from bonding failures. In some structures, this appears as a thickening or expansion, sometimes forming a bubble in an area.

### Rotor Blade Design

Wooden rotor blades have an infinite fatigue life provided that moisture is kept out of them. The metal attachment fittings on the wooden blade do have a fatigue life; therefore, a life limit is placed on the blade. Although wooden blades are used, they do not perform as well as modern metal or composite blades. Another comparison is composite rotor blades versus metal rotor blades. Composite rotor blades are closer to an "on condition" replacement status, potentially saving thousands of dollars for operations but are softer and at least require the leading edge abrasion strip to be replaced. Metal blades are generally less expensive to manufacture; composite blades may perform better and last longer with less maintenance, but at a higher initial cost.

### Powerplant Design

When discussing the powerplant, comparisons of reciprocating and turbine engines can be explained. A reciprocating engine uses much less fuel than a turbine engine does, but turbine engines can produce much more power from a very light powerplant for more flight hours. However, a turbine engine can easily be ten times the cost of a reciprocating powerplant. Turbine engines are usually much more reliable, but failures are often much more dramatic with high-speed shrapnel flying through other structures, causing tremendous damage throughout the powerplant.

### Antitorque System Design

Enclosed antitorque tailrotor systems are usually more resistant to foreign object damage but can be more complicated to build and heavier, which causes them to perform less well than open antitorque systems. Open tailrotors may perform better aerodynamically with less weight and less cost but are more vulnerable to damage and tend to produce more drag when flown at higher cruise airspeeds.

### Landing Gear System Design

Wheel type landing gear systems are easier to retract for increased cruise speeds with less fuel flow and are much easier to store and move when conducting maintenance. Wheel type landing gear systems are more expensive compared to skid type landing gear with many more parts to inspect, buy, and maintain. They usually have brakes, which also require inspection, service, repair, or replacement. Skid type landing gear is a simpler design, easier to manufacture, and reasonably light in weight. Skid landing gear is a parasitic drag source, which increases exponentially at faster airspeeds. Skid landing gear are less expensive overall but more difficult to move for maintenance or storage and always requires additional equipment on the ground.

Inform the student to use care when handling the actual helicopter, training aids, and/or training devices. Moving parts, sharp edges, protruding components, and/or hydraulic pressures may cause hazardous situations. Material Safety Data Sheets (MSDS) instruction should be reviewed during preflight and postflight if the student comes in contact with any of the fluids in or on the helicopter.

## Airframe

Airframe discussions should explain that the airframe, or structure, of a helicopter can be made of different types of material. *Figure 5-1* is an example of the many different materials that are used in the construction of a helicopter. The importance of learning the structures and construction materials of the helicopter is to help the student to determine the airworthiness of the helicopter and potential failures and hazards to be found on pre and post flight inspections. The goal is not to make the pilot into a helicopter aeronautical engineer but rather a safe pilot who can understand the full consequences of an unknown condition of a helicopter component. Helicopters can be made of metal, wood, or

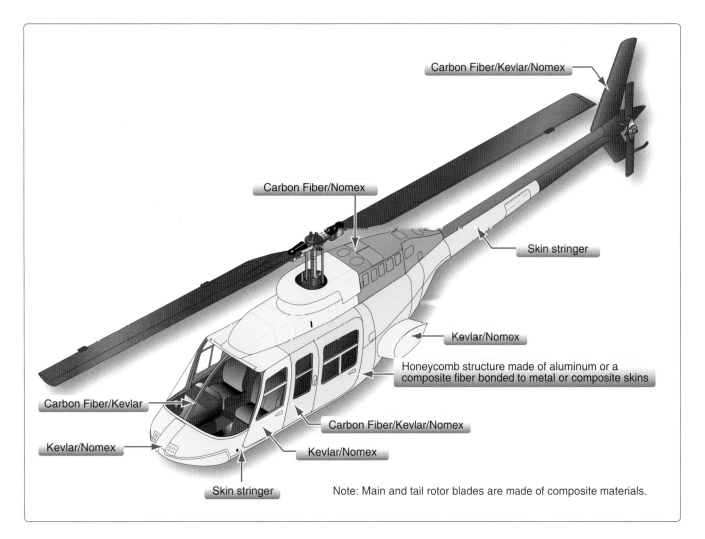

**Figure 5-1.** *Airframe materials.*

composite materials, or a combination of the two. A sample piece of the airframe from the manufacturer is a good prop for introducing the student to the different types of material(s) used. Point out some of the areas that may be made of a variety of materials. One such area is near the exhaust or other areas that must withstand extreme heat for which most manufacturers use titanium. The following is a list of advantages and disadvantages of aluminum and composites that help the student learn about each.

## Aluminum
### *Advantages*

- Predictable strength, which is certified by the manufacturer of the metal and which is recorded with each batch.

- The metal conductivity enables a single-wire electrical system to be used, which saves weight and complexity.

- The metal conductivity maximizes antenna reception and transmission.

- Recognized lightning strike properties and protection.

- Minimal UV degradation from sunlight.

- Controllable moisture problems, such as corrosion.

- Will transmit loads and can bend without failure.

- Good bonding techniques can eliminate P-static for better radio reception.

- Smoothness of construction is predetermined; no extensive sanding or filling is required.

- Paint chips do not materially affect the integrity of the underlying aluminum.

- The Federal Aviation Administration (FAA) readily accepts it as a construction medium and is knowledgeable about its physical properties, causing minimal delays in the certification process.

### *Disadvantages*

- Form blocks must be built to hydroform the metal in a soft state, which then has to be heat-treated to regain its strength.

- Due to the setup of the hydro-form process, there is a high per-unit-part cost unless large batches are produced at one time, in which case inventory carrying costs increase.

- Thin aluminum, as used in general aviation aircraft, cannot be compound curved and still carry structural loads, thereby increasing drag in some areas if improperly engineered.

- Antennas need to be exposed for proper operation, necessitating the use of very expensive, low drag antennas for high speed.

- It is almost impossible to build laminar flow wings with the skin thickness used in general aviation aircraft.

- Any flexing of the structure promotes metal fatigue and can suffer from structural failure.

- Operation under certain conditions (ocean salty air, corrosive chemicals found in agricultural operations) can lead to corrosion caused structural failures.

### Composite Construction
#### *Advantages*

- Ease of construction in small lots. These planes are typically assembled by individuals in their garages.

- Low cost for one-off projects as minimal tooling is required.

- Can be compound curved for maximum drag reduction and still carry structural loads.

- Electronic transparency means antennas can be hidden inside for streamlining without loss of reception.

- Easier to get smooth surfaces for laminar flow designs, which contributes to some additional speed.

- Cracks do not usually propagate in composite structures.

- Structures can be stronger for light weight components.

#### *Disadvantages*

- Strength varies from batch to batch. Difficult to detect voids.

- Since there is no metal frame, there is no common ground; a two-wire electrical system is required.

- Without any electrical conductivity, there is very poor lightning strike protection.

- Ultraviolet light degradation due to sunlight.

- Delamination problems due to moisture.

- Composites tend to break without warning at failure loads, unlike aluminum which can bend and still survive, and usually provide some warning prior to failure.

- Poor electrical bonding causes static interference with radios.

- Requires expensive paint maintained in perfect condition (without chips or scratches) to keep sunlight and moisture out; otherwise, composites degrade like an old fiberglass boat. Poor acceptance by the FAA due to unknown physical properties, such as aging and delamination.

- Very labor intensive to construct repeatable components.

- Composites do not afford any fire or heat protection and can be the source of deadly fumes in the case of an accident or fire.

- Composites require new tools and machines to repair.

## Fuselage

Perhaps one of the best teaching devices for the fuselage discussion is an actual helicopter. Better yet is a helicopter with all or most outside panels removed or undergoing major maintenance or inspection. This open view provides an opportunity for the instructor to point out (literally) to the student engine housing, transmission, avionics, flight controls, and the powerplant. Also, the student can view the seating arrangement for your particular helicopter as you identify where the pilot, crew, passengers, and cargo are seated or placed. *[Figure 5-2]*

## Main Rotor System

Rotor blade design and theory can be very complex to a new student. Discussion should begin with simpler designs and then move towards the more complex as the students understanding progresses. A cut-away of the main rotor assembly is a useful tool for instructors in a classroom environment. This allows students to see how each input from the cyclic and the collective affects the hub assembly. A static helicopter is also very helpful in demonstrating the moving parts of the hub assembly. Explain to the student that the purpose of the main rotor is to produce lift. Show the student the main parts of the mast, hub, and rotor blades.

The three basic classifications of main rotor system are rigid, semirigid, and fully articulated. Main rotor systems are classified according to how the main rotor blades are attached and their movement relative to the main rotor hub. Show which rotor system is installed on the student's particular helicopter and how it is identified.

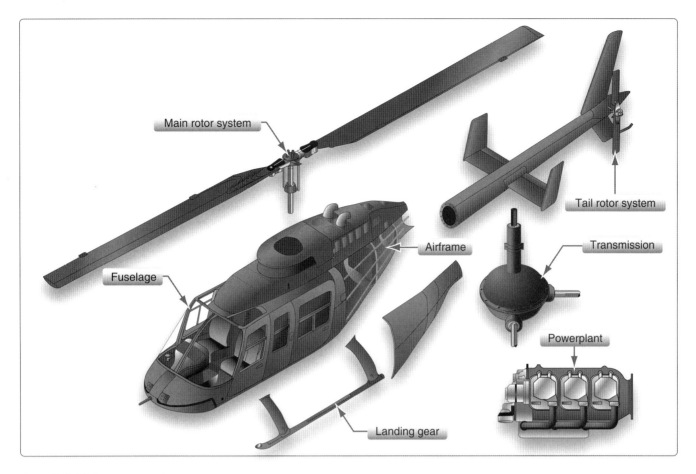

**Figure 5-2.** *Major components.*

Discussions with the student regarding the different types of main rotor systems can be accomplished with additional manufacturer drawings. Pay particular attention to the type of main rotor system that the student will be flying. By now, the student probably has read about the different types of rotor systems but may not fully understand the differences. Explain to the student identifiable characteristics of each rotor system that make it different from the other system types. Also, be ready to answer the student's questions about the advantages and disadvantages of each type of system, cost, and maintenance requirements versus ride quality, performance, reliability, and durability.

## Rigid Rotor System

When introducing the rigid rotor system, instructors should explain that the system is mechanically simple, but structurally complex because operating loads must be absorbed in bending rather than through hinges. *[Figures 5-3 and 5-4]* The rigid rotor was developed by Irven Culver (1911 to 1999) of Lockheed Aircraft Corporation to bring the simplicity of fixed-wing flight to helicopters. In a rigid rotor system, the blades, hub, and mast are rigid with respect to each other. The rigid rotor system is mechanically simpler than the fully articulated rotor system. There are no vertical or horizontal hinges so the blades cannot flap or drag, but they can be

feathered. Operating loads from flapping and lead/lag forces must be absorbed by bending rather than through hinges. By flexing, the blades themselves compensate for the forces that previously required rugged hinges. The result is a rotor system that has less lag in the control response because the rotor has much less oscillation. The rigid rotor system also negates the danger of mast bumping inherent in semirigid rotors. The rigid rotor can also be called a hingeless rotor.

Explain the other advantages of the rigid rotor system to the student (e.g., a reduction in the weight and drag of the rotor hub, higher control loads). Without the complex hinges, this rotor system is much more reliable and easier to maintain than the other rotor configurations. Rigid rotor systems require flexible rotor blades to produce a tolerable ride quality, but allow better maneuverability.

## Semirigid Rotor System

Discuss with the student the main parts of the semirigid rotor system. Explain that it was named for its lack of the lead-lag hinge that a fully articulated rotor system has. The rotor system can be said to be rigid in plane because the blades are not free to lead and lag; however, they are not rigid in the flapping plane (through the use of a teeter hinge). Therefore, the rotor is not rigid, but not fully articulated either; it is

**Figure 5-3.** *A Westland Lynx four-blade rigid main rotor.*

**Figure 5-4.** *Main rotor blade attachment joint on rigid main rotor system.*

semirigid. The parts of the semirigid rotor system that should be identified are teeter hinge, blade grip, blade pitch change horn, and pitch link. (NOTE: The swashplate assembly is described on page 5-12.)

Also, discus the difference between teetering (flapping) versus feathering. On any rotor system, flapping occurs when the blade moves up and down. On a rigid rotor system, this occurs when the blade bends. On an articulated system, the blade flaps up and down around a teetering hinge. On a two-bladed, semirigid teetering system, the blades flap in unison around the flapping hinge, such as in a Bell 206. The semirigid main rotor system is designed such that as the blades cone and flap for different airspeeds, the rotor blade center of gravity centers around the teetering hinge such that the flap down is mostly cancelled out by flap of the other side.

Examples of the semirigid rotor system are found on the Bell 230, the Bell 222 *[Figures 5-5, 5-6, and 5-7]*, and the Bell 206. *[Figure 5-8]* Point out to the student that the Bell 206 head does not include coning hinges. Instead, the rotor head is designed with a pre-cone angle to the blade retention system, and other coning forces are simply dealt with by bending of the blades.

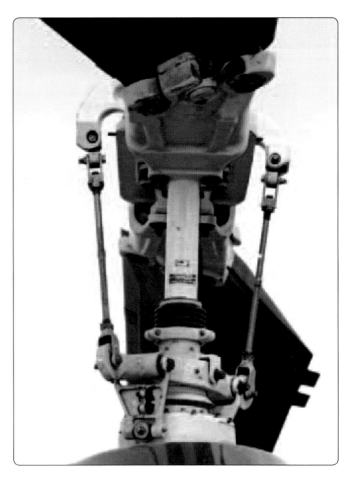

**Figure 5-5.** *Bell 230 semirigid rotor system and swashplate assembly.*

**Figure 5-6.** *Bell 230 semirigid rotor system.*

**Figure 5-7.** *Main rotor blade grip of the Bell 230 semirigid rotor system.*

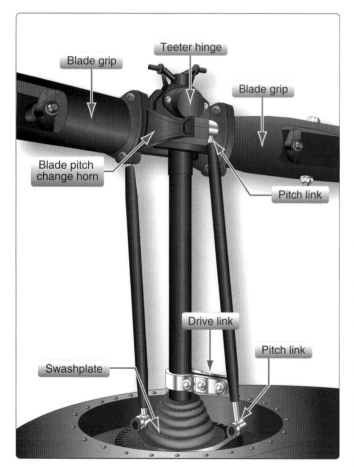

**Figure 5-8.** *Bell 206 semirigid rotor system.*

When discussing the semirigid rotor system, instructors should explain that some are designed with an underslung rotor system which mitigates the lead/lag forces by mounting the blades slightly lower than the usual plane of rotation so the lead and lag forces are minimized. As the blades cone upward, the center of pressure of the blades are almost in the same plane as the hub. Further explain that if the semirigid rotor system is an underslung rotor, the center of gravity (CG) is below the mast attachment point. This underslung mounting is designed to align the blade's center of mass with a common flapping hinge so that both blades' centers of mass vary equally in distance from the center of rotation during flapping. The rotational speed of the system tends to change, but this is restrained by the inertia of the engine and flexibility of the drive system. Only a moderate amount of stiffening at the blade root is necessary to handle this restriction. Simply put, underslinging effectively eliminates geometric imbalance.

## Fully Articulated Rotor System

Fully articulated rotor systems can accommodate larger loads and faster airspeeds with good ride quality. Because there are more blades, the load can be spread among them resulting in lower initial angle of attacks which allows the retreating blade more margin above stall which allows increased forward airspeed before $V_{NE}$. They have increased expenses due to the many parts that make up the rotor system, which also make preflight more complicated. The fully articulated rotor system is also susceptible of ground resonance if certain factors coincide. As with the other types of rotor systems, the student should have read about the fully articulated rotor system. *[Figure 5-9]* The student should be able to use the solidity ratio to explain how each blade carries only a portion of the total load. It is about the wing loading (in pounds) to the total area of the wing (in square feet). The instructor may need to review with the student basic aerodynamics of airfoils and airflows necessary to develop lift. Full articulation is also found on rotor systems with more than two blades. Using the rotor, show the student how the fully articulated system allows each blade to lead and lag, flap up and down, and feather. *[Figure 5-10]*

The purpose of the drag hinge and dampers is to absorb the acceleration and deceleration of the rotor blades caused by Coriolis Effect. *[Figure 5-11]* Older hinge designs relied on conventional metal bearings. By basic geometry, this precludes a coincidental flapping and lead/lag hinge and is cause for recurring maintenance. Newer rotor systems use elastomeric bearings, arrangements of rubber and steel that can permit motion in two axes. Other than solving some of the above-mentioned kinematic issues, these bearings are usually in compression, can be readily inspected, and eliminate the maintenance associated with metallic bearings.

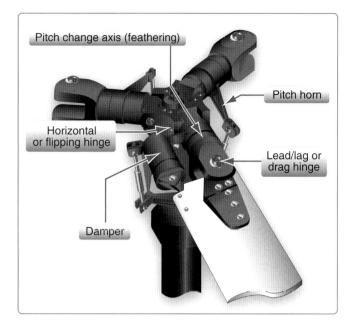

**Figure 5-10.** *Fully articulated rotor system.*

Elastomeric bearings are naturally fail-safe and their wear is gradual and visible. The metal-to-metal contact of older bearings and the need for lubrication is eliminated in this design.

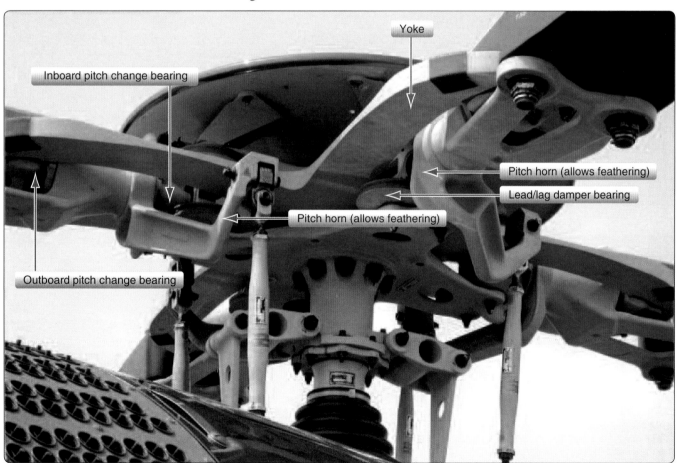

**Figure 5-9.** *Bell 427 fully articulated rotor system. This system is often referred to as soft in plane; each blade operates independently and leads, lags, and flaps in a controlled manner due to elastomeric construction.*

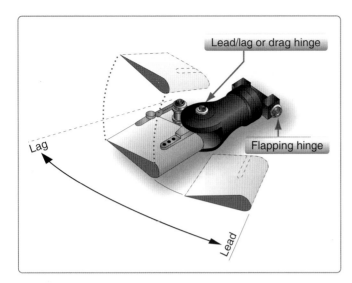

**Figure 5-11.** *Lead/lag hinge allows the rotor blade to move back and forth in plane.*

Coning or flapping hinges-allows the blades to flap up and as airspeed is increased, allows the main rotor blades to flap due to differences in the relative wind speeds. *[Figure 5-11]*

Feathering hinges allow the main rotor system blades to change pitch individually as they cycle around the rotor disk to allow for direction thrust control application.

This is a good time to reiterate to the student what was covered in the Helicopter Flying Handbook, Chapter 4, Helicopter Flight Controls, and how each input from the controls (cyclic and collective) independently or collectively affects the rotor system.

*Figures 5-10* and *5-11* depict how the blade acts in its rotation about the mast. Explain to the student that the blade is normally kept in a horizontal plane during its rotation by centrifugal force. However, high winds during runup or shutdown when the blades are turning at a low speed could affect this (and cause damage as well). The damage occurs when the blades flex up or down greater than normal. Another factor to consider is how the flapping force is affected by the severe rigor of the maneuver (rate of climb, forward speed, aircraft gross weight, hard landing, etc.).

Coning or flapping hinges allow the blades to flap up and as airspeed is increased, allows the main rotor blades to flap due to differences in the relative wind speeds. The feathering hinges allow the main rotor system blades to change pitch individually as they cycle around the rotor disk to allow for direction thrust control application.

Explain to the student that modern rotor systems may use the combined principles of the rotor systems mentioned above. Some rotor hubs incorporate a flexible hub, which allows the blade to bend (flex) without the need for bearings or hinges. These systems, called flextures, are usually constructed from composite material. Elastomeric bearings may also be used in place of conventional roller bearings. Elastomeric bearings are constructed from a rubber-type material and have limited movement that is perfectly suited for helicopter applications. Flextures and elastomeric bearings require no lubrication and, therefore, require less maintenance. They also absorb vibration, which means less fatigue and longer service life for the helicopter components.

## Bearingless Rotor System

When discussing the bearingless rotor system, explain to the student how the structures of the blades and hub are manufactured differently to absorb stresses. Bearingless rotor systems, such as the Eurocopter systems, have contact surfaces or load points made of elastomeric composite components that deform and twist to allow blade movement. Most of these components are "on condition" life items versus metal components which must be changed at certain times due to metal fatigue. The composite components are designed so that even if a portion fails, the aircraft can make a safe landing. *[Figure 5-12]*

The hingeless (bearingless) rotor system functions much as the articulated system does, but uses elastomeric bearings and composite flextures to allow flapping and lead lag movements of the blades in place of conventional hinges. Its advantages are improved control response with less lag and substantial improvements in vibration control. It does not have the risk of ground resonance associated with the articulated type unless the landing gear system needs servicing. The hingeless rotor system is also considerably a more expensive system.

## Tandem Rotor

On a tandem rotor helicopter, two rotors turn in opposite directions at opposite ends of a long hull. The rotors are usually synchronized through a transmission system so that the main rotor shafts can be little more than a blade length apart. Tandem rotor helicopters operate a little differently from the single rotor variety. Tandem rotor helicopters have no tail rotor, so there is no translating tendency to combat, but there are pedals for directional control at a hover. The cyclic control, which is used as it always has been in single rotor helicopters, has not changed either. *[Figure 5-13]*

One deviation to the tandem rotor system is the side-by-side twin rotor system. *Figure 5-14* shows an example of the Kamen K-Max intermeshing (side-by-side) rotor system, which dates back to the old H-4 Husky, and is a modified tandem rotor system. It is optimized for external load

**Figure 5-12.** *A Eurocopter EC-135 hingeless and bearingless rotor.*

**Figure 5-13.** *Tandem rotor, or dual rotor, helicopters have two large horizontal rotor assemblies, a twin rotor system instead of one main assembly, and a small tail rotor.*

operations, and is able to lift a payload of over 6,000 pounds, which is more than the helicopter's empty weight. The K-MAX relies on the two primary advantages of synchropter over conventional helicopters: 1) it is the most efficient of any rotor-lift technology, and 2) it has a natural tendency to hover. This increases stability, especially for precision work in placing suspended loads. At the same time, the synchropter is more responsive to pilot control inputs, making it easily possible to sling a load thus to scatter seed, chemicals, or water over a larger area.

## Coaxial Rotor System

Students should be shown the Coaxial rotor system, which consists of a pair of helicopter rotors mounted one above the other on concentric shafts, that is one shaft inside another with the same axis of rotation, but that turn in opposite directions. *[Figure 5-15]* Explain that this configuration is a feature of helicopters produced by the Russian Kamov helicopter design bureau. Coaxial rotors solve the problem of angular momentum by turning each set of rotors in opposite directions. The equal and opposite torques from the rotors

**Figure 5-14.** *Kaman K-Max with the modified tandem rotor system.*

upon the body cancel out. Rotational maneuvering, yaw control, is accomplished by increasing the collective pitch of one rotor and decreasing the collective pitch on the other. This causes a controlled differential of torque.

Coaxial rotors reduce the effects of dissymmetry of lift through the use of two rotors turning in opposite directions,

causing blades to advance on either side at the same time. One other benefit arising from a coaxial design include increased payload for the same engine power. A tail rotor typically wastes some of the power that would otherwise be devoted to lift and thrust; all of the available engine power in a coaxial rotor design is devoted to lift and thrust. Reduced noise is a second advantage of the configuration. Part of the loud slapping noise associated with conventional helicopters arises from interaction between the airflows from the main and tail rotors, which in the case of some designs can be severe. Also, helicopters using coaxial rotors tend to be more compact (occupying a smaller 'footprint' on the ground) and consequently have uses in areas where space is at a premium. Another benefit is increased safety on the ground; by eliminating the tail rotor, the major source of injuries and fatalities to ground crews and bystanders is eliminated.

The coaxial rotor system has the following disadvantages:

1. Mechanical complexity.

2. Poor hover performance characteristics of the smaller rotor disk in higher altitudes and warmer climates.

3. Heavier, stiffer blades required to prevent the blades from flexing into the other rotor rotating in the opposite direction.

4. Heavier rotor head and hub components to control and retain the heavier blades.

## Swashplate Assembly

Explain to the student that the rotating swashplate couples stationary cyclic motion with rotating cyclic control

**Figure 5-15.** *A Kamov Ka-32A-12 has a coaxial rotor system.*

movements. The drive link ensures that the rotating swashplate stays synchronized with the main rotor as it turns. The antidrive link and lever are attached to the aft side of the inner ring and swashplate support, preventing rotation of the inner ring. Point out to the student where these controls are connected. Also, point out (if installed) the stationary swashplate, rotating swashplate, pushrods, antidrive link, uniball, and pitch horns. [Figure 5-16] During preflight inspect for obvious damage, condition, and security of all components.

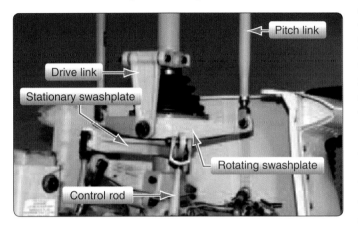

**Figure 5-16.** *Stationary and rotating swashplates.*

Explain to the student that there are several different mechanisms for transmitting cyclic and collective inputs to the main rotor system. The Robinson R22 and R44 have the swashplate mounted on a monoball. This allows the entire swashplate to slide up and down on the rotor mast (for collective inputs) and tilt (for cyclic inputs). *Figure 5-17* shows how the swashplate slides up and down to transmit a collective pitch change. *Figures 5-17* and *5-18*

should be used by the instructor as references. Demonstrating to the student the actual movements on the helicopter is a better option, if available.

*Figure 5-17* depicts how collective inputs affect the swashplate assembly. The red arrow is pointing to the bottom of the swashplate (A), B shows the entire swashplate has moved up the mast. Note the effect on the pitch links.

Small hashed lines show that in B, the pitch link has moved up along with the swashplate (compare the top of the pitch link and the left-hand coning hinge bolt in the two pictures). Since the entire swashplate has moved up without changing its tilt, the pitch links have all moved up a set amount, but continue to move up and down during rotation in response to the tilt of the swashplate.

*Figure 5-18* depicts how the cyclic inputs affect the swashplate assembly. Notice that the swashplate in A is basically level, while in B it has been tilted. The tilt forces the pitch link to move up as it travels to the right-hand side of the picture, and move back down as it travels to the left-hand side of the picture. As it moves up and down, the blade pitch increases and decreases.

NOTE: On some helicopters, the control rods were routed internally up through the main rotor mast to protect them. On those helicopters, the cyclic inputs come down from the top of the mast and the swashplate is under the transmission, where it is all covered and protected from wires (Enstrom).

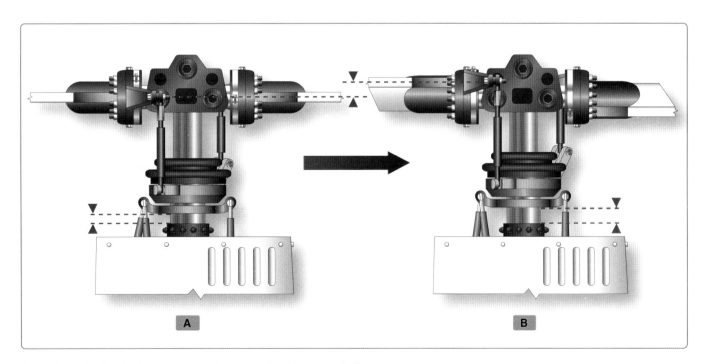

**Figure 5-17.** *Collective inputs on a stationary and rotating swashplate.*

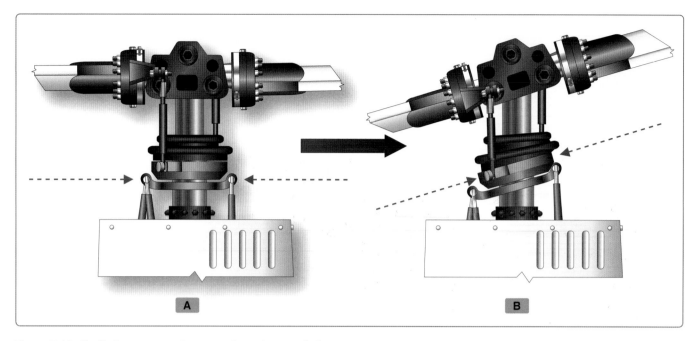

**Figure 5-18.** *Cyclic inputs on stationary and rotating swashplate.*

## Antitorque Systems

### Tail Rotor

Explain to the student that the tail rotor is required on a single rotor helicopter to overcome the torque effect. This torque effect is the result of the fuselage turning in the opposite direction of the main rotor system. *Figure 5-19* depicts the main rotor blades turning counterclockwise and the fuselage (torque direction) turning clockwise in order to compensate for the unwanted torque of the fuselage. On a static aircraft, show the student how the inputs of antitorque pedals effect the pitch change in the tail rotor. Discuss with the student the emergency procedures for loss of tail rotor authority, loss of tail rotor thrust, loss of tail rotor components (forward CG shift), a break in the tail rotor drive system, and fixed pitch settings.

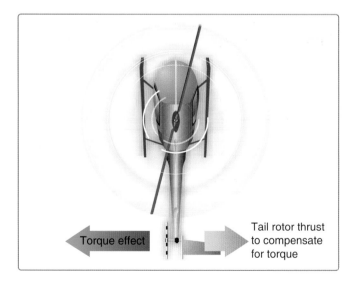

**Figure 5-19.** *Antitorque rotor produces thrust to oppose torque.*

Point out to the student the different parts of the tail rotor (if installed), including the pitch change tube, pitch change link, and the cross head assembly. *[Figure 5-20]* The Bell model 427 tail rotor assembly shown in Figure 5-20 has an internal control rod which is designed this way for protection. Demonstrate that as the crosshead assembly moves in and out, it will change the pitch angle of the tail rotor blades via the pitch change link and pitch horns. When left pedal is applied, control tubes are moved and the lever assembly retracts the control tube. As the control tube retracts, the crosshead moves closer to the yoke assembly; tail rotor blade pitch is increased.

Show how the tail rotor is much like the main rotor, except it is turned on its side and provides thrust instead of lift. Another way to describe the tail rotor is to compare it to an airplane propeller which also generates thrust and does not provide lift. Reinforce to the student the importance of keeping the antitorque pedals free of obstructions and having full range of movement. Emphasize that if a loose object fell during flight and were not retrieved, it could jam the pedals and reduce aircraft controllability.

### Other Types of Antitorque System

Explain to the student that there are several different types of anti-torque systems. One is the fenestron, or "fan-in-tail," design. A fenestron is a fully enclosed tail rotor. It is essentially a ducted fan. The housing is integral with the tail skin, and, like the conventional tail rotor it replaces, is intended to counteract the torque of the main rotor. Fenestrons have between eight and eighteen blades. These may have variable angular spacing so that the noise is distributed over different frequencies and, thus, is quieter. The housing allows a higher rotational speed than a conventional rotor,

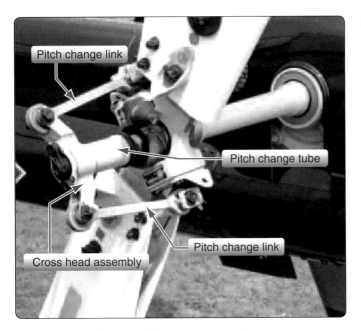

**Figure 5-20.** *Bell model 427 tail rotor assembly.*

allowing it to have smaller blades. *[Figure 5-21]* The smaller diameter allows use of higher fan speeds and sometimes requires higher fan rpm ranges to equal thrust from a much larger unducted system. The housing, although somewhat heavier, does offer some protection on the ground and is more streamlined in forward flight. Discuss with the student that propellers and rotors alike are designed to be less than transonic at the tips.

**Figure 5-21.** *Fenestron, or "fan-in-tail," antitorque system. This design provides an improved margin of safety during ground operations.*

The other type of tail rotor is the NOTAR® system (no tail rotor). The NOTAR system represents the first significant configuration change to conventional helicopters since 1939 when Igor Sikorsky flew the first conventional rotorcraft. "The new system uses the Coanda effect of air flowing over or around the surface of the tail boom to create lateral lift. This counteracts the torque of the main rotor. The NOTAR system shortens drive shafts, gearboxes, and the rotor unit

itself. This reduction in the parts count is a distinct advantage over conventional tail rotor craft. *[Figure 5-22]*

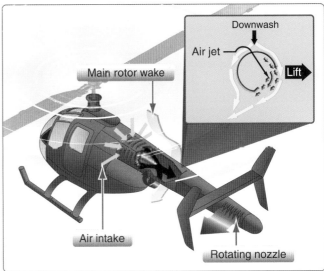

**Figure 5-22.** *The Coanda effect supplies approximately two-thirds of the lift necessary to maintain directional control in a hover. The remaining lift is created by directing the thrust from the controllable rotating nozzle.*

In operation, the NOTAR system draws low-pressure air in through an air intake located at the top of the airframe to the rear of the main rotor shaft. A variable-pitch fan pressurized the tail boom to a relatively constant 0,5 psi. The air is fed to two starboard side slots and a direct jet thruster. The slots provide the necessary antitorque force. The rotating jet thruster provides direction control. The two slots are located at 70 and 140 degrees and allow ejected air to mix with the main rotor downwash to establish the Coanda effect. The main rotor downwash is normally dissipated as essentially symmetric separation on both sides of the tail boom in a hover. The pressurized boom inject low-pressure air at 250 fps onto the Coanda surface (outer surface of the tail boom), which results in the deflection and produces about two-thirds of the required antitorque force. This force is predictable. It is controlled by the appropriate location of the slot and control of the air jet that exits from the slot.

In other words, the tail boom reacts like an airplane wing, only sideways. The increased air speed over the starboard side of the tail boom causes lateral lift, pushing against the torque forces trying to spin the helicopter clockwise. This is the same result that a tail rotor achieves when it propels the tail in a counterclockwise motion.

The main rotor downwash skews as velocity is increased, and the circulation control slot is uncovered, resulting in proportional loss of antitorque force. The vertical tail surface provides the directional stability with forward speed. In sideward flight, the effective angle of attack is changed as a function of the main rotor thrust and sideward velocity inflow effects. When the downwash is altered by motion other than hovering, the system reduces the Coanda effect, and the thruster picks up more of the load. This keeps the system forces balanced. The tail fin, which does not come into play when hovering, also becomes effective when flying forward.

The direct thruster provides the remaining one-third of the force needed to counter the torque of the main rotor. The thruster rotates, moving the opening either to the right or left. In this way, directional control is achieved.

## Engines

Discuss the different types of engine that may be found on most modern helicopters: reciprocating (or piston) and turbine. Discuss with the student the emergency procedures for engine related problems, such as loss of power (underspeed) or rapid increase in power (overspeed) while in flight. Authorized fuel types for a specific engine should also be topics of discussion at this time.

### Reciprocating Engine (Piston)

Explain that the reciprocating engine is the most widely used powerplant in light helicopters and is designed to specific standards of reliability. It must be capable of sustained high power output for long periods. Explain to the student the cycle of the reciprocating engine, as depicted in *Figure 5-23*. Discuss the intake cycle (induction stroke, fuel/air mixture), compression cycle (fuel/air mixture ignited by spark plug), power cycle (burning mixture expands), and the exhaust cycle (burned gases escape). The manufacturer may be able to provide a diagram of the internal components of the reciprocating engine. This allows further discussion with the student of the internal workings of the engine.

It is very important for instructors to teach the student to understand what the engine is supposed to do, how it works while flying and what happens when something breaks and how the pilot should react. The instructor should be able to discuss the octane requirements of a gasoline engine or jet fuel classifications if teaching in a turbine engine powered machine. Some engines require the settings to be changed for different fuels. The instructor must ensure the student can determine the difference between jet fuel and avgas when sumping the tanks. Explain that besides smell and the oiliness tests, there is the white paper test, where a drop is placed on a piece of white paper or paper towel. Avgas will leave a distinctive ring from the dye in the fuel whereas jet fuel tends to leave an oily yellow ring.

### Turbine Engine

Explain to the student that the turbine engine is also widely used today on larger and most all of the military helicopters. Because turbine engines have a continuous combustion process which allows more horsepower to be developed form a smaller unit. Since the power is developed from circular rotation instead of reciprocating motion, power is smoother and engine stresses are reduced which contributes to reliability. The expense comes from the high temperature tolerant materials and close-tolerance manufacturing processes needed to produce the turbine engine. A turbine engine provides a high power-to-weight ratio, which a reciprocating engine cannot provide. Some have a power-to-weight ratio three times that of the piston engine.

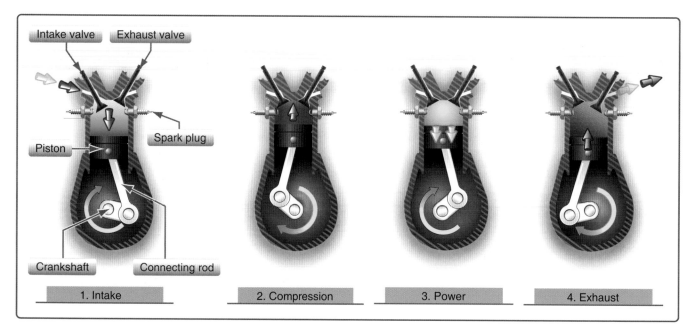

**Figure 5-23.** *The arrows indicate the direction of crankshaft and piston motion during the four-stroke cycle.*

Turbocharged or supercharged piston engines can operate well at high altitudes. Weight per horsepower and reliability are the main factors favoring the turbine engine. Explain that the working cycle of the turbine engine is similar to that of the piston engine (i.e., induction, compression, combustion, and exhaust). One other difference is the fact that the piston engine combustion (power) is intermittent; in the turbine engine, each process (cycle) is continuous. The manufacturer may be able to provide a diagram of the internal components of the turbine engine. This will allow further discussions with the student regarding the internal workings of the engine. *Figure 5-24* is an example of a turbine engine.

The instructor should also explain the increased fuel usage in a turbine engine is due to the continuous combustion process and the fact that approximately the first 50-60 percent of the engine's power is required to sustain the engine's induction and systems such as the oil system and electrical generator. This accounts for a turbine engines high idle speed. A turbine engine may idle at 16,000 rpm and generate maximum power at 38,000 rpm. Turbine engine power curves are very steep and it may need 6-10 seconds or longer to begin generating large increased power demands. There is very little extra power available at close to idle settings from turbines. Usually, turbine engines must be above 80-90% rpm to develop moderate power output. This is the reason to keep turbine engines under power loads to have the power available if needed.

Show the student the main parts of the turbine engine (compressor, combustion chamber, turbine, and accessory gearbox assembly). Then, discuss what each section is doing during flight or 100 percent power, as depicted in *Figure 5-24*. Many helicopters use a turboshaft engine to drive the main transmission and rotor systems. The main difference between a turboshaft and a turbojet engine is that most of the energy produced by the expanding gases is used to drive a turbine (turboshaft engine) rather than producing thrust through the expulsion of exhaust gases (turbojet engine). The instructor should fully understand and be able to explain that the turbine and four-stroke helicopter engines both have four cycles: intake, compression, power, and exhaust. This continuous combustion process is the main limitation due to material limitations. The extreme heat and centrifugal forces place tremendous stress on the rotating parts of the combustion section.

When operating helicopters with turbine engines, instructors should teach the student about starting batteries and the different characteristics of a lead acid and Ni-cad (nickel-cadmium) batteries. Lead acid batteries generally do not have the energy density per pound of the Ni-cad batteries, but cost much less and have much lower maintenance costs. Lead acid batteries also tend to have a sloping power output curve that can allow the operator to perceive impending failure and replace the battery; however, lead acid batteries must be specially designed to withstand the deep charge that happens during a turbine engine start. The student should be reminded of the differences between the start times of a reciprocating engine (a relatively short period of time) and the prolonged turbine starting sequence (lasting 30–60 seconds not counting a cooling period if the internal engine

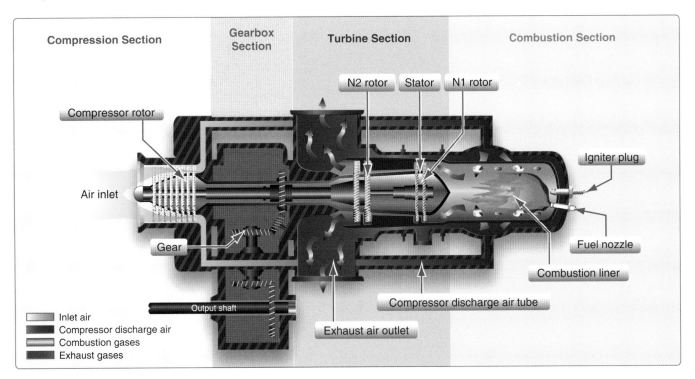

**Figure 5-24.** *Turbine engine.*

temperatures are initially too high). Additionally, the battery for a turbine engine installation must be designed with sufficient residual reserve to furnish cooling rotation in the event of an aborted or hot start.

Ni-cad batteries have much higher energy densities for their weight and, most significantly, can withstand the long, very high current drain necessary to start a turbine engine in cold temperatures. One advantage of Ni-cad batteries is the almost flat output power curve. The uniform output provides consistent turbine starter activation. Unfortunately, Ni-cad batteries produce a very flat consistent discharge power output which suddenly and rapidly decreases at the end of its charge, and this means that it can be very difficult to determine if the battery is at full capacity or towards the end of the charge curve.

To receive proper service and consistent turbine starts, battery voltage and battery charge indications must be closely and consistently monitored for long-term, gradual changes and be maintained in accordance with the manufacturer's recommendations. This usually requires completely discharging and charging the individual battery cells. Most manufacturers then require that batteries be reassembled with equal output cells for best results. For more information on starter batteries, the instructor should review chapter 10 of the Aviation Maintenance Technician—General Handbook.

As a reminder:

1. The compressor draws air into the plenum chamber and compresses it.

2. That air is directed to the combustion section where fuel is injected into it.

3. The fuel-air mixture is ignited and allowed to expand.

4. This combustion gas is then forced through a series of turbine wheels, causing them to turn.

5. Turbine wheels provide power to both the engine compressor and the accessory gearbox.

6. Power is provided to the main rotor and tail rotor systems through the freewheeling unit, which is attached to the accessory gearbox power output gear shaft.

7. During the starting process, follow the manufacturer's requirements closely for hot or slow starting procedures. A fully charged battery will help in most cases.

8. Always follow the manufacturer's cool down procedures to allow internal parts to settle to cooler uniform temperatures as much as possible before engine shut off.

Now, briefly explain what each section comprises and any emergency actions related to each one.

## Compressor

The compressor is similar to a fan. As air is drawn inward, stator vanes act as a diffuser at each stage, decreasing air velocity and increasing air pressure. The high pressure air then passes through the compressor manifold where it is distributed to the combustion chamber via discharge tubes.

Discuss with the student the phenomenon of a compressor stall (engine surges). Explain to the student how reducing the airflow might correct the condition. This is accomplished by activating the bleed air system, which vents excess pressure to the atmosphere and allows a larger volume of air to enter the compressor to unstall the compressor blades. Help the student understand the compressor air control system installed in the helicopter and explain the probable failure modes. If the inlet guide vanes fail closed or if a bleed air valve fails open, the pilot will notice much higher engine combustion temperatures at lower power settings with the maximum power available being very limited. If the guide vanes fail open or the bleed air valves fail closed, high power operations will probably be normal but compressor stalling and possible flameouts may occur as the power demand is reduced.

## Combustion Chamber

The combustion chamber is where the fire takes place anytime the engine is running properly. An igniter plug connected to the combustion chamber ignites the fuel/air mixture only when starting the engine. If installed with an auto-relight, the igniter may attempt to automatically relight the fuel/air mixture in an engine flame-out condition. Discuss with the student what is done if the engine should flame out during flight. Altitude and time available should be mentioned as well.

## Turbine

Discussions about the turbine need to be tailored to the specific helicopter that is being flown as each there are differences in how the two sections of the turbine are coupled to the drive line. For example, the Rolls-Royce (formerly Allison) (Bell JetRanger and BO-105), Lycoming engines (Astars), and Pratt and Whitney (BH-212) are free turbine engines with separate shafts for compressors and power turbines. Older Gazelle and Alouettes use a single-shaft turbine with a centrifugal clutch to allow starting, much like the older and often larger reciprocating-engine-powered helicopters. Turbines will always have the two sections of compression and combustion. What varies is how the sections are coupled to the drive line. Common in most helicopters now is the free turbine design, which uses one inner shaft from the combustion section turbine to drive the accessory gearbox, oil pump, fuel pump, starter/generator and the compressor to sustain the engine. This is typically called the N1(NG). A separate outer shaft around the inner shaft driven by the power output turbine wheel usually goes through the gearbox to be reduced in rpm and support the output drive shaft. This is typically called the N2 and is dedicated to driving the main rotor, tail rotor, drive system, and other accessories such as generators, alternators, and air conditioning (if installed).

Help the student correlate possible emergencies, such as NR/NG overspeeds, to what is happening in the turbine and ensure that the student understand why the steps being taken for the emergency procedures help alleviate or control the problem so that they can safely land the helicopter. Memorizing emergency procedures is part of the beginning learning process for students, but the ultimate goal should be to help them recognize the onset of a system/component failure and then how to properly react to ensure a safe landing.

## Transmission System

Explain to the student the purpose of the transmission system. It transfers the work done by the engine to the main rotor, tail rotor, and other components of the helicopter that rely on engine propulsion. Discuss the main components of the transmission and where they are located on the helicopter:

1. Main rotor transmission
2. Antitorque drive system
3. Clutch
4. Freewheeling unit
5. Rotor brake (if installed)

Point out the location of the oil level sight gauge. Also, point out the location of chip detectors that are associated with the transmission and engine (if detectors are installed). Identify the location of the warning lights on the pilot's instrument panel.

Chip detectors give advance warning of possible excessive engine or transmission wear, which could prevent an impending failure. This early warning can also greatly reduce the cost of engine and transmission overhaul. The chip detectors illuminate warning light(s) when metal chips bridge the gap in the magnetic probe of the chip detector.

NOTE: Some helicopters use chip detectors that have burn-off capability (fuzz burners). When a metal chip(s) bridge the gap in the magnetic probe a warning light is illuminated on the instrument panel. The chip(s) are automatically charged with an electrical current with the ability to eliminate most small particles.

## Main Rotor Transmission

Explain to the student that the main rotor transmission is designed as a gear reduction, reducing engine power to rotor revolutions per minute (rpm). With a horizontally mounted engine, the transmission changes the axis of rotation from the horizontally mounted engine to the vertical axis of the rotor shaft. In many helicopters, the transmission also supports or carries the entire weight of the helicopter. Because of this, the transmission brackets should be checked on preflight for stability and condition.

Explain to the student that the rotor rpm is kept at a predetermined setting during normal flight. During autorotation, the rotor rpm must be maintained by the pilot to continue a normal rate of descent. Remember, very low rotor rpm is unrecoverable as the blades will fold up and airflow will not increase the rpm.

Discuss with the student that a high rotor rpm during autorotation increases the rate of descent. Low rotor rpm initially slows the rate of descent; however, if rpm is allowed to decrease excessively, the helicopter may fall almost vertically. Little or no collective is available at the bottom of the autorotation. Maintaining the autorotation rpm that is set by the helicopter manufacturer is important. *Figure 5-25* depicts various types of tachometer used to maintain/monitor the rotor rpm.

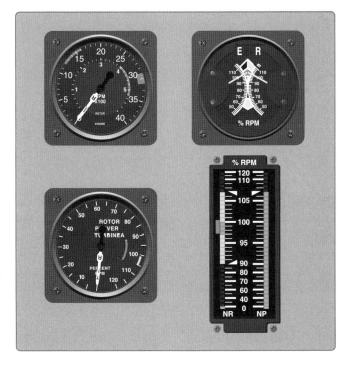

**Figure 5-25.** *There are various types of duel-needle tachometers; however, when the needles are superimposed or joined, the engine rpm ratio is the same as the gear reduction ratio.*

## Antitorque Drive System

Explain to the student that the drive system may be exposed or placed inside of a covered tail boom depending on the type of helicopter. Point out to the student the different parts of the tail rotor drive system, (if installed) such as the hanger bearings, flex couplings, input seal, and output seal. Also, point out what to look for during preflight (leaks, loose fittings, or obvious damage). The instructor should ensure the student understands the common failure modes and weak links. For example, the witness pins on the shafts at the couplings, coupling packs, slippage marks, and metal particles indicating a movement between the surfaces (around a loose rivet). The tail rotor gear box should also be covered at this time. Fluid levels and attaching hardware are important preflight items to check. *[Figure 5-26]*

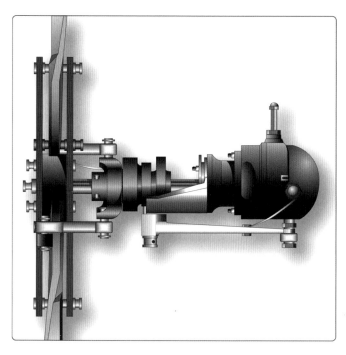

**Figure 5-26.** *Tail rotor and tail gearbox of a Robinson 22.*

## Clutch

Explain to the student the purpose of the clutch (if installed) on the helicopter, and how a centrifugal clutch works. A centrifugal clutch is a clutch that uses centrifugal force to connect two concentric shafts, with the driving shaft nested inside the driven shaft. The input of the clutch is connected to the engine crankshaft while the output may drive a shaft, chain, or belt. As engine rpm increases, weighted arms in the clutch swing outward and force the clutch to engage. The most common types have friction pads or shoes radially mounted that engage the inside of the rim of a housing. On the center shaft, there are assorted extension springs, which connect to a clutch shoe. When the center shaft spins fast enough, the springs extend, causing the clutch shoes to engage the friction face. It can be compared to a drum brake

in reverse. When the engine reaches a certain rpm, the clutch activates, working. This results in waste heat but, over a broad range of speeds, it is much more useful than a direct drive in many applications. Those using the belt clutch system must be very careful to ensure full engagement and engagement procedures. Excessive throttle can quickly ruin an engine because there is no load during the initial starting, so the engine can speed past its rpm redlines very quickly. Those events require expensive teardowns and overhauls. Most large helicopters use a clutch during the start sequence and then gradually engage the rotor system to normal operating rpm.

Free-turbine engines do not need a clutch because there is little load from the rotor system. The rotor slowly starts turning during the start sequence and gradually achieves normal operating rpm.

Explain to the student that there are three main types of clutch found on reciprocating helicopters.

1. Centrifugal clutch—briefly explain how the centrifugal clutch operates and how to determine if the clutch is operating normally, using the rotor tachometer.

2. Belt drive clutch—briefly explain how the belt drive clutch operates and how to determine if the clutch is operating normally using the rotor tachometer. Show the student the location of the pulley belts and that the pilot must check for frays, tears, or cracks on the belt(s) during preflight of the helicopter.

3. Sprag clutch—explain how the sprag clutches have inclined ramps and rollers. If the drive shaft is faster than the driven shaft, the rollers are forced against the ramps and the clutch locks up and transmits full power. If the driven shaft is turning faster than the driving shaft, the rollers retreat down the incline and allow the driven shaft to rotate freely, hence the freewheeling clutch.

NOTE: A clutch is used to disconnect the engine from the rotor load to enable a starter motor to turn the engine for starting. Some turbine helicopters have a centrifugal clutch (Gaszelle) that engages the rotor system above about 28,000 rpm. Also, the R-22, R-44, and HU-269 series use belt clutches to allow the engine to be started without excessive loading. The older Hillers and Bell 47 series machines used centrifugal clutches mounted above the engines.

## Freewheeling Unit

Explain to the student that all helicopters are fitted with a form of freewheeling unit. Also, explain the purpose of the freewheeling unit and where it is located on the helicopter. The freewheeling unit makes autorotations possible by disconnecting the dead or failed powerplant from the transmission and removing the drag from the rotor system. One of the most popular types is the sprag clutch. The freewheeling unit allows the engine to drive the rotors but does not allow the rotors to turn the engine. When the engine(s) fail, the main rotor still has a considerable amount of inertia and still tends to turn under its own force and through the aerodynamic force of the air through which it is flying. The freewheeling unit is designed to allow the main rotor to rotate now on its own regardless of engine speed. This principle is the same as being in a car and pushing the clutch in, or putting it into neutral while the car is still moving—the car coasts along under its own force. This occurs regardless of what is done to the accelerator pedal.

## Fuel System

Explain to the student the parts and functions of the fuel system. *Figure 5-27* illustrates a typical gravity feed fuel system.

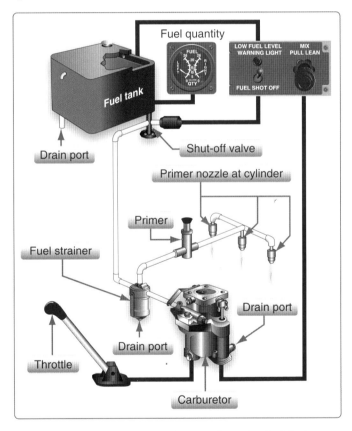

**Figure 5-27.** *The components of a typical gravity feed fuel system in a helicopter with a reciprocating engine.*

Show the student how to properly check the fuel for water or other contaminants. Also point out to the student the location of the fuel shutoff valve in case of an emergency. If installed, show the student how to operate the hand-operated fuel primer and why a primer if installed for a carburetor engine must be closed and locked for proper engine operation.

Explain to the student the purpose and part(s) of the engine fuel control system and the location of the system. Each type of helicopter (reciprocating or turbine engine) requires a different type of fuel control, and each one also has a different type of delivery for the fuel control.

1. Reciprocating engines have a carburetor or are fuel injected.

2. Turbine engines have several types of fuel control systems:

   a. Full Authority Digital Engine Control (FADEC)—engine is electronically controlled with no mechanical connections. Requires electricity to fully operate and function.

   b. Mechanical Units—no power is needed, it is all mechanical and is reliable but not as efficient.

   c. Hydro/Mechanical hybrid units have some characteristics of both. Usually older versions of early attempts at FADEC type systems. Many had a manual reversion capability.

Show the student the major components of the fuel control system, if installed. Two types of fuel control system that are used today by most modern turbine helicopters are the FADEC and analog electronic engine control (EEC).

True FADECs have no form of manual override available (in case of FADEC failure), giving the computer full authority over the operating parameters of the engine. If a total FADEC failure occurs, the engine fails.

If the engine is controlled digitally and electronically but allows for manual override, it is considered an EEC or electronic control unit. An EEC allows the pilot to continue to operate the engine with the throttle while in emergency mode (manual mode). Electronic supervisory control allows the pilot to override the digital side of the fuel control and operate in the analog mode during emergency mode of operations.

NOTE: Many turbines still utilize the older type electronic fuel control systems, which may not be quite as efficient as the newer systems, but operate without electrical power and are quite reliable. Manual operation is easily possible.

## Engines

### Reciprocating Engines

#### *Carburetor*

Explain to the student the need to make adjustments to the carburetor ("full rich" to "leaning the mixture") and why. Refer the student to the FAA-approved Rotorcraft Flight Manual (RFM) and point out the specific procedures for a particular helicopter.

Discuss with the student what the indications are if the fuel mixture is too rich (engine seems rough/reduced power) or leaned out too much (high engine temperature, possibly damaging). The mixture in most cases should be adjusted on the ground because an overly lean mixture can cause the engine to stop, resulting in a forced autorotation and attempt to restart the helicopter in flight.

### Carburetor Ice

*Figure 5-28* depicts how ice affects the carburetor. Discuss with the student why ice may form on the internal surfaces of the carburetor. Carburetor ice has two sources: 1) Venturi cooling from air expansion and 2) fuel vaporization absorbing heat. Both effects combine to cool moisture in the air to below freezing. In some installations, the Venturi effect can cause icing around the butterfly in fuel injection systems, but it is a rare instance. Recommend reviewing FAA Advisory Circular (AC) 20-113, Pilot Precautions and Procedures To Be Taken in Preventing Aircraft Reciprocating Engine Induction System and Fuel System Icing Problems. Also, discuss the indications of carburetor icing (e.g., decrease in engine rpm or manifold pressure, carburetor air temperature gauge outside the safe operating range, and engine roughness) and how to correct for the icing condition.

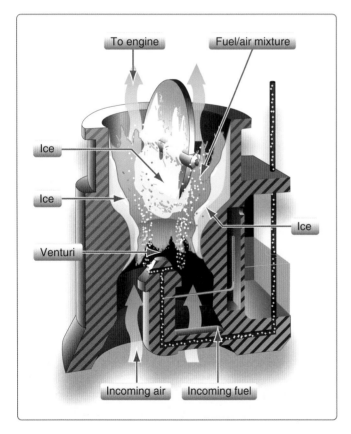

**Figure 5-28.** *Carburetor ice reduces the size of the air passage to the engine, restricting the flow of the fuel/air mixture and reducing power.*

Point out to the student the FAA-approved RFM procedure for carburetor heat. Engine rpm should decrease as hot air is introduced into the engine because hot air is less dense. If the engine rpm does not decrease, the flight should be canceled until the defect is corrected and ensure that a deficiency entry is made into the helicopter's logbook or maintenance tracking sheet. Explain to the student that *Figure 5-29* is a depiction of how a typical carburetor heat system functions. Remind the student at this time that if too little carburetor heat is applied and ice kills the engine, the freewheeling unit will prevent restarts of the engine without use of the starter.

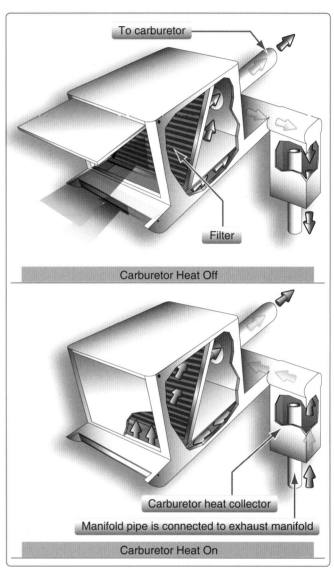

**Figure 5-29.** *When the carburetor heat is turned ON, normal air flow is blocked, and heated air from an alternate source, usually from the exhaust manifold, flows through the filter to the carburetor.*

### Fuel Injection

Explain to the student how the fuel injection differs from the carburetor system and why the system eliminates carburetor icing. When there is no carburetor, airflow is controlled by butterfly but no need for venture because the fuel is injected under pressure which reduces the cooling effect. Also if the fuel is infected at the intake port of the engine, the fuel vaporization temperature drop doesn't enter into the situation at all. Even if the fuel is injected at the butterfly, it vaporizes en route to the cylinder so the temperature drop occurs inside the warm engine where there is plenty of heat.

## Electrical Systems

Show the student the electrical diagram that is provided by the helicopter manufacturer and discuss with the student the major components and functions of the electrical system.

Explain how each system works with one another from the start sequence through the power off sequence (shutdown). At a minimum, show the student the location of the following items (if installed) and most importantly explain the function failure modes of the various components and enough about the locations for a thorough preflight:

1. Battery
2. Battery switch
3. Starting vibrator
4. Ammeter (discuss how to read it and what the numbers indicate)
5. Starter switch
6. Starter
7. Alternator
8. Alternator switch
9. All circuit breakers and switches (Note: FAA policy states that if a nonessential circuit breaker pops up or opens, do not reset in flight. If it is an essential circuit breaker, allow one reset only. Resetting circuit breakers could result in an in-flight fire. For more information, refer to the Special Airworthiness Information Bulletin, CE-10-11R1.)

For a student flying a turbine-powered helicopter, point out the starter/generator. A starter/generator load indicator is often located on the pilot's instrument panel to indicate the condition of the starter/generator system. A turbine helicopter pilot should fully understand the difference between a loadmeter and an ammeter and what the indications really mean in order to understand what the real failure is and the correct procedure to follow.

Flight is still possible during a total loss of electrical power, and students should be taught to remain calm and safely land the helicopter. The engine continues to operate normally without electrical power. The battery, if fully charged, provides a limited time of power for items such as radio(s). Also, discuss the steps to take in the event of electrical circuit breakers tripping or fuses burning out. Electrical fire in flight should be covered as well.

## Hydraulics

Hydraulic systems vary slightly with different helicopter designs. Pilots must understand the system on the specific helicopter that is being flown. Not all helicopters rely on hydraulic assist for the control inputs, and smaller helicopters usually do not use hydraulics in an effort to keep total weight of the airframe down. Larger helicopters (light to heavy) incorporate hydraulics to overcome high control forces. The discussion should begin with showing the student the manufacturer's hydraulic schematic and indicating where the pressure and return lines are located. Walk through the entire hydraulic system, showing the student the location of components and explain what the basic functions are of each component.

### Hydraulic System Components

Always refer to the proper Rotorcraft Flight Manual for the specific hydraulic system that the helicopter is equipped with. The following is a list of hydraulic system components and their functions with which the student should be familiar.

Hydraulic reservoir—The reservoir has three lines: overboard scupper drain, systems return line, and the pump supply line. The pump supply line uses both gravity feed and pump suction to keep the hydraulic pump supplied with fluid. The reservoir may be pressurized to prevent cavitations for helicopters that are capable of higher altitudes and to ensure positive control pressure. The hydraulic reservoir is usually located higher than the hydraulic pump to ensure adequate fluid gravity flow to the pump and is mounted on a bracket, which is located near the transmission. A window is provided on the cowling for inspection of the sight glass. A sight glass is provided to determine when the reservoir needs servicing. Normal fluid level is indicated when hydraulic fluid completely fills sight glass or on some helicopters, is filled to a set level on the sight glass.

Hydraulic pump—provides the pressure to operate the servos and the entire system pressure is regulated by the pump via the pressure line. If the pump is driven off the transmission and it fails, there is usually a shear shaft which breaks to allow the transmission to keep rotating so that the helicopter can be landed safely. Other systems drive the pump from the engine. An engine failure will also include a hydraulic failure. The pilot must understand the system on the specific helicopter being flown.

Quick disconnects—usually seen from the cabin roof and need to be checked for security. What is important is that the student understands how to ensure that the quick disconnect fittings are fully seated and locked together. The quick disconnects located on the pressure side are where fluid flows through from the hydraulic pump. On the return side, the quick disconnect is the last component through which the fluid flows before returning to the hydraulic reservoir. These components allow maintenance to isolate the hydraulic reservoir and pump from the hydraulic system.

Filter bypass indicator—the pressure and return filters both have a pressure indicator that should be checked during preflight. When the indicator is in the reset position, it will be flush and not seen. An extended red indicator indicates an impending filter stoppage. The system is also affected by low temperature, pressure surges and excessive vibration. The red indicator pops out when a set differential pressure across the filter is exceeded. The difference in pressure is not the same for all helicopters; therefore, instructors should teach students what the pressure differentials are for the helicopter being flown. Once the red indicator pops it will remain extended until it is reset manually. Refer the student the proper Rotorcraft Flight Manual for reset procedures.

Bypass check valve—the helicopter is equipped with a bypass system and there is an obstruction in the filter causing a pressure differential (the differential point will be different for each hydraulic system), the bypass valve will open and allow unfiltered fluid to flow directly to the reservoir. This feature allows the pilot to safely land the helicopter with the hydraulic system still working.

Relief valve—part of the hydraulic system, and located between the pressure and return portions of the hydraulic system. The unit protects the system from overpressurization in the event of a hydraulic pump malfunction.

Solenoid valve—designed to provide pressure to the system when it is deenergized. The solenoid valve is de-energized when the HYD SYS switch is in the HYD SYS position or in the event of loss of electrical power to the valve. Placing the HYD SYS switch to the OFF position will energize the valve and pump pressure will be blocked with the system pressure connected back to the reservoir.

Hydraulic system switch—located inside the cockpit. When the hydraulic switch is placed in the HYD SYSTEM position, the solenoid is deenergized. The solenoid is then energized when the hydraulic switch is placed in the OFF position. This system is a "fail safe" system which requires power to disable. Therefore, pulling the circuit breaker for the hydraulics might restore the system if it happens to be an electrical control

problem. If the hydraulics have failed, it is most important for the hydraulics to be switched off to ensure the hydraulic system does not come back online when large control forces are being applied. A gross over controlling situation could result which could lead to damaging the helicopter.

Pressure manifold—a distribution point that permits hydraulic pressure to evenly flow to all actuators.

Pressure switch—the switch opens if the hydraulic pressure ever becomes low. The pilot should see an indication in the cockpit that the hydraulic pressure is low.

External check valves—prevent reverse flow of the hydraulic fluid from the actuator when pressure is lost. The return line check valve permits the return fluid from the directional control actuator to flow out of the actuator, and then pack into the inlet port. When hydraulic pressure is lost, this type of design permits the directional control actuator to remain full of fluid and prevents feedback forces in the flight controls.

Servo actuators—varies with each helicopter design, but a common hydraulic system will have four servo actuators: one directional, one collective, and two cyclic.

Return manifold—fluid leaves the actuators and travels through the return manifold and recycles through the return filter. The fluid then passes through the quick disconnect coupling to the hydraulic.

### Hydraulic System Failure

Explain to the student what the procedures are if a hydraulic system failure occurs. Discuss with the student the difference in control after a hydraulic system failure while at a hover or in forward flight. Hover is difficult because of the tendency of overcontrolling the helicopter and the stiffness of the controls. A run-on landing is a suitable option during a hydraulic system failure.

NOTE: Some hydraulic systems operate at pressures exceeding 1,000 pounds per square inch (psi). Students should be cautioned about searching for hydraulic leaks while the system is still under pressure. The system accumulator can have high system pressure for long periods of time after shutdown and if any part of the human body is exposed to such high pressure streams, those streams can act like a needle and puncture the skin injecting the toxic fluid into the body.

## Stability Augmentation Systems (SAS)

The stability augmentation system (SAS) was developed from an earlier method which prevented the cyclic from flopping around, force trim, which would hold the cyclic control only in the position at which it was released. Force trim was a passive system that simply held the cyclic in a position which gave a control force to transitioning airplane pilots who had become accustomed to such control forces. Students should learn that SAS is an active stabilization system that helps the helicopter track the position of the cyclic relative to the horizon. Some systems are designed to use as much as 10% of the total servo travel to control the helicopter. This is achieved automatically without inputs from the pilot; with this type of system installed, the pilot work load is reduced. The helicopter is a bit more stable with SAS installed, and it dampens unwanted helicopter movement during flight and at a hover. Instructors should show the student the SAS actuators, which are mounted on the hydraulic servos and are fed information from gyros that sense the: pitch, roll, and yaw axes of rotation. Important information to relay to the student is that the SAS requires power, both for the stabilization platform and for the actuators. Like any other helicopter system, they are subject to failure and instructors need to discuss emergency actions that may be required if the system were to fail.

## Autopilot

Explain to the student that the more sophisticated SASs have additional features, such as an autopilot. As suggested, the autopilot can perform certain duties as selected by the pilot. Some of the basic systems perform only basic functions, such as heading and altitude.

Some of the advanced systems perform certain functions, such as climb/descent rate, navigation capabilities that track to points and some fly instrument approaches to a hover without any additional pilot input.

Autopilot is widely used by the United States Coast Guard to assist in search and rescue and to recover the helicopter during adverse weather conditions, as well as in many turbine powered helicopters which allows for single pilot IFR operations.. NOTE: It is important to refer to the autopilot operating procedures located in the RFM, if autopilot is installed.

## Environmental Systems (Heating/Cooling)

Explain to the student that many smaller helicopters only have doors as part of their environmental systems. Show where the doors will be stored and how to properly store all loose equipment and seat belts. Once the doors are removed, stress the importance of a clean and secure cabin. Many accidents have occurred when objects have blown from the cabin and damaged both the mainrotor and tailrotor. Pilots have lost maps, charts, sunglasses, cushions, jackets etc. from the cabin or cockpit. Flapping seatbelts can also cause unnecessary damage to the side of the helicopter in flight.

Show how the ram air functions and the location of any levers that are used to control ram air. If installed, show the location and controls of the air conditioning unit. The pilot should be well versed in the operation and restriction of use of the air conditioning unit. Many units are restricted from use during takeoffs and landing due to power demands. Ensure the student refers to the RFM for the proper operating procedures.

Discuss the cabin heating system with the student and locate the heater ducts and switches that control them. Piston-powered helicopters use a heat exchanger shroud around the exhaust manifold, and turbine-powered helicopters use a bleed air system for cabin heat. Any other systems that use forced air or heat should be discussed at this time, such as defog blowers for the main windscreen.

## Anti-Icing Systems

First and foremost, students need to understand that anti-icing is the process of protecting against the formation of frozen contaminant, snow, ice, or slush on a surface. Icing can occur as the helicopter sits over night or during flight. In either case, icing becomes a hazard and if not attended to can be disastrous. Include the following topics when discussing anti-icing systems: engine anti-ice, carburetor icing, preflight, and deicing.

### Engine Anti-Ice

Discuss with the student the importance of using the engine anti-ice system if certain conditions are encountered and the loss of performance when the system is in use. The instructor should be able to explain why the engine anti-ice decreases power so much. The following information should be explained to the student:

1. Engine anti-ice uses bleed air to heat inlet.

2. The bleed air exits the inlet area into the inflow which decreases the air density due to the high temperature air.

3. Although the anti-ice may keep the engine operating, everything else is still subject to icing. Real icing conditions dictate an immediate exit from those conditions.

4. The windshield is subject to icing on the exterior and fogging on the interior from the crew and occupants breathing. Rarely do helicopters have windshield anti-icing or deicing certification.

### Carburetor Icing

Carburetor icing can occur during any phase of flight, and it is particularly dangerous when you are using reduced power, such as during a descent. Explain that the pilot may not notice it during the descent until trying to add power. Teach the student about the possible indications of carburetor icing: decrease in engine rpm or manifold pressure, the carburetor air temperature gauge indicating a temperature outside the safe operating range, and engine roughness. Because changes in rpm or manifold pressure can occur for a number of reasons, closely check the carburetor air temperature gauge when in possible carburetor icing conditions. Show the student that the carburetor air temperature gauges are marked with either a yellow or green caution operating arc. Instructors should refer the student to the FAA-Approved RFM for the specific procedure regarding when and how to apply carburetor heat. In most cases, you should keep the needle out of the yellow arc or in the green arc. This is accomplished by using a carburetor heat system, which eliminates the ice by routing air across a heat source, such as an exhaust manifold, before it enters the carburetor.

### *Preflight and Deicing*

Instructors should stress the importance of checking for fuselage and component icing when doing the preflight with the student. Rotorblades, pitot tubes, and engine parts are all susceptible to icing and should be checked thoroughly before starting the helicopter. Explain that de-icing is the process of removing frozen contaminant, snow, ice, slush, from a surface. Deicing of the helicopter fuselage and rotor blades is critical prior to starting. If possible, show the student a helicopter that has been sheltered from the elements and then compare it to one that has not. Helicopters that are unsheltered by hangars are subject to frost, snow, freezing drizzle, and freezing rain all of which can cause icing of rotor blades and fuselages, rendering them unairworthy until cleaned. Asymmetrical shedding of ice from the blades can lead to component failure and shedding ice can be dangerous because it could hit any structures or people that are around the helicopter. The tail rotor is vulnerable to shedding ice damage. Thorough preflight checks should be made before starting the rotor blades. If any ice was removed prior to starting, ensure that the flight controls move freely. While inflight, deicing systems (i.e., helicopters so equipped) should be activated immediately after entry into an icing condition.

## Instructor Tips

- Always supervise the student when first introduced to the helicopter. *[Figure 5-31]*

- Point out to the student the danger areas and the sharp portions of the helicopter.

- Show the student the "No Step" areas, if present.

- Allow the student to touch each component of the helicopter as it introduced and say the name of the part. On the next preflight, the student should begin to describe the function of each part, and on every preflight after that, the student should be asked the next component or components in order of the checklist until the student has learned the functions of each component of that specific helicopter.

- If appropriate, tell the student well ahead of time what will be covered during the next lesson and what the student should study or reference. The student should always be briefed in a quiet area so there are no distractions that can take their attention away from the discussion.

## Chapter Summary

The components, sections, and systems that were covered in this chapter were described so you as the instructor can convey this information to your student. This is your guide to further create a lesson plan and teach the student the "whys" of the helicopter components, sections, and systems. What was not covered in this chapter was the responsibility of the instructor to also introduce the student to the helicopter's service reports, what they mean, and how to obtain them.

## Helicopter Components, Sections, and Systems

**Objective**

The purpose of this lesson plan is to introduce the student to the basic components, sections, and systems of the helicopter. The student will demonstrate a basic knowledge of the location of components, sections, and systems (if installed). The student will have a basic understanding of how the systems operate (if installed).

**Content**

1. Preflight discussion:
   a. Discuss lesson objective and completion standards.
   b. Normal checklist procedures coupled with introductory material identifying the location of components, sections, and systems.

2. Review basic helicopter components, sections, and systems.

3. Instructor actions:
   a. Use preflight as introduction to the basic helicopter components, sections, and systems.
   b. Have the student identify the components, sections, and systems. Cover at a minimum (if installed): airframe, fuselage, main rotor, tail rotor, swashplate assembly, freewheeling unit, engine, transmission, fuel system, electrical system, hydraulics, and anti-icing system. (Note: avionics and navigation systems could be included in this list of systems.)
   c. Introduce only a few components at a time. Too much information too soon overwhelms students.
   d. Perform a check on learning once the student has had the opportunity to see and identify the components, sections, and systems. Have the student point out and explain each of the systems.

4. Student actions:
   Study the helicopter components, sections, and systems in the appropriate operator's manual and, if applicable, any FAA-approved RFMs.
   a. Be able to identify the location of selected helicopter components, sections, and systems.
   b. Be prepared to discuss with the instructor, an understanding of the selected helicopter components, sections, and systems.

**Postflight Discussion**

1. Review what was covered during this phase of training.
2. Preview and assign the next lesson. Assign *Helicopter Flying Handbook*, Chapter 5, Rotorcraft Flight Manual.

**Figure 5-31.** *A sample lesson plan.*

# Chapter 6
# Rotorcraft Flight Manual

## Introduction

An essential component of flight instruction is ensuring the student becomes familiar with publications that provide information required for flight. One of those publications, the Rotorcraft Flight Manual (RFM), not only contains information critical to safe flight, it is also required to be carried in the aircraft. Introduce students to this manual and the material it contains at the beginning of the training program. *[Figure 6-1]*

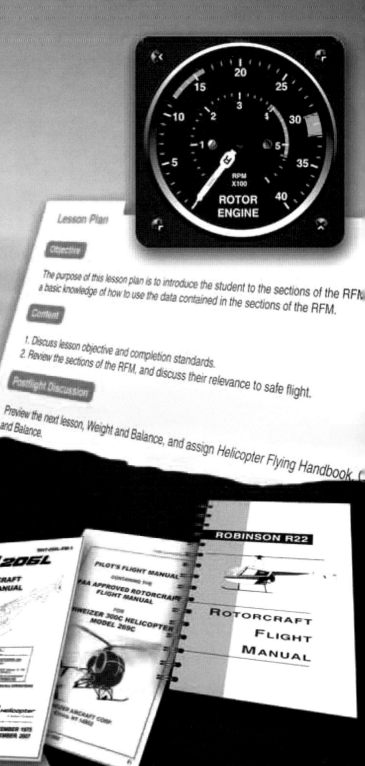

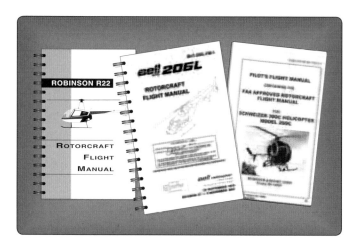

**Figure 6-1.** *Rotorcraft Flight Manuals.*

## Introducing the Manual

When introducing the manual for a particular helicopter, a good place to begin is the certification basis of the helicopter as defined by Title 14 of the Code of Federal Regulations (14 CFR) part 27, Airworthiness Standards: Normal Category Rotorcraft, or Part 29, Airworthiness Standards: Transport category rotorcraft. The certification establishes the framework for development of the helicopter itself as well as the performance requirements for that certification level.

Another method for introducing the RFM to a student is to review the manual section by section. Discuss the information contained in each section and why that information is important to a pilot. Explain that some manufacturers use the title Pilot's Operating Handbook (POH). If POH is used, the title page must include a statement to the effect that the document is a Federal Aviation Administration (FAA) approved RFM.

Emphasis must be placed on the fact that no two RFMs are exactly alike. Explain to the student that although the manual format is standardized for the make and model of an aircraft, some of the information contained therein relates to a specific aircraft. The title page indicates the registration number and serial number of the aircraft to which the manual applies. Ensure the student understands that an after-market POH does not meet the requirements of the RFM.

Another piece of information to include would be the subject of supplements to the RFM. Students need to be advised of the need for RFM supplements when equipment not included in the type certificate data sheet (TCDS) is installed into the aircraft. The supplement often changes procedures and limitations stated in the RFM and would be more restrictive than the more general factory information.

The TCDS is a formal description of the aircraft, engine, or propeller. It lists limitations and information required for type certification, including airspeed limits, weight limits, thrust limitations, etc. Assist the student in finding the TCDS on the FAA website and show the student the basis for the building of that helicopter and where the limitations are stated. *[Figure 6-2]*

## Sections of the Manual

### General Information (Section 1)

The general information section provides basic descriptive information on the helicopter and the powerplant and is a good place for the instructor to familiarize the student with the aircraft being flown. Since this section provides the basic dimensions of the aircraft, examples of how the student might use the information can be incorporated into the lesson plan. Discussion of the aircraft's overall dimensions can be used to calculate hangar area required. The diameter of the rotor disk can be used to determine the altitude of aircraft for ground effect.

The dimensions of the helicopter primarily define the amount of space needed for the helicopter's landing and takeoff areas. For example, the hover pad should be the main rotor diameter plus the tailrotor's diameter and some margin for maneuvering error and enough distance to be able to clear obstructions during takeoff and landings. Review the takeoff and landing charts with the student so they have a better understanding of the distances involved. Review the aerodynamics so the student remembers why vertical takeoffs and landings, even if possible, are not advised, and probably are not the safest maneuvers usable.

The instructor should teach the student that for part 27 helicopters, the charts are advisory. Whereas, the charts for a part 29 helicopter may be limiting depending on the verbiage in the limitations section of the RFM. However, should an accident occur, operating outside given acceptable parameters may be grounds for a careless and reckless determination.

### Operating Limitations (Section 2)

This section contains those limitations necessary for the safe operation of the aircraft and should be thoroughly reviewed with the student. Divide this section into subsections for discussion purposes: instrument markings, operating limits, loading limits, and flight limitations; explain each one to the student. Show the student how the information related to instrument markings is depicted on the instruments in the aircraft. Make the information relevant by explaining the markings and how exceeding the limits affects flight.

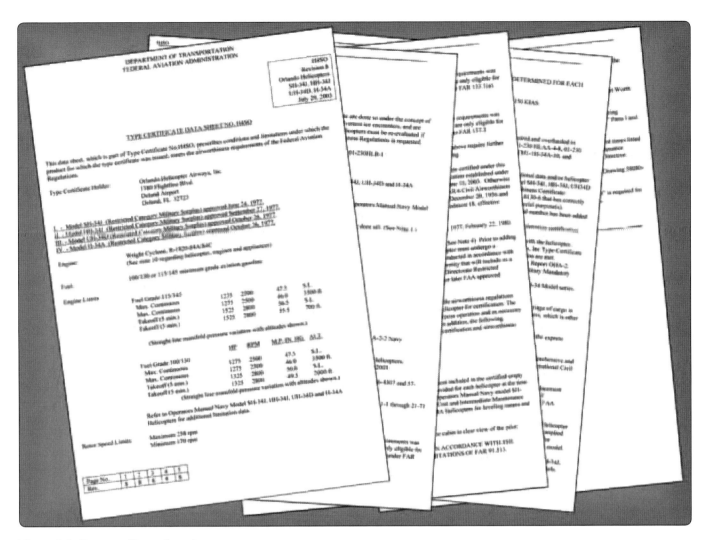

**Figure 6-2.** *Type certificate data sheet.*

Show the student the flight limitations section and discuss how the limitations relate to safe flight. Review the weight and loading distribution section, emphasizing the information concerning maximum certificated weights, as well as the center of gravity (CG) range. Point out any prohibited maneuvers or restrictions to flight.

Do not assume a student knows what a placard is—show an actual placard on the aircraft. Always relate what is written in the RFM to what and where it is on the actual aircraft. First, identify the tachometer and explain the markings. Then, point out which needle indicates engine revolutions per minute (rpm) on which scale and where to read the rotor rpm. *[Figure 6-3]* The student should be well aware of the importance of maintaining rotor rpm at all costs. For example, in dealing with emergency procedures not necessarily documented in the RFM, if the rotor tachometer fails in flight, maintaining powered flight and monitoring the engine tachometer should keep the rotor within the limitations. However, this should never be used to allow takeoff with an inoperative main rotor tachometer. The student should appreciate that if the rotor is in the low green range, then the glide distance will be somewhat farther, but too little

**Figure 6-3.** *Identify and explain all placards to students when instructing on limitations.*

rotor rpm allows the blades to fold or bend upward in flight with no chance of recovery, or little to no cushioning energy during the touchdown phase of an autorotation. Excessively high main rotor rpm can overstress the blade retention parts leading to immediate blade loss or begin stresses which lead to blade loss in the future.

## Emergency Procedures (Section 3)

Remind the student that while flight is generally safe, it does include an element of risk. One of the risks associated with flight is equipment malfunction or failure. These include engine, tail rotor, or system failure, or fire. Explain that the manufacturer has developed procedures for coping with emergencies and includes those procedures in this section. *[Figure 6-4]*

Accidents and helicopter operating history proves that training in emergency procedures in helicopters is beneficial, although sometimes expensive. Experience indicates that helicopter training is much more demanding than airplane training due to the differences in the machines and failure modes. However, with the proper training, helicopter flight can as safe as airplane flight.

Although the FAA does not encourage memory items for checklist procedures, due to the complexity and aerodynamics of the helicopter, the helicopter pilot must be better trained, more attentive, and more responsive than the average airplane pilot. In many situations, the helicopter pilot must react almost instantly and accomplish the necessary items without the aid of a checklist since the emergency may end before a checklist can be located, read, and followed.

For example, helicopters commonly fly at an altitude of approximately 1,000' above ground level (AGL). If an engine or driveline component fails at this altitude, the time remaining aloft is only about 44 seconds for common 4 or 5 place reciprocating-engine powered helicopters with a power off descent rate of 1,350 feet per minute (fpm).

Remind the student that, in those 44 seconds (or less with the entry into autorotation), the pilot must:

1. Achieve autorotation airspeed,
2. Control the rotor rpm,
3. Confirm the wind direction on the surface,
4. Find a suitable landing area,
5. Maneuver into the suitable landing area while missing any obstructions (e.g., wires, fences, trees, and towers),
6. Indentify the actual landing spot during a low, quick, reconnaissance and then

7. Complete the autorotative landing while:
   a. Thinking about a mayday call and
   b. Accomplishing checklist cleanup items (e.g., fuel valve closure, battery switch to off, and activation of the ELT).

All of these tasks must be accomplished while using both hands to fly and keeping the aircraft in trim with the pedals. Imagine the helicopter descending faster than the power-off descent rate of 1,350 fpm, and that most helicopter pilots train in aircraft much lighter than gross weight. In the rare event something fails, that emergency landing may be the pilot's first at that weight.

Using the RFM, show the student how the manufacturer distinguishes the emergency procedures section and makes it easy to find. Review the various types of emergencies included in the RFM and the manufacturer's recommended procedures. Explain that manufacturers identify the immediate actions that aid the pilot in maintaining aircraft control when coping with an emergency. These actions must be completed instinctively, without reference to a checklist. If the situation permits, immediate action should always be followed by completing the full emergency checklist to ensure all items have been covered.

More in-depth discussion of helicopter emergencies can be found in the Helicopter Flying Handbook, Chapter 12, Helicopter Emergencies. Emphasize to the student that procedures in the RFM take precedence when there are differences between the RFM and other publications regarding pilot actions.

## Normal Procedures (Section 4)

A student should understand that the normal procedures section of the RFM is used more than any other section since it includes checklists for all phases of the flight from the preflight inspection of the aircraft to the postflight inspection. Explain to the student how proper use of a checklist ensures manufacturer's procedures are followed. Demonstrate the proper use of the checklist for preflight, engine start through shutdown, and postflight. Remember, an instructor serves as an important role model for the student. Never complete a checklist from memory. Reviewing accidents in which checklist items were missed is a good technique to help the student understand the importance of checklists in day-to-day helicopter operations.

More information concerning normal flight operations and procedures can be found in Chapter 9, Basic Flight Maneuvers, and Chapter 10, Advanced Flight Maneuvers, as well as the Helicopter Flying Handbook.

## EMERGENCY PROCEDURES

**EMER ENG SHUTDWN**
1. ENG PWR CONT lvrs - OFF
2. ENG FUEL SYS sel - OFF
3. FUEL BOOST PUMP switch(es) - OFF

**EMER APU START**
1. FUEL PUMP sw - APU BST
2. APU CONTR sw - ON

**SINGLE ENG FAIL**
1. Collective - Adj to keep RPM
2. Ext cargo/stores - Jett (if reg)
   *(if continued fit is not possible)*
1. Land ASA Poss
   *(if continued fit is possible)*
1. Establish SE airspeed
2. Land ASA Prac

**DUAL ENG FAIL**
1. AUTOROTATE

**DECREASING % RPM R**
1. Collective - Adj to keep RPM
2. ENG PWR CONT lever - LOCKOUT low pwr engine Maint TRQ 10% < good eng
3. Land ASA Prac

**INCR % RPM R**
1. ENG PWR CONT lvr - Retard high eng. Maint TRQ 10% < other eng
2. Land ASA Prac
   *(if affected eng not respond)*
1. Establish SE airspeed
2. EMER ENG SHTDN (aff eng)
3. Refer to SE failure emer proc

**% RPM OSCILLATION**
1. Slowly retard ENG PWR CONT on bad engine
   *(if oscillation stops)*
1. LOCKOUT on bad eng
2. Control eng manually
3. Land ASA Prac
   *(if oscillation continues)*
1. ENG PWR CONT lever beck to FLY
2. Retard ENG PWR CONT lvr on other eng
   *(when oscillation stops)*
1. Place eng in LOCKOUT
2. Control eng manually
3. Land ASA Prac

**% TRQ SPLIT BETWEEN ENGs**
1. Appropriately use ENG PWR CONT lever of bad engine
2. Land ASA Prac

**ENG COMPRS STALL**
1. Collective - Reduce
   *(if stall continues)*
1. ENG PWR CONT lvr - Retard (aff eng)
2. ENG PWR CONT lvr - FLY (aff eng)
   *(if stall recurs)*
1. EMER ENG SHTDN (aff eng)
2. Refer to SE failure emer proc

**ENG OIL FIL BYPASS, ENG CHIP, ENG OIL PRESS HI or LOW, ENG OIL TEMP HIGH, ENG OIL TEMP, ENG OIL PRESS Caution Light On**
1. ENG PWR CONT lvr - Retard to reduce torque aff eng
   *(if oil press low or temp high)*
1. EMER ENG SHTDN (aff eng)
2. Refer to SE failure emer proc

**ENG HIGH SPEED SHAFT FAIL**
1. Collective - Adjust
2. EMER ENG SHTDN (aff eng)
3. Refer to SE failure emer proc

**LIGHTNING STRIKE**
1. ENG PWR CONT lvrs - Adj
2. Land ASA Poss

**LOSS OF T/R THRST**
1. AUTOROTATE
2. ENG PWR CONT lvrs - OFF
   *(when landing point is assured)*

**LOSS OF T/R THRST at LOW A/S or HOVER**
1. Collective - Reduce
2. ENG PWR CONT lvrs - OFF at 5–10 ft

**T/R QUADRANT Light On with Loss of T/R Cont**
1. Collective - Adj
2. Land ASA Prac

**T/R QUADRANT Light On with No loss of T/R**
1. Land ASA Prac

**PEDAL BIND-No Light**
1. Use pedal force to oppose it
2. TRIM sw - Off
   *(if not restored)*
1. BOOST sw - Off
   *(if not restored with boost off)*
1. BOOST sw - ON
   *(if not restored)*
1. TAIL SERVO sw - BACKUP
2. Land ASA Prac

**# 1 T/R RTR SERVO Light On and Back-UP PUMP ON Light Off or #2 T/R SERVO ON Light OFF w/no auto switchover**
1. TAIL SERVO sw - BACKUP
2. BACKUP HYD PUMP sw - ON
3. Land ASA Prac

**MAIN XMSN OIL PRESS Light On/XMSN OIL PRESS LOW/XMSN OIL TEMP HIGH or XMSN OIL TEMP Light On**
1. Land ASA Poss
   *(if time permits)*
1. Slow to 80 kts
2. EMER APU START
3. GEN 1 and 2 switches - OFF

**CHIP INPUT MDL LH or RH Light On**
1. ENG PWR CONT lvr (aff eng) - IDLE
2. Land ASA Poss

**Figure 6-4.** *Example of an emergency procedures checklist.*

## Performance (Section 5)

The performance section contains all the information required by the regulations and any additional performance information the manufacturer determines may enhance a pilot's ability to operate the helicopter safely. When discussing this section, emphasize the importance of the correct use of charts in preflight planning to determine fuel consumption or power or torque available for the given flight conditions. Explain the differences between the in ground effect (IGE) and out of ground effect (OGE) hover charts. The most important function of discussing OGE versus IGE hover charts is the possible performance restriction(s) if the helicopter cannot meet OGE requirements. In those instances, instructors should show the student how to plan approaches more carefully and maneuver the helicopter to maintain ETL at all times. The instructor should ensure the student understands the performance limitations, even if the information is "advisory" and not regulatory limiting.

Ensure the student understands how to read the height/velocity diagram and the importance of the information obtained from the chart, as well as how not to fly in the avoid areas of the chart unless the flight task specifically demands that profile. Again, make the use of the charts meaningful by giving examples of how the information contained in this section is used in determining the parameters of a flight. For example, if the student plans the flight using the highest temperature forecast and the highest altitude expected, then a practical expectation of the helicopter's performance can be derived. If IGE hover is not possible by the charts, then the loss of translational lift will probably result in a landing—desired or not! Always plan to hover IGE. If IGE hover is not possible according to the charts, is the flight really necessary at that weight? The student should remember that a helicopter needs power to stop, come to a hover, and take off to clear obstructions. Maybe less fuel can be carried, or more than one trip can be flown to move the people or weight in a safe manner. Direct the student to the Helicopter Flying Handbook, Chapter 8, Helicopter Performance.

## Weight and Balance (Section 6)

The weight and balance section should contain all the information required by the FAA to calculate weight and balance. Explain to the student the importance of determining the weight and balance of the aircraft for each flight. Weight balance is very important to a helicopter with its limited CG range. More important than the structural maximum gross weight may be the maximum gross weight for the altitude and temperature. Since the helicopter uses almost maximum power for hovering, the loss of power due to thinner air at higher altitudes and temperatures can be especially critical to helicopter operations above cool sea level points.

Show the student how the information provided in this section is used to make weight and balance calculations. Make the information relevant by demonstrating how to complete a weight and balance computation for a training flight. Refer the student to the Helicopter Flying Handbook, Chapter 7, Weight and Balance, or the Pilot's Handbook of Aeronautical Knowledge, Chapter 9, Weight and Balance, for more information regarding weight and balance.

## Aircraft and Systems Description (Section 7)

The Helicopter Flying Handbook, Chapter 5, Helicopter Components, Sections, and Systems, contains a general description of the various systems found on a helicopter. Explain to the student that this section of the RFM contains a specific description of the systems on the aircraft. Emphasize that a good pilot becomes familiar with the systems on the aircraft because detailed knowledge of the systems of an aircraft is essential for determining whether flight is advisable. Explain to the student the best way to become familiar with the systems is to study the information in this section.

## Handling, Servicing, and Maintenance (Section 8)

The handling, servicing, and maintenance section describes the maintenance and inspections recommended by the manufacturer, as well as those required by the regulations, and airworthiness directive (AD) compliance procedures. Explain that an AD is a notification to owners and operators of certificated aircraft that a known safety deficiency with a particular model of aircraft, engine, avionics, or other system exists and must be corrected. If a certificated aircraft has outstanding ADs with which the operator has not complied, the aircraft is not considered airworthy. Thus, it is mandatory for an aircraft operator to comply with an AD. ADs usually result from service difficulty reported by operators or from the results of aircraft accident investigations. They are issued either by the national civil aviation authority of the country of aircraft manufacture or of aircraft registration. When ADs are issued by the country of registration, they are almost always coordinated with the civil aviation authority of the country of manufacture to ensure that conflicting ADs are not issued.

In detail, the purpose an AD is to notify aircraft owners that:

- The aircraft may have an unsafe condition, or

- The aircraft may not be in conformity with its basis of certification or of other conditions that affect the aircraft's airworthiness, or

- There are mandatory actions that must be carried out to ensure continued safe operation, or

- In some urgent cases, the aircraft must not be flown until a corrective action plan is designed and carried out.

ADs are mandatory in most jurisdictions and often contain dates or number of additional aircraft flying hours by which compliance must be completed. ADs may be divided into two categories; those of an emergency nature requiring immediate compliance prior to further flight, and those of a less urgent nature requiring compliance within a specified period of time. The student should know where the AD compliance record is located in the maintenance logs of the helicopter and the proper way to check it. Owner/operators of an aircraft must now request to receive ADs electronically as the ADs are no longer mailed to owners of record; however, the owner/operators are still responsible for AD compliance.

Acquaint the student with how this section outlines the service, maintenance, and inspection intervals for the different components of the aircraft. Point out that it also establishes time-between-overhaul (TBO) limits for components, such as rotor blades and gear boxes.

Explain to the student that this section also describes the preventive maintenance required, as well as ground handling and storage procedures. The pilot should always know what type and brand of oils are being used. The RFM may specify what is acceptable, but it is best to not mix brands in engines and transmissions/gearboxes.

## Supplements (Section 9)

When acquainting the student with the supplements section, explain that this section provides pertinent information necessary for the installation and operational considerations of optional equipment, such as floats or spray equipment. *[Figure 6-5]* Stress that the information may be provided by the aircraft manufacture or the optional equipment manufacturer, and that in either case, it becomes part of the RFM when the equipment is installed. Other important information located in the supplement section includes any changes required in the normal, abnormal, and emergency checklists; and servicing, preflight or maintenance requirements. Additions or supplements should always be placed in the Supplement Section of the flight manual so that the owner/operator maintains all additions and changes in

**Figure 6-5.** *A helicopter using spray equipment.*

one place rather than creating alternate checklists that could cause important information to be misplaced or forgotten.

## Instructor Tips

- Much of the information in this chapter lends itself to oral assessment of fact questions (discussed in Chapter 5, Assessment, from the Aviation Instructor's Handbook) based on memory or recall.

- To engage the interest of the student, make the information as relevant as possible to the flying experience. *[Figure 6-6]*

---

### Rotorcraft Flight Manual

**Objective**

The purpose of this lesson plan is to explain the relationship between the RFM and safe procedures and introduce the sections of the RFM.

**Content**

1. Discuss lesson objective and completion standards.
2. Review the sections of the RFM, and discuss their relevance to safe flight.
3. Discuss that all flights and ground procedures should remain within established parameters.
4. Explain that any questions to the written procedures should not be discussed while flying but on the ground with legitimate and authoritative sources.
5. Discuss the importance of advisory information to safe flights

**Postflight Discussion**

Preview and assign the next lesson. Assign *Helicopter Flying Handbook,* Chapter 6, Weight and Balance.

---

**Figure 6-6.** *Sample lesson plan.*

## Chapter Summary

This chapter provided discussion points for the instructor to help the student learn the important limitations stated in the RFM. It also explained how to assist the student in understanding the importance of preflight planning in regard to safe flights, as well as recognizing the critical steps in the published emergency procedures. The applicable 14 CFR regulations were discussed and how the relationships integrate to satisfy safety regulations and safe flight practices.

# Chapter 7
# Weight and Balance

## Introduction

This chapter is intended to help the instructor to teach the student the basics of helicopter weight and balance and how to compute the weight and balance forms. Definitions are abbreviated slightly, as the instructor's experience is recognized when working with the weight and balance form and records. It is not intended to instruct the student on the actual weighing of aircraft.

The student must understand the reasons for weight and balance control. Listed in *Figure 7-1* are some of the important topics of discussion. Discuss the list with the student. Refer the student to the Aircraft Weight and Balance Handbook, FAA-H-8083-1, as well as the Rotorcraft Flight Manual for that specific helicopter.

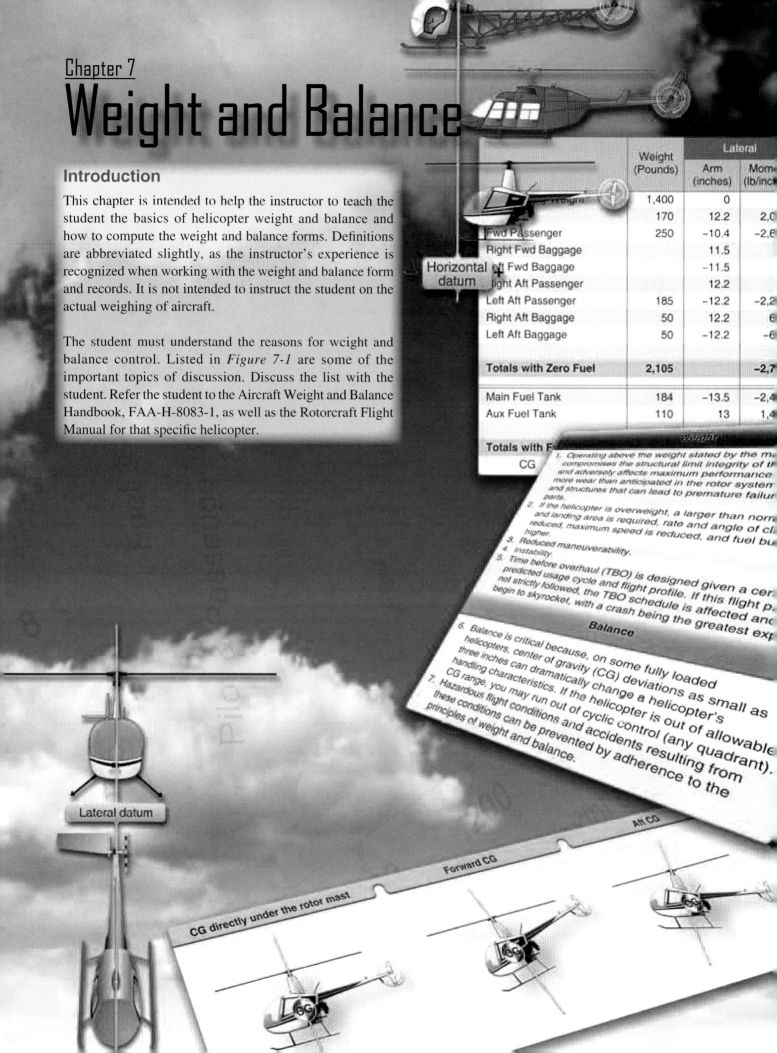

| | Weight (Pounds) | Lateral | |
| --- | --- | --- | --- |
| | | Arm (inches) | Moment (lb/inch) |
| | 1,400 | 0 | |
| | 170 | 12.2 | 2,0 |
| Fwd Passenger | 250 | −10.4 | −2,6 |
| Right Fwd Baggage | | 11.5 | |
| Left Fwd Baggage | | −11.5 | |
| Right Aft Passenger | | 12.2 | |
| Left Aft Passenger | 185 | −12.2 | −2,2 |
| Right Aft Baggage | 50 | 12.2 | 6 |
| Left Aft Baggage | 50 | −12.2 | −6 |
| **Totals with Zero Fuel** | **2,105** | | **−2,7** |
| Main Fuel Tank | 184 | −13.5 | −2,4 |
| Aux Fuel Tank | 110 | 13 | 1,4 |

Horizontal datum

Lateral datum

CG directly under the rotor mast

Forward CG

Aft CG

**Weight**

1. Operating above the weight stated by the ma compromises the structural limit integrity of th and adversely affects maximum performance more wear than anticipated in the rotor system and structures that can lead to premature failur parts.
2. If the helicopter is overweight, a larger than norm and landing area is required, rate and angle of cli reduced, maximum speed is reduced, and fuel bu higher.
3. Reduced maneuverability.
4. Instability.
5. Time before overhaul (TBO) is designed given a cer predicted usage cycle and flight profile. If this flight p not strictly followed, the TBO schedule is affected and begin to skyrocket, with a crash being the greatest exp

**Balance**

6. Balance is critical because, on some fully loaded helicopters, center of gravity (CG) deviations as small as three inches can dramatically change a helicopter's handling characteristics. If the helicopter is out of allowable CG range, you may run out of cyclic control (any quadrant).
7. Hazardous flight conditions and accidents resulting from thesu conditions can be prevented by adherence to the principles of weight and balance.

| Weight |
|---|
| 1. Operating above the weight limit stated by the manufacturer compromises the structural integrity of the helicopter and adversely affects maximum performance. It causes more wear than anticipated in the rotor system, driveline, and structures that can lead to premature failure of a part or parts. |
| 2. If the helicopter is overweight, a unknown larger than normal takeoff and landing area is required, rate and angle of climb are reduced by an unknown value, maximum speed is reduced by an unknown value, and fuel burn rate is higher. |
| 3. Reduced maneuverability. |
| 4. Instability. |
| 5. Time before overhaul (TBO) is designed given a certain predicted usage cycle and flight profile. If this flight profile is not strictly followed, the TBO schedule is affected and costs begin to skyrocket, with a crash being the greatest expense! |

| Balance |
|---|
| 6. Balance is critical because, on some fully loaded helicopters, center of gravity (CG) deviations as small as three inches can dramatically change a helicopter's handling characteristics. If the helicopter is out of allowable CG range, you may run out of cyclic control (any quadrant). |
| 7. Hazardous flight conditions and accidents resulting from these conditions can be prevented by adherence to the principles of weight and balance. |

Note: The responsibility for proper weight and balance control begins with the helicopter engineers and designers and extends to the aircraft mechanics who maintain the aircraft and the pilots who operate them. Discuss with the student the importance of staying below the maximum weight of the helicopter. Explain that the Laws of Physics are unbreakable; attempts to break these laws through poor decisions such as overloading or simply an error in calculations will instead result in the breakage of the helicopter. Weight is important because it is the basis for determining the maneuvering load with the G loading factored onto the rotor system and transmission. Explain to the student that the manufacturer, through the engineering and testing process, has developed limitations to be applied for the safe operation of the helicopter. Adherence to these limitations will avoid placing excessive stress and fatigue on weak points throughout the helicopter. The student should not assume what the weak point or limiting factor is to exceed selected limitations. Due to the distance and complexity of the drive train, the antitorque system in many helicopters may be the actual unstated weak point. Too much antitorque thrust over a period of time has led to failures of the tail rotor blades, gearboxes, pylon, and tail boom attachment points.

**Figure 7-1.** *Weight and balance topics for an instructor to use in discussions with students.*

# Weight

The weight of a helicopter is critical during any and all flight maneuvers. This chapter primarily considers the weight of the loaded helicopter while at rest.

## Definitions

Discuss all the weight and balance terms with the student which can be found in the Aircraft Weight and Balance Handbook, FAA-H-8083-1. The student should have an understanding of the terms in order to consistently and successfully monitor weight and balance and complete weight and balance forms.

Never intentionally exceed the load limits for which a helicopter is certificated. Operating above the maximum weight could result in structural deformation or failure during flight if excessive load factors, strong wind gusts, or turbulence were encountered, as well as operating below a minimum crew weight could adversely affect the handling characteristics of the helicopter. Operations at or below the minimum weight of the helicopter can also affect the autorotational characteristics of the helicopter.

Other factors to consider when computing weight and balance are high altitude, high temperature, and high humidity conditions, which result in a high density altitude. As density altitude increases, more power is required. Any adjustment to gross weight by varying fuel, payload, or both, affects the power required. For this reason, the maximum operational weight may be less than the maximum allowable weight. In-depth performance planning is critical when operating in these conditions.

Most small helicopters have some type of seating limitations. Two examples are the Robinson R-22 and Robinson R-44. The R-22 has a 240-pound seat limit (includes any cargo below the seat). The R-44 has a 300-pound limit with a 50-pound cargo limit included within the 300 pounds. Therefore, if 50 pounds of cargo were placed under a seat, the passenger or pilot could weigh no more than 250 pounds. Also, the external load limit includes the weight of the load and the lifting slings and hardware.

## Determining Empty Weight

A helicopter's weight and balance records contain essential data, including a complete list of all installed optional equipment. Use these records to determine the weight and balance condition of the empty helicopter. Lead the student through a weight and balance problem, as well as preflight planning exercise before most flights. Once the student is proficient at weight and balance, then he or she should be allowed to reuse the previous planning information (provided that the conditions did not change).

When a helicopter is delivered from the factory, the basic empty weight, empty weight CG, and useful load are recorded on a weight and balance data sheet included in the Federal Aviation Administration (FAA)-approved Rotorcraft Flight Manual (RFM). If equipment is removed, replaced, or additional equipment installed, these changes must be reflected in the weight and balance records. Major repairs or alterations must be recorded by a certificated mechanic. When the revised weight and moment are recorded on a new form, the old record is marked with the word "superseded" and dated with the effective date of the new record.

## Balance

Some helicopters may be properly loaded for takeoff, but near the end of a long flight with almost empty fuel tanks, the CG may have shifted enough for the helicopter to be out of balance laterally or longitudinally. Before making any long flight, the CG with the fuel available for landing must be checked to ensure it is within the allowable range. It is essential to load the aircraft within the allowable CG range specified in the RFM's weight and balance limitations.

### Center of Gravity (CG)

The CG is defined as the theoretical point where all of the aircraft's weight is considered to be concentrated. Improper balance of a helicopter's load can result in serious control problems.

The allowable range in which the CG may fall is called the CG range. The exact CG location and range are specified in the RFM for each helicopter. In addition to making a helicopter difficult to control, an out-of-balance loading condition also decreases maneuverability since cyclic control is less effective in the direction opposite to the CG location.

Changing the CG changes the angle at which the aircraft hangs from the rotor. When the CG is directly under the rotor mast, the helicopter hangs horizontally; if the CG is too far forward of the mast, the helicopter hangs with its nose tilted down; if the CG is too far aft of the mast, the nose tilts up. *[Figure 7-2]* Discuss with the student some of the reactions of the helicopter and problems associated with a forward CG and an aft CG.

### CG Forward of Forward Limit

Forward CG may occur when a heavy pilot and passenger take off without baggage or proper ballast located aft of the rotor mast. This situation becomes worse if the fuel tanks are located aft of the rotor mast because as fuel burns the CG continues to shift forward.

Teach the student to assess the helicopter control response prior to each flight. Have the student get the helicopter light on the skids/gear and ensure helicopter is free of surface obstructions or attachments and will ascend to a hover in a nearly level attitude. Ensure that there is enough cyclic control to continue. Once at a low hover, ensure the helicopter remains nearly level. Point out to the student what a normal hover attitude looks like, and that if things do not feel or look right, then to lower the collective slowly and land the helicopter. Attempt to determine why the helicopter is responding in such a way. Adjustment or reduction of load may be necessary.

Do not continue flight in this condition. Decelerating the helicopter in this condition may be very difficult or impossible as well. In the event of engine failure and the resulting autorotation, there may not be enough cyclic control to flare properly for the landing.

### CG Aft of Aft Limit

Exceeding aft CG may occur when:

- A lightweight pilot takes off solo with a full load of fuel located aft of the rotor mast.

- A lightweight pilot takes off with maximum baggage allowed in a baggage compartment located aft of the rotor mast.

- A lightweight pilot takes off with a combination of baggage and substantial fuel, both of which are aft of the rotor mast.

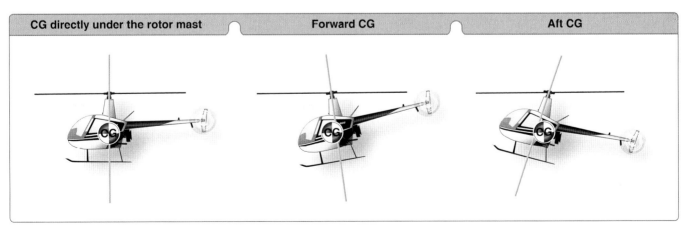

**Figure 7-2.** *The location of the CG strongly influences how the helicopter handles.*

Easily recognized when coming to a hover, the helicopter has a tail-low attitude, and needs excessive forward displacement of cyclic control to maintain a hover in a no-wind condition. If there is a wind, even greater forward cyclic is needed.

NOTE: One technique used in smaller helicopters is to show the student what a tail-low (nose high) attitude looks like. To do this, seat the student (pilot seat) in a parked helicopter (engine off). Pull the tail boom down or lift the front skids until the tail stinger or guard touches ground. Tell the student to commit this helicopter position to memory and never allow the helicopter to achieve this attitude while in flight.

Gusty or rough air could accelerate the helicopter to a speed faster than that produced with full forward cyclic control. In this case, dissymmetry of lift and blade flapping could cause the rotor disk to tilt aft. With full forward cyclic control already applied, a pilot might not be able to lower the rotor disk, resulting in possible loss of control or rotor blades striking the tail boom.

### Lateral Balance

Discuss with the student the problems associated with lateral balances.

With small helicopters, it is generally unnecessary to determine the lateral CG for normal flight instruction and passenger flights. The cabins are relatively narrow and most optional equipment is located near the centerline. If there is an unusual situation that could affect the lateral CG, such as a heavy pilot and a full load of fuel on one side of the helicopter, its position should be checked against the CG envelope. Certain types of helicopters require a pilot to verify the lateral balance often, such as external hoist operations (hoist located on the side of the helicopter).

Helicopter engineers determine the amount of cyclic control that is available. They establish both the longitudinal and lateral CG envelopes. This allows the pilot to load the helicopter and still have sufficient cyclic control for all flight conditions. Operating the helicopter outside these limits is inadvisable and control of the helicopter may not be possible in some conditions.

### Ballast

Ballast is some form of weight placed in a specific location in an aircraft intended to maintain the CG within limits by compensating for unfavorable weight and balance conditions, thereby ensuring correct control margins. Two types are permanent ballast and temporary ballast.

1. Permanent ballast—the removal or addition of equipment to the helicopter has an effect on aircraft weight and balance. It may be necessary to install

ballast weights to maintain the CG position within the CG limits.

2. Temporary ballast—such weights that may be necessary to compensate for missing crewmembers or equipment in order to maintain CG position. The amount and location of temporary ballast required to maintain safe flight is determined by the pilot or the airframe and powerplant (A&P) mechanics. Ensure that any ballast is properly secured or strapped down to prevent movement. When ballast moves, it compounds the CG problem instead of relieving it.

## Weight and Balance Calculations

To determine whether a helicopter is properly loaded, ask the following questions:

1. Is the gross weight less than or equal to the maximum allowable gross weight?

   Answer: Add the weights of the items comprising the useful load to the basic empty weight of the helicopter (pilot, passengers, fuel, oil (if applicable), cargo, and baggage).

2. Is the CG within the allowable CG range, and will it stay within the allowable range throughout the duration of flight including all loading configurations that may be encountered?

   Answer: Use CG or moment information from loading charts, tables, or graphs in the RFM. Then using one of the methods described below, calculate the loaded moment and/or loaded CG and verify that it falls within the allowable CG range shown in the RFM.

   NOTE: Discuss with the student which direction the CG shifts as fuel is burned. On some helicopters, the fuel tanks are behind the CG, causing it to shift forward as fuel is used. Under some flight conditions, the balance may shift enough that there is not sufficient cyclic authority to flare for landing. For these helicopters, the loaded CG should be computed for both takeoff and landing weights.

   Use CG or moment information from loading charts, tables, or graphs in the Rotorcraft Flight Manual (RFM). Calculate the loaded moment and/or loaded CG and verify that it falls within the allowable CG range shown in the RFM.

It is important to note that any weight and balance computation is only as accurate as the information provided. Therefore, determine passenger weight and add a few pounds to account for the additional weight of clothing, especially during the winter months. The baggage weight should be weighed on a scale, if practical. If a scale is not available, compute personal loading values according to each individual estimate.

## Weight Versus Aircraft Performance

Overloading an aircraft may cause structural failure or result in reduced engine and airframe life. An increase in gross weight has the following effects on aircraft performance:

a. Increases takeoff distance

b. Reduces hover performance

c. Reduces rate of climb

d. Reduces cruising speed

e. Reduces maneuverability

f. Reduces ceiling

g. Reduces range

h. Increases landing distances

NOTE: Ensure that the student understands the importance of not overloading the helicopter (operating at maximum weights) and also understands how the helicopters weight versus performance is also affected. Accurate performance planning based on current and forecast environmental conditions is critical in determining the helicopter's actual maximum operational weight for any given set of conditions

*Figure 7-3* indicates the standard weights for specific operating fluids.

| | |
|---|---|
| Aviation Gasoline (Avgas) | 6 lb/gal |
| Jet Fuel (JP-4) | 6.5 lb/gal |
| Jet Fuel (JP-5) | 6.8 lb/gal |
| Reciprocating Engine Oil | 7.5 lb/gal*(1.875 lb/qt) |
| Turbine Engine Oil | |
| ... Varies between 7.5 and 8.5 lb/gal*(1.875 and 2.125 lb/qt) | |
| Water | 8.35 lb/gal |

*Outside of this chart, oil weight is generally given in pounds per gallon (lb/gal) while oil capacity is usually given in quarts; therefore, convert the amount of oil to gallons before calculating its weight. Remember, four quarts equal one gallon.

**Figure 7-3.** *When making weight and balance computations, always use actual weights if they are available, especially if the helicopter is loaded near the weight and balance limits.*

## Reference Datum

*Figure 7-4* indicates some of the different locations of the horizontal reference datum. Balance is determined by the location of the CG, which is usually described as a given number of inches from the reference datum. The horizontal reference datum is an imaginary vertical plane or point arbitrarily fixed somewhere along the longitudinal axis of the helicopter from which all horizontal distances are measured for weight and balance purposes. There is no fixed rule for its location; it may be located at the rotor mast, the nose of the

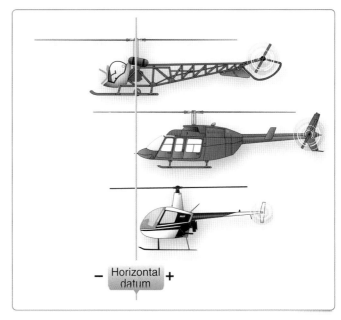

**Figure 7-4.** *While the horizontal reference datum can be anywhere the manufacturer chooses, some manufacturers choose the datum line at or ahead of the most forward structural point on the helicopter, in which case all moments are positive. This aids in simplifying calculations. Other manufacturers choose the datum line at some point in the middle of the helicopter, in which case moments produced by weight in front of the datum are negative and moments produced by weight aft of the datum are positive.*

helicopter, or even at a point in space ahead of the helicopter. The lateral reference datum is usually located at the center of the helicopter. The location of the reference datum is established by the manufacturer and is defined in the RFM. *[Figure 7-5]*

The lateral CG is determined in the same way as the longitudinal CG, except the distances between the scales and butt line zero (BL 0) are used as the arms. Arms to the right of BL 0 are positive and those to the left are negative. The BL 0 (sometimes referred to as the buttock) is a line through the symmetrical center of an aircraft from nose to tail. It serves as the datum for measuring the arms used to find the lateral CG. Lateral moments that cause the aircraft to rotate clockwise are positive (+), and those that cause it to rotate counterclockwise are negative (–).

## Arm (Station)

The horizontal distance from the datum to any component of the helicopter or to any object located on the helicopter is called the "arm" or "station."

## Moment

If the weight of an object is multiplied by its arm, the result is known as its moment. Think of a moment as a force that results from an object's weight acting at a distance. Moment

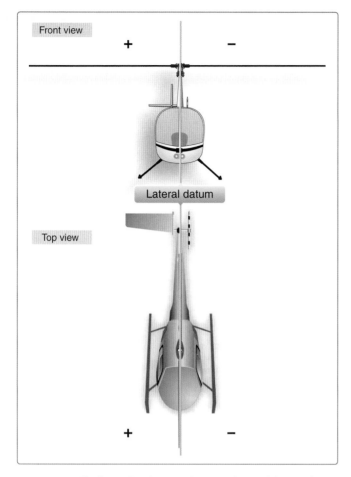

**Figure 7-5.** *The lateral reference datum is located longitudinally through the center of the helicopter; therefore, there are positive and negative values.*

is also referred to as the tendency of an object to rotate or pivot about a point. The farther an object is from a pivotal point, the greater its force.

## Weight and Balance Methods

Discuss with the student the two different methods of obtaining the weight and balance of the helicopter: computational and loading chart.

### *Computational Method*

The first method is the computational method, which uses simple mathematics to solve weight and balance problems.

1. Ascertain the total weight and total moment of the helicopter and ensure it does not exceed the maximum allowable weight under existing or forecast conditions. In this case, the total weight of the helicopter is under the maximum gross weight of 3,200 pounds.

   *Tip:* The empty weight CG can be considered the arm of the empty helicopter. Use care in recording the weight of each passenger and baggage. Recording each weight in its proper location is extremely important for the accurate calculation of a CG.

2. Now, divide the total moment by the total weight. This gives the CG of the loaded helicopter.

The bottom of the chart in *Figure 7-6* lists the CG range for this particular helicopter. Take the CG figure and compare it to the allowable limits of 106.0 inches to 114.2 inches. The CG's location of 109.9 inches is within the acceptable range. NOTE: If the CG falls outside the acceptable limits, adjust the loading of the helicopter.

| | Weight (pounds) | Arm (inches) | Moment (lb-in) |
|---|---|---|---|
| Basic Empty Weight | 1,700 | 116.5 | 198,050 |
| Oil | 12 | 179.0 | 2,148 |
| Pilot | 190 | 65.0 | 12,350 |
| Forward Passenger | 170 | 65.0 | 11,050 |
| Passengers Aft | 510 | 104 | 53,040 |
| Baggage | 40 | 148 | 5,920 |
| Fuel | 553 | 120 | 66,360 |
| **Total** | **3,175** | | **348,918** |
| CG (loaded) | | 109.9 | |
| Max Gross Weight = 3,200 lb  CG Range = 106.0–114.2 inches | | | |

**Figure 7-6.** *In this example, the helicopter's weight of 1,700 pounds is recorded in the first column, its CG or arm of 116.5 inches in the second, and its moment of 198,050 lb-in in the last. Notice that the weight of the helicopter multiplied by its CG equals its moment.*

If adjustments need to be made to the loading, refer the student to the Aircraft Weight and Balance Handbook, FAA-H-8083-1.

### *Loading Chart Method*

The second method is the loading chart method. Three sample problems are given for this method. Take time to ensure the student fully understands the weight and balance concept and computations.

*Figure 7-7* is an example of a loading chart. To use this chart, a pilot must:

1. Subtotal the empty weight, pilot, and passengers. This is the weight at which to enter the chart on the left.

2. The next step is to follow the upsloping lines for baggage and then for fuel to arrive at a final weight and CG.

NOTE: Any value on or inside the envelope is within the range.

*Sample Problem 1*

The student may not understand how or why the numbers are entered on the chart. (Explain the numbers to the student. Most manufacturers provide samples with their charts;

figuring the charts out is a matter of inputting your own numbers.) The loading chart used in sample problems 1 and 2 is designed to calculate the loaded CG graphically and show whether it is within limits, all on a single chart.

Determine if the gross weight and CG are within allowable limits under the following loading conditions for a helicopter based on the loading chart in *Figure 7-7*. To use the loading chart for the helicopter in this example, add up the items in a certain order. The maximum allowable gross weight is 1,600 pounds.

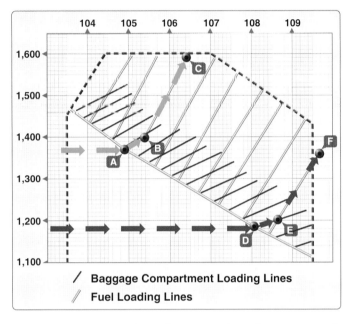

**Figure 7-7.** *Loading chart illustrating the solution to sample problems 1 and 2.*

NOTE: In *Figure 7-7*, the helicopter weights are located on the left side of the chart and the CG limits are located on the top of the chart.

| ITEM | WEIGHT (LB) |
| --- | --- |
| Basic empty weight | 1,040 |
| Pilot | 135 |
| Passenger | 200 |
| Subtotal | 1,375 (point A) |

1. Follow the green arrows in *Figure 7-7*. Enter the graph on the left side at 1,375 lb., the subtotal of the empty weight plus the pilot and passenger weights. Move right to Point A.

| Baggage compartment load | 25 lb |
| --- | --- |
| Subtotal | 1,400 lb (Point B) |

2. Move up and to the right, parallel to the baggage compartment loading lines to 1,400 lb (Point B).

| Fuel load (30 gallons) | 180 lb |
| --- | --- |
| Total weight | 1,580 lb (Point C) |

3. Continue up and to the right, this time parallel to the fuel loading lines, to the total weight of 1,580 lb (Point C).

Point C is within allowable weight and CG limits.

*Sample Problem 2*

Assume that the pilot in sample problem 1 discharges the passenger after using only 20 lb of fuel.

| ITEM | WEIGHT (LB) |
| --- | --- |
| Basic empty weight | 1,040 |
| Pilot | 135 |
| Subtotal | 1,175 (Point D) |

1. Follow the purple arrows in *Figure 7-7*, starting at 1,175 lb on the left side of the graph, then to Point D.

| Baggage compartment load | 25 lb |
| --- | --- |
| Subtotal | 1,200 lb (Point E) |

2. Continue to Point E.

| Fuel load | 160 lb |
| --- | --- |
| Total weight | 1,360 lb (Point F) |

3. Continue to Point F.

Note the total weight of the helicopter is well below the maximum allowable gross weight; however, point F falls outside the aft allowable CG limit. As standard practice, compute the weight and balance with zero fuel to verify that a helicopter remains within the acceptable limits as fuel is used. It is imperative that weight and CG be within allowable limits for all phases of the flight or successive flights. Weight and CG change as loading configurations change; therefore, more than one weight and balance calculation needs to be accomplished. Initial calculations are based upon a fully loaded aircraft for the first takeoff; subsequent calculations address loading configuration changes, such as passenger drop off, fuel consumption and refueling. Addressing calculations for all projected loading configurations ensures that weight and CG remain within limits during all phases of the flight.

Discuss with the student some options he or she could use in order to fix the out-of-CG condition.

*Sample Problem 3*

Calculate moments for each station.

| ITEM | WEIGHT (LB) | MOMENT |
|------|-------------|--------|
| Basic empty | 1,102 | 10.8 |
| Pilot and front passenger* | 340 | 28.3 |
| Fuel* | 211 | 22.9 |
| Baggage | 0 | 0 |
| Total | 1,653 | 162.0 |

*Use chart in *Figure 7-8*.

1. Record the basic empty weight and moment.

2. Record the weights of the pilot, passengers, fuel, and baggage on a weight and balance worksheet.

3. Determine the total weight of the helicopter.

4. Determine if the helicopter weight is within limits. (load limit on bottom of chart in *Figure 7-9*)

5. Determine the moments for a pilot and passenger (340 lb) and fuel (211 lb). *[Figure 7-8]*

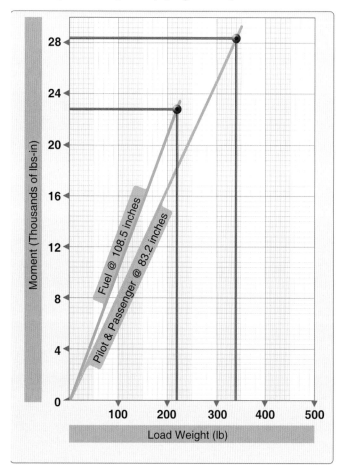

**Figure 7-8.** *Moments for fuel, pilot, and passenger.*

6. Using *Figure 7-8*, start at the bottom scale labeled load weight. Draw a line from 211 lb up to the line labeled "Fuel—station 108.5." Draw a line to the left

|  | Weight (lb) | Moment (lb-in/1,000) |
|--|------|--------|
| Basic Empty Weight | 1,102 | 110.8 |
| Pilot and Front Passenger | 340 | 28.3 |
| Fuel | 211 | 22.9 |
| Baggage | | |
| Total | 1,650 | 162.0 |

**Figure 7-9.** *CG/Moment chart.*

to intersect the moment scale and read the fuel moment (22.9 thousand lb-in).

7. Do the same for the pilot/passenger moment. Draw a line from a weight of 340 lb up to the line labeled "Pilot & passenger—station 83.2." Go left and read the pilot/passenger moment (28.3 thousand lb-in).

8. After recording the basic empty weight and moment of the helicopter and the weight and moment for each item, total and record all weights and moments.

9. Plot the calculated takeoff weight and moment on the sample moment envelope graph. Based on a weight of 1,653 lb and a moment/1,000 of 162 lb-in, the helicopter is within the prescribed CG limits.

NOTE: Reduction factors are often used to reduce the size of large numbers to manageable levels. In *Figure 7-8*, the scale on the loading graph gives moments in thousands of pound-inches. In most cases, when using this type of chart, a pilot need not be concerned with reduction factors because the CG/moment envelope chart normally uses the same reduction factor. *[Figure 7-9]*

### Combination Method

The combination method usually uses the computation method to determine the moments and CG. Then, these figures are plotted on a graph to determine if they intersect within the acceptable envelope.

The example in *Figure 7-10* illustrates that with a total weight of 2,399 pounds and a total moment of 225,022 lb-in, the CG is 93.8. Plotting this CG against the weight indicates that the helicopter is loaded within the longitudinal limits (Point A).

## Calculating Lateral CG

Some helicopter manufacturers require that pilots also determine the lateral CG limits. Lateral balance of an airplane is usually of little concern and is not normally calculated. But some helicopters, especially those equipped for hoist operations, are sensitive to the lateral position of the CG, and the Pilot's Operating Handbook (POH) includes both longitudinal and lateral CG envelopes, as well as information on the maximum permissible host load.

These calculations are similar to longitudinal calculations. However, since the lateral CG datum line is almost always defined as the center of the helicopter, a pilot is likely to encounter negative CGs and moments in the calculations. Negative values are located on the left side while positive stations are located on the right.

When completing the steps below to calculate lateral CG, refer to *Figure 7-11*.

1. When computing moment for the pilot, 170 lb is multiplied by the arm of 12.2 inches, resulting in a moment of 2,074 lb-in. As with any weight placed right of the aircraft centerline, the moment is expressed as a positive value.

| | Weight (pounds) | Lateral Arm (inches) | Lateral Moment (lb-in) |
|---|---|---|---|
| Basic Empty Weight | 1,400 | 0 | 0 |
| Pilot | 170 | 12.2 | 2,074 |
| Fwd Passenger | 250 | –10.4 | –2,600 |
| Right Fwd Baggage | | 11.5 | 0 |
| Left Fwd Baggage | | –11.5 | 0 |
| Right Aft Passenger | | 12.2 | 0 |
| Left Aft Passenger | 185 | –12.2 | –2,257 |
| Right Aft Baggage | 50 | 12.2 | 610 |
| Left Aft Baggage | 50 | –12.2 | –610 |
| **Totals with Zero Fuel** | **2,105** | | **–2,783** |
| Main Fuel Tank | 184 | –13.5 | –2,484 |
| Aux Fuel Tank | 110 | 13 | 1,430 |
| **Totals with Fuel** | **2,399** | | **–3,837** |
| CG | | –1.6 | |

**Figure 7-11.** *Computed lateral CG.*

2. The forward passenger sits left of the aircraft centerline. To compute this moment, multiply 250 lb by –10.4 inches. The result is a moment of –2,600 lb-in.

3. Once the aircraft is completely loaded, the weights and moments are totaled and the CG is computed. Since more weight is located left of the aircraft centerline,

| | Weight (pounds) | Longitudinal Arm (inches) | Longitudinal Moment (lb/in) |
|---|---|---|---|
| Basic Empty Weight | 1,400 | 107.75 | 150,850 |
| Pilot | 170 | 49.5 | 8,415 |
| Forward Passenger | 250 | 49.5 | 12,375 |
| Right Forward Baggage | | 44 | 0 |
| Left Forward Baggage | | 44 | 0 |
| Right Aft Passenger | | 79.5 | 0 |
| Left Aft Passenger | 185 | 79.5 | 14,708 |
| Right Aft Baggage | 50 | 79.5 | 3,975 |
| Left Aft Baggage | 50 | 79.5 | 3,975 |
| **Totals without fuel** | **2,105** | | **194,298** |
| Main Fuel Tank | 184 | 106 | 19,504 |
| Auxiliary Fuel Tank | 110 | 102 | 11,220 |
| **Totals with fuel** | **2,399** | | **225,022** |
| CG | | 93.8 | |

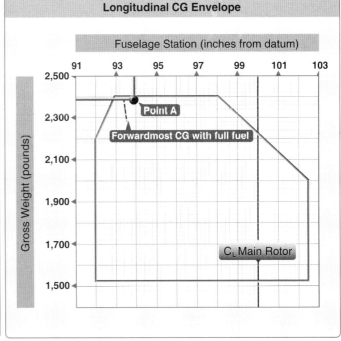

**Figure 7-10.** *Use the computed longitudinal CGs (left) and the longitudinal CG envelope (right) to determine if the helicopter is loaded properly.*

the resulting total moment is –3,837 lb-in. To calculate CG, divide –3,837 lb-in by the total weight of 2,399 lb. The result is –1.6 inches, or a CG that is 1.6 inches left of the aircraft centerline.

Lateral CG is often plotted against the longitudinal CG. *[Figure 7-12]* In this case, –1.6 is plotted against 93.8, which was the longitudinal CG determined in the previous problem. The intersection of the two lines falls well within the lateral CG envelope.

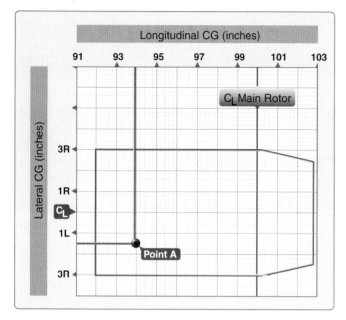

**Figure 7-12.** *Use the lateral CG envelope to determine if the helicopter is properly loaded.*

## Instructor Tips

•   Review all the terms associated with weight and balance with the student.

•   Review the sample weight and balance problems with the student.

•   Ensure that the student understands weight and balance must remain within limits after all loading, unloading, refueling, and fuel consumption.

•   Ensure that the student understands current and forecast environmental conditions affect the performance of the aircraft, thereby affecting weight and loading limitations.

•   Review and practice with the student the actual weight and balance forms for the helicopter to be flown.

•   Ensure the student understands the charts by using weights and/or balances that may put the helicopter in an out-of-weight or out-of-balance condition.

•   The more the student works the weight and balance figures, the more proficient the student will become at computing weight and balance. *[Figure 7-13]*

## Chapter Summary

This chapter explained how to introduce weight and balance to the student, how to read the charts necessary for computing weight and balance of the helicopter, and definitions of common terms associated with weight and balance.

## Weight and Balance

### Objective

The purpose of this lesson plan is to give the student the solid understanding of weight and balance concepts (including computations) necessary for safe flight.

### Content

1. Preflight discussion: discuss lesson objective and completion standards.

2. Review terms associated with weight and balance.

3. Instructor actions:
   a. In a classroom environment, review all terms associated with weight and balance.
   b. Discuss with the student the difference between the computational method and the combinational method.
   c. Have the student review and practice the computations for sample problems 1, 2, and 3.
   d. Once the student has a understanding of the contents of chapter 7, introduce that student to the actual weight and balance computations for his or her helicopter using the FAA-approved Rotorcraft Flight Manual.
   e. Perform a check on learning by giving the student known figures that may place the helicopter in a out of weight or balance condition.
   f. Ensure the student knows what to do if the helicopter is in an out-of-weight or out-of-balance condition.

4. Student actions:
   a. Study the terms associated with weight and balance.
   b. Be able to compute weight and balance for the helicopter that is being operated.
   c. Be prepared to discuss with the instructor your understanding of the helicopter weight and balance forms.
   d. Be prepared to discuss with the instructor what actions you can take if your helicopter is overweight/underweight or in an out-of-balance condition.

### Postflight Discussion

- Review what was covered during this phase of training.
- Preview and assign the next lesson. Assign *Helicopter Flying Handbook*, Chapter 7, Helicopter Performance.

**Figure 7-13.** *Sample lesson plan.*

# Chapter 8
# Helicopter Performance

## Introduction

It is imperative for the student to realize safety correlates directly to a comprehensive understanding of performance planning. Performance planning incorporates engine power available and rotor system efficiency to establish helicopter performance values. The student must be able to recognize the impact of density altitude, helicopter weight, and other environmental factors in order to maximize each helicopter's unique capabilities, while understanding its limitations. Failure to conduct performance planning properly may lead to disastrous consequences.

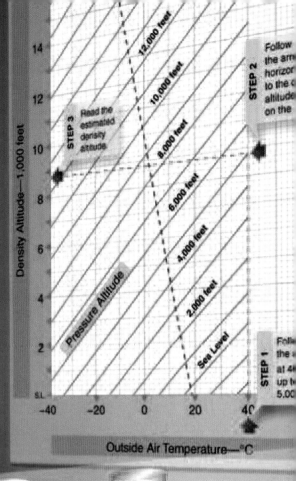

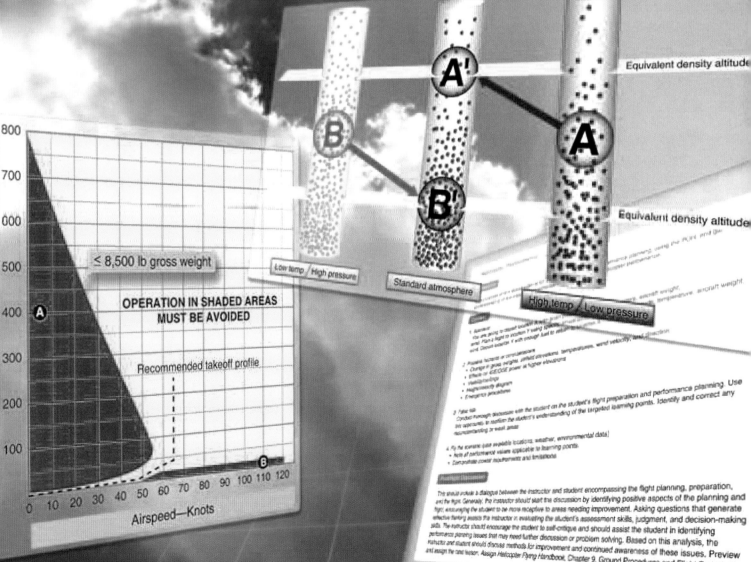

# Factors Affecting Performance

Instructors should ensure the student has a firm grasp on the four major factors affecting helicopter performance: density altitude, weight, loads, and wind. Emphasis must be placed on the importance of proper and thorough performance planning prior to each flight.

## Density Altitude

The instructor should convey to the student the need for a comprehensive knowledge of density altitude. The student must understand what combinations of high elevations, low atmospheric pressure, high temperatures, and high humidity directly impact density altitude and its affect on helicopter performance.

Explain that density altitude is pressure altitude corrected for nonstandard temperature. High density altitude refers to thin air while low density altitude refers to dense air. *[Figure 8-1]* Air density is affected by changes in altitude, temperature, and humidity. Conditions that result in a high density altitude are high elevations, low atmospheric pressures, high temperatures, high humidity, or some combination of these factors. Lower elevations, high atmospheric pressure, low temperatures, and low humidity are more indicative of low density altitude.

Using the building block concept, explain the factors that may increase or decrease density altitude. The predominant factors are atmospheric pressure, altitude, temperature, and moisture. Use examples of each factor's impact on the performance planning used for the aircraft to be flown. Stressing the impact that each factor has on density altitude leads to a greater awareness of the student to these ambient conditions. The lift equation is an effective tool that can be used to show the affect of an increase or decrease of density altitude on aircraft performance.

$$Lift = \frac{coefficient\ of\ lift \times density \times velocity^2 \times lifting\ surface\ area}{2}$$

Once the student has a clear understanding of how density altitude affects helicopter performance, you can advance to actual flight demonstrations and hands-on practice. The instructor can set artificial limitations on the amount of available torque or temperature to simulate operations in environments that cause actual reductions in power margins. While this provides a demonstration to the student, it is not a substitute for actual conditions.

The instructor could ask the student to determine performance for a maneuver, such as OGE hover, for a low or field elevation, and then for one at a much higher altitude normally found in the area of flight. Then, the instructor could discuss

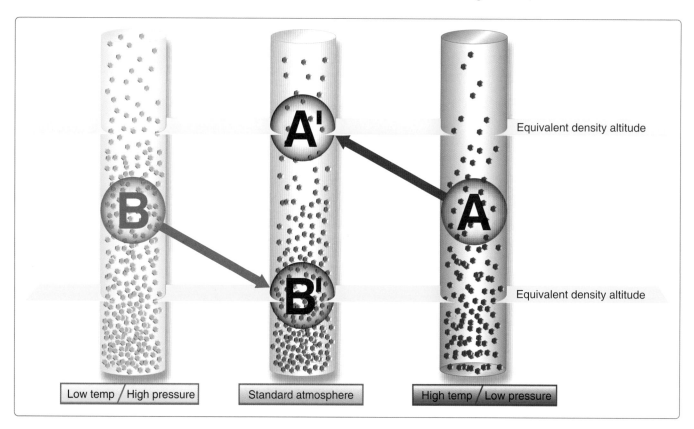

**Figure 8-1.** *Three atmospheres are illustrated. The Standard Atmosphere (29.92 "Hg and 15 °C) is shown in the middle in gray. A less dense atmosphere (A) (lower pressure and/or higher temperature) is shown on the right in red. A denser atmosphere (B) (higher pressure and/or colder temperature) is illustrated on the left in blue.*

the difference with the student and ask the student to describe the practical response differences of the helicopter in the two situations. The instructor may be able to simulate higher elevation power of the helicopter performance by artificially limiting the power the student can use (e.g., limit the manifold pressure or torque to be used). However, the instructor cannot limit the lift and efficiency of the rotor system. The instructor must ensure the student understands the lack of fidelity of the simulation and the hazards awaiting the pilot at high density altitudes, such as lower $V_{NE}$, less antitorque available, less lift from the rotor system, slower engine response, and higher collective pitch settings for cruise and hovering.

If the student makes errors during the simulated high-altitude practice, the increased power is still there, whereas the power is simply not there at actual higher altitudes. At higher altitudes, helicopters do not have the margins of performance necessary to correct errors in planning or judgment. Due to the retreating blade stall characteristic of helicopters, increasing altitudes yield no performance improvements over 5,000 feet to 6,000 feet depending on the design and powerplant(s). Therefore, when operating in higher terrain, any altitude gains above takeoff elevation usually results in decreasing performance. By the very nature of a helicopter, a pilot must be versed in the performance limitations imposed by the change in the environment. The pilot must understand the limitations of increased altitudes and plan accordingly. Therefore, the instructor must include that planning in the syllabus.

Refer to Chapter 10 of the Pilot's Handbook of Aeronautical Knowledge (FAA-H-8083-25) for a detailed explanation of density altitude as it relates to aircraft performance and Chapter 3 for more information on the structure of our atmosphere.

## Weight

The student should possess a basic understanding of the aerodynamics of weight opposing lift. As the weight of the helicopter increases, the power required to produce lift also must increase. The instructor should use the helicopter's performance charts in the relevant Pilot's Operating Handbook (POH) to demonstrate minimum and maximum weight configurations and how that correlates to required power with a given set of environmental conditions. A beneficial technique is to show various power/torque requirements at graduated weight and altitude increments, culminating with high gross weight/high altitude condition. This demonstrates the additional power required as weight and/or density altitude increase. Validation of this exercise can be accomplished using similar data in performance charts while flying at various weights and altitudes and noting the corresponding torque values, or using fuel weight at various stages of the flight. As fuel (i.e., weight) is burned off, power

requirements decrease. This gives the student a practical application of his planning.

## Loads

The strength of the helicopter is measured by the total load the rotor blades are capable of carrying without causing permanent damage. The load imposed upon the rotor blades depends largely on the type of flight. The blades must support not only the weight of the helicopter and its contents (gross weight), but also the additional loads imposed during maneuvers.

In straight-and-level flight, the rotor blades support a weight equal to the helicopter and its contents. So long as the helicopter is moving at a constant altitude and airspeed in a straight line, the load on the blades remains constant. When the helicopter assumes a curved flightpath—all types of turns (except hovering turns utilizing pedals only), flares, and pullouts from dives—the actual load on the blades is much greater because of the centrifugal force produced by the curved flight. This additional load results in the development of much greater stresses on the rotor blades.

## *Load Factor*

The load factor is the actual load on the rotor blades at any time, divided by the normal load or gross weight (weight of the helicopter and its contents). Any time a helicopter flies in a curved flightpath, the load supported by the rotor blades is greater than the total weight of the helicopter. The tighter the curved flightpath, that is, the steeper the bank, or the more rapid the flare or pullout from a dive, the greater the load supported by the rotor; therefore, the greater the load factor. The load factor and, hence, apparent gross weight increase is relatively small in banks up to 30°. Even so, under the right set of adverse circumstances, such as high-density altitude, gusty air, high gross weight, and poor pilot technique, sufficient power may not be available to maintain altitude and airspeed. Above 30° of bank, the apparent increase in gross weight soars. At 30° of bank, the apparent increase is only 16 percent, but at 60°, it is 100 percent.

If the weight of the helicopter is 1,600 pounds, the weight supported by the rotor in a 30° bank at a constant altitude would be 1,856 pounds (1,600 + 256). In a 60° bank, it would be 3,200 pounds; and in an 80° bank, it would be almost six times as much or 8,000 pounds.

One additional cause of large load factors is rough or turbulent air. The severe vertical gusts produced by turbulence can cause a sudden increase in angle of attack, resulting in increased rotor blade loads that are resisted by the inertia of the helicopter.

To be certificated by the Federal Aviation Administration (FAA), each helicopter must have a maximum permissible limit load factor that should not be exceeded. As a pilot, you should have the basic information necessary to fly a helicopter safely within its structural limitations. Be familiar with the situations in which the load factor may approach maximum and avoid them. If you meet such situations inadvertently, you must know the proper technique.

## Wind

Much like density altitude, awareness of the wind's influence plays a large part in performance planning. To avoid the potential for wind-induced incidents, the student must understand the impact of wind on the handling of the aircraft, as well as performance planning.

One simple demonstration can be conducted while completing hover checks. Caution must be taken not to jeopardize controllability while performing this demonstration. In some cases, the instructor may need to fly so the student can focus more on the engine instruments. Position the aircraft into the wind, note the power required, then conduct pedal turns at 90° increments. At each subsequent heading change, note the variation in power required and difficulty in maintaining heading control. Depending on the wind velocity, moderate to sizable increases in power will be noticed. This will facilitate understanding of the impact that directional wind has on power requirements and the importance of wind direction awareness. Additionally, this demonstrates to the student the changes in power required by the tail rotor to overcome the tendency of the aircraft to weather vane into the wind. A comparison of hover power into the wind versus with a tail wind is very effect in demonstrating this flight characteristic. In gusty wind conditions, it is also important to note the momentary spikes in torque while attempting to maintain a stationary hover.

Discussing the impact of wind on translational lift best illustrates contrasting effects of a takeoff or landing into a headwind or with a tailwind. If, in a no-wind situation, translational lift occurs when airspeed reaches approximately 16 to 24 knots, then the impact of directional wind will increase or decrease that range. Noting this, the student can see the advantage of using the headwind to more quickly depart the vortices caused by in ground effect conditions. Conversely, the student will understand that a prolonged in-ground-effect condition (and the need for greater power) exists during a takeoff with a tailwind conditions, because the aircraft must accelerate more to outrun the wind and pass through translational lift.

Also discuss with the student that during a landing, if there is a headwind present, it may help prevent a settling with

power situation from developing. Conversely, if there is a tailwind, settling with power conditions are encountered earlier as the helicopter slows for a landing. A crosswind is much more preferable to a tailwind. The instructor should work with the student to ensure an understanding of the apparent groundspeed versus airspeed factors and differences. If a pilot begins a landing approach with a 10 knot tailwind, at some time in the approach, the helicopter experiences a zero knot airspeed, which means a total loss of translational lift and thrust. In order to maintain the approach angle, more power must be added. If the conditions of less than 10 knots of airspeed, more than 300 feet per minute rate of descent and more than 20 percent power is applied to the rotor system exist, the helicopter is prone to encounter settling with power.

## Height/Velocity Diagram

Ensure the student understands the information in the height/velocity diagram in the applicable POH/RFM and knows how to fly to avoid those unsafe areas in the height/velocity diagram. *[Figure 8-2]* Referring to the aerodynamics of autorotation, explain that this chart shows those heights above ground and airspeeds which, in the event of engine or drive train failure, an experienced pilot should be able to make a safe autorotational landing. The instructor must stress that the conditions provided in the chart are ideal conditions with an experienced pilot.

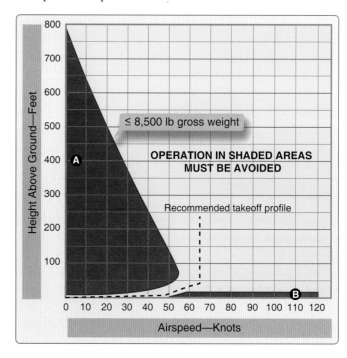

**Figure 8-2.** *Height/velocity diagram.*

The instructor may wish to revisit the aerodynamic theory of autorotation. Specifically, discuss that altitude equates to potential energy; therefore, during autorotational descent the unpowered rotor system maintains kinetic energy as the descending helicopter loses potential energy in the form of altitude. Through the use of turns, flares, and collective control, the pilot can regulate the amount of available kinetic energy within the rotor system. This rotational kinetic energy in the rotor system is used during the deceleration and landing to slow and cushion the landing.

If power is lost in situations of higher altitudes [*Figure 8-2, area A*] and low to no airspeed, such as OGE operations, the helicopter may not be able to maintain enough kinetic energy (rotation of the rotor system) to establish the minimum rate of descent airspeed necessary for a successful landing. Also, in flight profiles with higher airspeed [*Figure 8-2, area B*] and extremely low altitudes, engine failure can cause loss of altitude that will not allow time to take appropriate action to establish an autorotational profile. An effective demonstration of this phenomenon is to have the student note the amount of altitude lost during a high-altitude entry into an autorotational profile and then relate this loss to what would happen in the same situation at low altitudes.

It is important to stress that reaction time and immediate response are critical for an experienced pilot to land the helicopter safely. Avoiding flight profiles in the shaded areas of the height/velocity diagram is not always possible. Some jobs require flight maneuvers or tasks with prolonged OGE operations, such as utility/power line flight, aerial photography, logging and other occupations. Have the student think through scenarios in which potential emergencies occur in these profiles. Discuss possible response options the student may have available in these scenarios.

As instructors, we should be familiar with primacy; what is first learned is often the first to be retained and practiced. Take time to discuss proper takeoff and landing profiles that minimize unnecessary exposure to these shaded areas of the height/velocity diagram. These techniques may include flying higher into the wind and minimizing excessive aircraft loading. The student must consistently demonstrate these maneuvers during each flight.

## Performance Planning

To make the discussion of performance planning relevant to the student, develop a series of exercises that are scenario based and require use of performance graphs from the POH/RFM. Each scenario based lesson plan must have a targeted learning point. Different scenarios can demonstrate the effect of each environmental factor (Atmospheric Pressure, Altitude, Temperature and Moisture) affecting density altitude.

*Figure 8-3* provides a scenario in which the student pilot departs a near sea level location and arrives at a substantially higher elevation. This scenario demonstrates the impact of a greater density altitude as well as the increased IGE and OGE power requirement to operate at that high altitude. Discuss with the student scenarios in which OGE power may not be available and the related consequences. Discuss aircraft system or component limitations, such as torque versus altitude-induced limitations or reductions in $V_{NE}$ airspeeds commonly found at higher elevations.

At the arrival location, ensure the student accurately prepares departure performance planning with added fuel to return to the initial departure location (plus required reserve fuel).

When possible, the scenarios should use high gross weight values that best support the learning objective. If applicable, combine the effects of external loads at higher altitudes, temperatures and wind velocities.

Use scenarios that provide significantly higher IGE/OGE power requirements to demonstrate helicopter capabilities or limitations and to determine go/no-go decisions based on the charted values.

## Instructor Tips

- Take every opportunity to advance the student's awareness of environmental conditions. Accurate and thorough performance planning are essential to safe and successful flight operations.

- Require the student to complete performance planning prior to each training flight. This instills the habit and highlights the importance of performance planning.

- The student must use the appropriate POH/RFM for the helicopter being flown for all performance planning. This allows the student to become familiarized with the capabilities and limitations of the particular helicopter.

- Remember, as an instructor, you should capitalize on opportunities throughout training to correlate the academic training to practical flight applications.

- Tailor scenarios to the student's reason for helicopter training. Students are better able to learn when the training takes on a meaningful role and seems more relevant to the student's goals.

- Use techniques such as artificially limiting available aircraft power. This allows the student to experience operations in reduced performance environments

## Helicopter Performance

**Objective**

The objective of this lesson plan is for the student to conduct performance planning, using the POH/RFM, and gain a deeper understanding of the impact environmental conditions have on helicopter performance.

**Content**

1. Scenario:
   You are going to depart location X with given pressure altitude, temperature, aircraft weight, wind. Plan a flight to location Y using specific arrival conditions of pressure altitude, temperature, aircraft weight, wind. Depart location Y with enough fuel to return to location X.

2. Possible hazards or considerations
   • Change in gross weights, airfield elevations, temperatures, wind velocity, and direction
   • Effects on IGE/OGE power at higher elevations
   • Visibility/ceilings
   • Height/velocity diagram
   • Emergency procedures

3. Preflight briefing
   Conduct thorough discussion with the student on the student's flight preparation and performance planning. Use this opportunity to reaffirm the student's understanding of the targeted learning points. Identify and correct any misunderstanding or weak areas.

4. Fly the scenario (use available locations, weather, environmental data)
   • Note all performance values applicable to learning points.
   • Demonstrate power requirements and limitations.

**Postflight Discussion and Debriefing**

This should include a dialogue between the instructor and student encompassing the flight planning, preparation, and the flight. Generally, the instructor should start the discussion by identifying positive aspects of the planning and flight, encouraging the student to be more receptive to areas needing improvement. Asking questions that generate reflective thinking assists the instructor in evaluating the student's assessment skills, judgment, and decision-making skills. The instructor should encourage the student to self-critique and should assist the student in identifying performance planning issues that may need further discussion or problem solving. Based on this analysis, the instructor and student should discuss methods for improvement and continued awareness of these issues. Preview and assign the next lesson. Assign *Helicopter Flying Handbook*, Chapter 8, Ground Procedures and Flight Preparations.

• Advise on progress towards ultimate goal by starting to praise positive performacnce
• Discuss items that need improvement
• Discuss ways to improve newly learned skills and reinforce previously learned skills
• Explain what to expect next in the training progression

**Figure 8-3.** *Sample scenario-based lesson plan.*

while allowing for an easier escape for the instructor while the student builds additional experience. If using this technique, the instructor must ensure the student is aware that it is only a simulation, and that actual conditions can be much more challenging.

• Allow the student to progress to more challenging scenarios and simulations only after mastering the basic fundamentals of the elements being taught.

## Chapter Summary

This chapter discussed the effects ambient weather, weight and environmental conditions have on performance planning and flight operations. Additionally, this chapter provided instructional methods to apply practical, in-flight application of academic knowledge.

# Chapter 9
# Preflight and Postflight Procedures

## Introduction

Ground operations, inspections, and checks must be accomplished prior, during, and at the completion of any flight. Procedures for accomplishing many of these tasks have been developed by the manufacturer and are contained in the Rotorcraft Flight Manual (RFM) or Pilot's Operating Handbook (POH).

# Checklists

It is essential that the student understand that the purpose of the checklist is not to provide a comprehensive procedure, but rather to provide a systematic, sequential directory of those tasks to be accomplished for the safe operation of the helicopter. The student should be shown that the checklist is derived from those procedures found in the applicable RFM/POH. [Figure 9-1]

It cannot be stressed enough that the use of the checklist ensures that items necessary for safe operation are not overlooked or forgotten. Checklists are of no value if the pilot is not committed to their use.

Instructor fundamentals, such as the Law of Primacy, play a very important role in helping the student develop routine use of checklists. While maintaining a positive attitude, discuss the use of checklists with the student. Demonstration and meaningful repetition establish positive habit patterns that the student will carry forward in all aviation applications.

Also, demonstrate techniques that reinforce good practices. Turning the blades to a 90° position during preflight or prior to start, although not necessarily listed as a step for some checklists, ensures the blades are untied and also allows for checking correlation of flight control movement.

As an instructor, use personal experiences to demonstrate why steps are performed. Perhaps, while conducting a preflight check of a fuel sample, it was learned that the sample contained water or contaminants. This example reaffirms the student's need for checklist use.

Illustrate situations where failure to follow checklist procedures may lead to catastrophic consequences, such as leaving the blades tied down during start, covers left on the aircraft during start, failure to place switches in the required position, failure to remove mooring equipment, and failure to conduct power checks before takeoff. Each of these examples and the resulting hazards can be discussed, showing the impact of not using the checklist or lack of understanding for that particular step.

At a minimum, a well designed checklist should encompass the following phases of flight:

- Preflight inspection
- Before engine start
- Engine start
- Before taxi
- Before takeoff
- After takeoff
- Cruise
- Descent
- Before landing
- After landing
- Engine shutdown and securing

## Required Documents

Emphasize to the student that a safe flight begins with a careful inspection of the helicopter. The purpose of the preflight inspection is two-fold:

- Determine if the helicopter is legally airworthy, and
- Determine if the helicopter is in condition for safe flight.

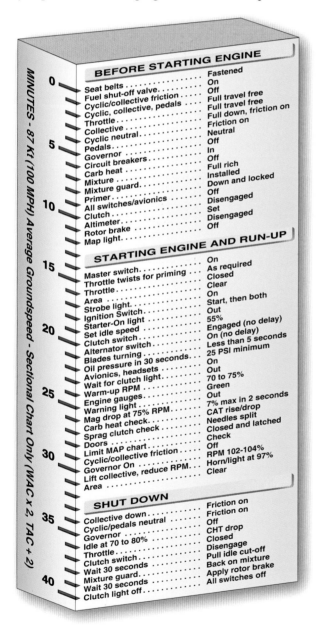

**Figure 9-1.** *Sample of a Robinson R22 checklist.*

The airworthiness is determined, in part, by the certificates and documents that must be on board the helicopter during operation. *[Figure 9-2]* Educate the student on potential liabilities and subsequent ramifications of not having verified these documents. An excellent reference publication for a student pilot is the Plane Sense handbook. This publication explains the fundamental information on the requirements of owning, operating, and maintaining a private aircraft.

Show the student each of the following documents and discuss where they can be found and what purpose they serve:

- Airworthiness certificate (registration number, manufacturer, serial number, category, etc.)

- Registration certificate (eligibility, application requirements, expiration date, etc.)

- Federal Communications Commission (FCC) radio station license, if required by the type of transmitters on board. Refer to Title 47 of the Code of Federal Regulations (47 CFR) part 87.18

- Operation limitations (which may be in the form of a Federal Aviation Administration (FAA) approved RFM and/or POH placards, instrument markings, or any combination thereof)

- Helicopter logbooks and inspection records

It is imperative to take the time to discuss what maintenance inspections entail and, when listed "as required" or "if installed," that the student understands the applicability of those statements. Explain calendar inspections versus hourly inspections. If an inspection has both calendar and hourly criteria, clarify which criteria has precedence. While the intent is not to produce an Airframe and Powerplant (A&P) mechanic, a pilot should demonstrate basic aviation maintenance knowledge. Further information regarding required documentation and inspections can be found in Title 14 of the Code of Federal Regulations (14 CFR) part 91.

## Preflight Inspection

Preflight inspections should only be accomplished after the instructor has conducted a thorough preflight briefing with the student which should be done in a place free of distractions to maximize learning and retention. If the student has not been told exactly what to expect step by step of the preflight inspection, they will probably be overwhelmed by the observations and information presented to them when conducting the inspection. A good initial preflight in a hangar is preferable to one that is conducted on the flightline. If a similar helicopter is disassembled for maintenance, this could serve as a good training aid and allow the student to see the actual part or piece of the helicopter that they may not be able to see during an actual preflight.

The preflight inspection should be performed in accordance with the printed checklist provided by the manufacturer for the specific make and model helicopter. Emphasize that the preflight inspection of the helicopter begins while approaching the helicopter on the ramp. Tell the student to note the general appearance of the helicopter, looking for obvious discrepancies, such as a landing gear out of alignment, structural distortion, skin damage, and dripping fuel or oil leaks. Once at the helicopter, inspection items should be followed in the order delineated in the checklist. Caution the student to be careful when preflighting components and to be aware of safety wire and cotter pins which can cause cuts

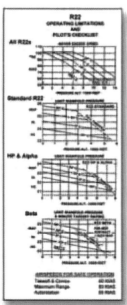

**Figure 9-2.** *Samples of important documents used before, during, and after flight.*

and puncture wounds. Particular attention should be paid to the fuel quantity, type and grade, and quality.

Ensuring freedom of movement of the flight controls prior to flight is essential to safety. Instruct the student not only to physically check for freedom of movement, but also to visually check and ensure there is nothing that could fall and wedge against a control linkage and restrict movement.

Explain to the student that the helicopter should have the required equipment for the type of flight to be flown. Required equipment can be found in 14 CFR section 91.205. Refer the student to 14 CFR section 91.213 regarding the minimum equipment list (MEL) for the helicopter. Discuss the purpose of the MEL, what items may be on the list, and the applicable limitations for instruments and equipment.

## Cockpit Management

While discussing cockpit management, it is important for the student to understand that aircraft control is always the first priority. In doing so, continue to instruct the basics of cockpit management. Ensure the student has all required equipment, documents, checklists, performance data, and navigation charts for the flight. Show the student how to organize these items to ensure they remain secured throughout the flight, as well as in the order of use.

Discuss the advantages and disadvantages of kneeboards and other document retention devices. (It is important for the instructor to note that the physical placement of such items can impact the student's flight dexterity or control. For example, a kneeboard on the right leg may interfere with the student's right hand and cyclic grip; on the left leg it may interfere with the collective.) Some document collection devices are bulky and may become hazardous if not properly stowed and/or secured. Another concern is operations with the doors removed, a luxury of helicopter flight that can potentially turn into a hazard. Secure all items before starting engines.

If a portable intercom, headset, or a hand-held global positioning system (GPS) is used, ensure that the routing of wires and cables does not interfere with the operation of flight controls. The pilot must retain the necessary freedom of movement to reach for radios, etc. Emphasis should be placed on the necessity for a pilot to be thoroughly familiar with all aircraft systems, switch functions, switch locations, control locations, and control functions, particularly for night operations. This is especially critical in helicopters because the pilot normally has both hands and both feet engaged in controlling the helicopter. Discuss procedures for applying friction to the controls to allow changing radio frequencies and other functions that require removing a hand from a flight control.

In a static aircraft, have the student conduct a "dry run" of sequencing through the material he or she has organized. This can be incorporated into practical lessons involving Before Engine Start, Engine Start, Before Taxiing, and Hover Checks. Have the student conduct mock radio calls and performance planning checks; all these tasks require organizational forethought of needed items, checklists, and publications. This ground instruction of cockpit management allows the student to focus on controlling the aircraft in flight rather than diverting attention to organizing in flight.

Once training has progressed to in-flight tasks, stress the importance of when and where to make calls and when and where to change frequencies, always focusing on controlling the aircraft first. Give situational examples, such as, "Do not wait until after you have entered Class B, C, or D airspace to organize the required publications, maps, checklists, and radio frequencies." The student should strive to think several minutes ahead of the aircraft.

During flight training, there must always be a clear understanding between the student and flight instructor of who has control of the aircraft. Prior to any dual training flight, a briefing should be conducted that includes the procedure for the exchange of flight controls. The following three-step process for the exchange of flight controls is highly recommended. When a flight instructor wishes the student to take control of the aircraft, he or she should say to the student, "You have the flight controls." The student should acknowledge immediately by saying, "I have the flight controls." The flight instructor confirms by again saying, "You have the flight controls." Only when the other pilot has confirmed he or she has the controls, will you relinquish the controls. *[Figure 9-3]*

**Figure 9-3.** *Exchanging flight controls.*

Part of the procedure should be a visual check to ensure the other person actually has the flight controls. When returning the controls to the flight instructor, the student should follow the same procedure the instructor used when giving control to the student. The student should stay on the controls until the instructor says: "I have the flight controls." There should never be any doubt as to who is flying the helicopter at any time. Numerous accidents have occurred due to a lack of communication or a misunderstanding about who actually had control of the aircraft, particularly between student and flight instructor. It is imperative that the student understands that the term "flight controls" is used in this procedure. A common point of confusion occurs when the pilot not flying announces a hazard to flight. For example, "I have the traffic." is construed to mean, "I have the traffic and the controls." The pilot flying the aircraft relinquishes the flight controls, misunderstanding the statement, and now is probably focused outside, looking for the traffic. At this point, no one is flying the aircraft! Establishing the above three-step process during initial training ensures the formation of a very beneficial habit pattern.

## Ground Operations

Instructors should describe in detail the process for ground handling and movement of helicopters on the ground in preparation for flight. Show the student where the manufacturer's recommended procedures for ground handling are found in the RFM. This discussion should include removal from typical storage locations, such as trailers and hangars, towing options, taxiing options, and choice of a launch area.

The rotor downwash from a helicopter can cause considerable damage to persons, property, and other aircraft. Ensure the student understands the importance of coordinating with all interested parties, including airport management, landowners, and other aircraft operators prior to conducting helicopter operations.

It is important a student understands the importance of operating a helicopter safely on the ground. This includes being familiar with standard hand signals that are used by ramp personnel and traffic control light signals from the control tower. [Figures 9-4 and 9-5]

Instructors should direct the student to the AIM for more information on hand signals and task the student to practice proper hand signals with another student. Hands-on experience and practice better reinforces the material and increases retention. To familiarize the student with control

**Figure 9-4.** *Hand signals.*

| Signal Type and Color | Aircraft on the Ground | Aircraft in Flight | Movement of Vehicles, Equipment, and Personnel |
|---|---|---|---|
| Steady green | Cleared for takeoff | Cleared to land | Cleared to cross Proceed Go |
| Flashing green | Cleared to taxi | Return for landing (to be followed by a steady green at the proper time) | Not applicable |
| Steady red | Stop | Give way to other aircraft and continue circling | Stop |
| Flashing red | Taxi clear of landing area or runway in use | Airport unsafe DO NOT land | Clear the taxiway or runway |
| Flashing white | Return to starting point on airport | Not applicable | Return to starting point on airport |
| Alternating red and green | General warning signal—exercise extreme caution. | | |

**Figure 9-5.** *Traffic control light signals.*

light signals from the control tower, the instructor can request the different signals from the tower over the radio so the student can see the real light signals. Tower windows are often tinted, which causes light colors to appear a slightly different color. As the student progresses, ask the student what signal to expect and arrange with the tower to delay the light signal some time after the radio transmission to allow the student to respond to the instructor.

The pilot needs to brief passengers and ground personnel. In addition to the required information in the regulations, the student should learn to tell ground personnel where the helicopter would go in case of malfunction, always to approach from a forward side of the helicopter, and approach only after acknowledgment by the pilot.

Many rotors can dip to as low as 4 feet off the surface. Usually, the forward portion of the main rotor dips the lowest to the ground, so approaching from the very front of the helicopter is usually not a safe route. Pilots should be taught to teach personnel never to come to the rear of a helicopter, except in the case of the BV-234/KV-107 to which rear access is safer. Special precautions must be taught for the BO-105 and BK 117 due to their rear access. Tail rotors have killed and injured as many people as airplane propellers.

Students should learn to keep their hands, arms, and hats down, and to be careful of long poles, tripods, survey targets, antennas, etc. No one should ever chase a hat that has blown away.

It is safest to stay in the helicopter until the blades stop or stand by the operations area until the blades stop, but often impractical due to other constraints. By the same token, the pilot should ensure that passengers are briefed to stay inside the airframe until the blades stop in the unlikely case that an unplanned event occurs.

Unlike airplanes, it is quite common for a passenger to be sitting in the front seat and be able to place a camera or other type obstruction around the collective, even when the dual controls are removed. The pilot should always ensure that any doors are closed and completely latched. Accidents have occurred when objects fell out of aft cabins or cargo bays and struck tail rotors.

If the helicopter is being flown with doors removed, this precaution is especially necessary.

## Engine Start

Discuss in detail the process of engine starting and initial operation. Stress the importance of using the engine start procedure recommended by the manufacturer in the RFM. Emphasize the hazards associated with engine start and blade rotation or rotor engagement.

Take time to show the student the reason for certain steps, such as electrical sequencing or throttle positioning. Most helicopters have steps in place to prevent possible damage to electrical systems or to prevent inadvertent fuel flow to the engine. Spend time with the student, enhancing his or her knowledge of these systems through explanation of these steps.

Unlike a fixed-wing aircraft, the helicopter has a 360° hazard area due to the main and tail rotors. Remember, it takes many flight hours and years of aviation exposure for the student to learn what an instructor already knows. Some things are learned from observation. As an instructor, take the time to demonstrate the flexibility of the main rotor and the minimum height it may droop. Also, use this opportunity to discuss approaching a running helicopter, the dangers of a drooping main rotor, and the effects of sloping or rising terrain. When possible, show the student the "invisible" tail rotor in operation. The high rpm of the tail rotor makes it virtually invisible to those unaware of its hazard.

Explain to the student that, whenever possible, the pilot must attempt to park the helicopter so the tail rotor is away from the most common access path to and from the helicopter. In addition to the invisible aspect of the tail rotor, the student should be advised to remember the tail rotor intake side (typically on the right side) does not warn the person by blowing wind in their face or make as much noise as the thrusting side of the tail rotor.

Prior to engine start, and while the student is sitting in the cockpit, have him or her look 90° to both sides and point out the distances needed for the helicopter to clear objects. Reinforce the clearances and visual appearances needed for safe flight along with the limited field of view from the cockpit.

Ensure the student understands engine overspeed or hot start procedures prior to engine start. Prior to engaging the starter or rotor system, caution the student to ensure:

- The helicopter is cleared 360°
- A call out of "CLEAR" is made
- A response has been made from anyone who might be near the helicopter

After starting procedures, conduct a thorough discussion and demonstration of operational checks in accordance with the checklist. *[Figure 9-6]*

## Taxiing

Instructors should place emphasis on the fact that a pilot should look outside the helicopter to the sides, as well as to the front, any time the helicopter is moving under its own power on the surface. The student must be aware of the entire area around the helicopter to ensure it remains well clear of all obstructions and other aircraft.

Explain the effects of wind on power requirements and the tail rotor and include a discussion of loss of tail rotor effectiveness (LTE). Students should be thoroughly familiar with the terms taxi, hover taxi, and air taxi.

A helicopter's unique abilities allow the pilot flexibility in maneuvering the helicopter. Do not allow that flexibility to lead to poor judgment. Think ahead of the helicopter movement. For example, if there are strong winds, it would be a good time for the instructor to instill the future planning

requirements by asking the student where the tail rotor is and what the effects of the strong winds are. Another example is system malfunction. Instructors should constantly be asking the student, "What would you do if the system malfunctioned right now?" Talking through these types of scenarios and emergencies helps the student prepare for an actual situation by rehearsing reactions, which could save a life. If in a controlled airspace environment, ensure the student pilot understands the taxi instructions prior to acknowledging them. Too frequently, pilots acknowledge instructions without fully understanding those instructions, potentially leading to confusion and/or mishaps.

Additional information related to taxiing helicopters can be found in the Helicopter Flying Handbook, Chapter 10, Basic Maneuvers, and the Aeronautical Information Manual (AIM), Chapter 4, Aerodynamics.

## Before Takeoff

Explain to the student that the before-takeoff check is a systematic procedure for making a check of the engine, controls, systems, instruments, and avionics prior to flight. *[Figure 9-7]* It is normally performed before taxiing to a position near the takeoff point. Emphasize that while performing the before-takeoff check, the student divides his or her attention between the inside and outside of the helicopter. It is at this point that the instructor can determine if the student has retained any discussion in cockpit management. Does the student have a plan for the many actions required while conducting the before takeoff checks? Remind the student that each helicopter has different features and equipment, which is why the takeoff checklist provided by the manufacturer or operator is used to perform the checks.

## After Landing

Impress upon the student that after-landing checks, if required by the RFM, should be completed using the manufacturer's established procedures. Movement of the helicopter to the parking area is accomplished according to instructions received from the tower or ground control. Remind the student that, at uncontrolled landing areas, it is the pilot's responsibility to ensure the helicopter does not present a safety hazard to persons or property on the ground. Helicopter pilots must always be aware of the extent of damage that the rotorwash can cause if aircraft and property are not secured properly. If the pilot is ever in doubt, request an alternate route or detour around the airfield if there appears to be loose items that will be affected by the rotorwash.

## Parking

Advise the student that whenever possible, the helicopter should be parked 90° from the actual or forecast winds.

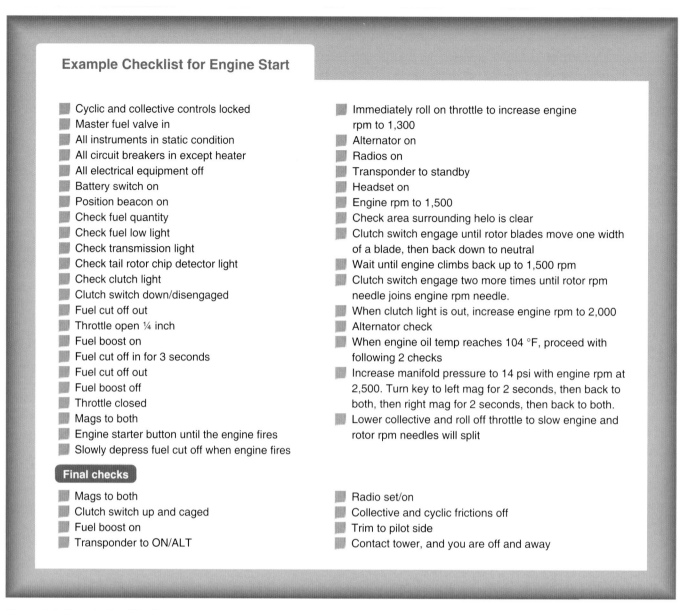

## Example Checklist for Engine Start

- Cyclic and collective controls locked
- Master fuel valve in
- All instruments in static condition
- All circuit breakers in except heater
- All electrical equipment off
- Battery switch on
- Position beacon on
- Check fuel quantity
- Check fuel low light
- Check transmission light
- Check tail rotor chip detector light
- Check clutch light
- Clutch switch down/disengaged
- Fuel cut off out
- Throttle open ¼ inch
- Fuel boost on
- Fuel cut off in for 3 seconds
- Fuel cut off out
- Fuel boost off
- Throttle closed
- Mags to both
- Engine starter button until the engine fires
- Slowly depress fuel cut off when engine fires

**Final checks**

- Mags to both
- Clutch switch up and caged
- Fuel boost on
- Transponder to ON/ALT

- Immediately roll on throttle to increase engine rpm to 1,300
- Alternator on
- Radios on
- Transponder to standby
- Headset on
- Engine rpm to 1,500
- Check area surrounding helo is clear
- Clutch switch engage until rotor blades move one width of a blade, then back down to neutral
- Wait until engine climbs back up to 1,500 rpm
- Clutch switch engage two more times until rotor rpm needle joins engine rpm needle.
- When clutch light is out, increase engine rpm to 2,000
- Alternator check
- When engine oil temp reaches 104 °F, proceed with following 2 checks
- Increase manifold pressure to 14 psi with engine rpm at 2,500. Turn key to left mag for 2 seconds, then back to both, then right mag for 2 seconds, then back to both.
- Lower collective and roll off throttle to slow engine and rotor rpm needles will split

- Radio set/on
- Collective and cyclic frictions off
- Trim to pilot side
- Contact tower, and you are off and away

**Figure 9-6.** *Sample checklist for engine start.*

## Before-Takeoff Checks

Prior to takeoff, the pilot shall consider the following items, and brief the passenger(s), as appropriate:

1. Hydraulics – ON
2. GOV – ON/RPM – 100%
3. Clear area (left, right, overhead)
4. Check gauges (engine, transmission, fuel)
5. Transponder – ON/Alt

**Figure 9-7.** *Sample guidelines for a takeoff briefing.*

Ensure the student understands that exceptions may be necessary under strong wind conditions (particular attention must be observed to avoid potential damage to the rotor head and/or tail boom). If possible, demonstrate alternate options.

Unless parking in a designated, supervised area, the student should be instructed to select a location and heading that prevents the propeller or jet blast of other airplanes from striking the helicopter broadside. Do not allow the student to become complacent. The flight is not complete until completion of all tasks and the pilot is walking away from the aircraft.

## Engine Shutdown

It is essential to stress the importance of following the checklist during engine shutdown. As the training flight nears completion, it is human nature to relax, becoming inattentive. The instructor must take time to go through the shutdown with the student, noting reasons for certain steps. For example, depending on the type of engine (reciprocating or turbine), different cool-down periods are necessary to prevent material damage to the internal components. What is learned first is most often retained; educate the student on the specifics and purpose of steps pertaining to the powerplant in use.

Additionally, certain steps are performed to identify possible system faults. For instance, if the helicopter uses a step to check battery voltage, explain why the check is performed. Another very common cooling time check is to turn off the fuel boost pumps and ensure the engine keeps running. A line leak or faulty component will kill the engine then, requiring repair prior to the next flight. The student should not be allowed to conduct shutdown procedures from memory.

## Postflight

Point out that a flight is never complete until the engine is shut down, rotors have stopped, and the helicopter is secured. When the rotors have stopped, the pilot then carries out a postflight inspection to include checking the general condition of the aircraft. Stress to the student that the postflight procedure is an essential part of any flight. Although not meant to be a thorough inspection, the postflight is no less important. Discrepancies noted on postflight allow maintenance personnel more time to make appropriate repairs and prepare the aircraft for subsequent flights.

Using experience, point out common or frequent trends found during postflight. Ask the student or have the student question what he or she may see. Are the fluid levels within limits? Are they at the same level as checked on preflight? Point out the concern over possible foreign object damage (FOD), specifically on the main and tail rotor blades and undercarriage. Observe any fluid or debris on the engine or transmission decks. Are fluids present on the ground or along the fuselage?

Show the student how to make appropriate entries in the aircraft logbook. Explain how vague entries cause confusion and to be descriptive in noting discrepancies. Also, discuss what policies or duties may be the pilot's responsibility, such as cleaning the cockpit area or windows. Demonstrate the correct method and materials to be used, if applicable.

## Securing and Servicing

When a flight is completed, the aircraft should be hangared or tied down and the rotor blades secured. Tying the blades down in the same manner each time instills positive habit transfer and awareness of tie-down and cover stowage and location. Ensure that the battery is off and have the student note the fuel level at the completion of the flight and ensure refuel is accomplished if necessary. Refer the student to the RFM for the procedures used to service and secure the helicopter.

## Instructor Tip

Remember, instructors are role models for student pilots. Make the use of checklists meaningful to the student by demonstrating the importance and reasons for each step. By using checklists on a regular basis, the instructor's actions underscore the importance of relying on checklists rather than memory. [Figure 9-8]

## Chapter Summary

This chapter described those flight preparations and ground procedures normally associated with helicopter flight. In this discussion, particular emphasis was placed on the use of the manufacturer's procedures and checklists in accomplishing various functions. For additional information in any of these areas, refer to the RFM for the helicopter being flown.

## Using Checklists

### Objective

The purpose of this lesson is demonstration by the student of the proper use of checklists during preflight inspection. The student will demonstrate the ability to perform a preflight inspection using the checklist.

### Content

1. Preflight discussion: lesson objective and completion standards
2. Instructor actions: review use of checklists
3. Student actions: select correct checklist and use it to perform a preflight inspection

### Postflight Discussion

Instructor critiques student performance, previews next lesson, and assigns the *Helicopter Flying Handbook*, Chapter 9, Basic Flight Maneuvers.

**Figure 9-8.** *Sample lesson plan.*

# Chapter 10
# Basic Flight Maneuvers

## Introduction

This chapter provides information to help the instructor explain and demonstrate basic flight maneuvers to students. Since instructors often forget the difficulties they encountered in mastering various flight maneuvers, this chapter is designed as a teaching aid to refresh the instructor's memory of the difficulties involved in learning various helicopter maneuvers.

## Basic Maneuvers

Basic flight maneuvers consist of five fundamental modes of flight: straight-and-level, turns, climbs, descents, and hovering. A student should understand that all maneuvers are based on one or a combination of these fundamental modes. Inform the student that the Practical Test Standards (PTS) establish the minimum standards or acceptable limits for the performance of each maneuver.

Prior to flight, a briefing should be conducted that includes the training to be accomplished, crew responsibilities in the event of an emergency, and the procedure for the exchange of flight controls. A positive three-step process in the exchange of flight controls between pilots is a proven procedure and one that is strongly recommended.

Note: Beginning with hover training as the first helicopter maneuver may be unwise. The student has not had the opportunity to learn how the helicopter reacts to control inputs. Learning control inputs for the first time three feet above the ground is unsettling for most. One option is for the instructor to take the student up to altitude and allow that student to become comfortable with the helicopter flight controls. Once the student is comfortable with the flight controls, proceed to lower and slower flight until hovering is finally achieved.

## Straight-and-Level Flight

It is important that the student be able to maintain a constant heading, altitude, and airspeed. Explain and demonstrate that straight-and-level flight is actually a series of small corrections needed to maintain the original attitude and heading following natural deviations caused by inadvertent control inputs or turbulence. The attitude required to maintain straight-and-level flight should be clearly defined using all available visual aids. One aid is the distance between the horizon and the tip-path plane of the rotor system. *[Figure 10-1]*

**Figure 10-1.** *Maintain a straight-and-level altitude by keeping the tip-path plane parallel to and a constant distance above or below the natural horizon.*

Roll attitude can be determined by using the tip-path plane or a canopy crossbar and its relation to the natural horizon.

Prominent objects on the ground should be used for heading reference. This encourages the student to look outside instead of concentrating too much on the instruments. While the student is gaining proficiency in straight-and-level flight, power is usually not adjusted once it is set, in order to maintain the desired cruise power.

## Instructional Points

It is important that the student learn to recognize the correct attitude for various flight maneuvers. The attitude of the helicopter usually indicates the rate of acceleration or deceleration of the helicopter and the airspeed, and is controlled by the cyclic. Altitude is primarily controlled by the use of the collective. To maintain forward flight, the rotor tip-path plane must be tilted forward to obtain the necessary horizontal thrust component from the main rotor. This usually results in an initially nose-low attitude. Due to the horizontal stabilizer, once the helicopter stabilizes in flight, the helicopter's fuselage will tend to return to a neutral, level attitude. The attitude of the helicopter should not be confused with the position of the rotor disk relative to the horizon. The lower the nose rotor disk is, the greater the power is that is required to maintain altitude, and the higher the resulting airspeeds. Conversely, the greater the power used, the lower the rotor disk must be to maintain at altitude. Since the helicopter is suspended beneath the rotor system, the angle of attack of the wings is not determined by the airframe's pitch as in an airplane. The horizontal stabilizer streamlines the helicopter airframe for reduced drag by applying more down force as airspeed increases, thereby raising the nose to a level (or almost level) cruising attitude.

Show the student, while in straight-and-level flight, any increase in the collective also increases the airspeed and altitude, due to an increase in lift and thrust. A decrease in the collective while holding airspeed constant causes a helicopter to descend. A change in the collective requires a coordinated change of the throttle to maintain a constant rpm. (In this handbook, all throttle discussions refer to helicopters without a governor or correllator.) Additionally, the antitorque pedals need to be adjusted to maintain heading and keep the helicopter in longitudinal trim.

To increase airspeed in straight-and-level flight, instruct the student to apply forward pressure on the cyclic and raise the collective as necessary to maintain altitude. To decrease airspeed, the student needs to apply rearward pressure on the cyclic and lower the collective as necessary to maintain altitude. The student should be guided to notice the yawing resulting from the changing of the torque and airflows over the vertical stabilizer as equipped, and make sufficient changes to maintain the heading and trim.

Once the student has maintained a specific altitude and airspeed with little deviation, point out that the helicopter's altitude and airspeed remain constant with a constant power setting. Small adjustments may be required to compensate for turbulence, but ensure the student uses outside references and does not focus on instruments alone. Assess the position of the rotor disk relative to the natural horizon. The attitude of the helicopter due to the influence of the horizontal stabilizer will approximate a consistent "level attitude" when not accelerating or decelerating. Also, heading is easier to maintain if the student at some point looks outside in line with the intended flightpath. Looking outside fulfills another very important task—scanning for other traffic and obstructions.

To prevent overcontrolling, teach control pressures and not movements. This is especially true in the first few lessons when the student is concentrating on control input and how the helicopter reacts.

## Common Student Difficulties

### Visualizing Attitude

The forward seating position and the excellent visibility in most helicopters may make it difficult for a student to visualize the attitude of the helicopter. It is important that the instructor provide all the assistance possible to ensure the student can determine an attitude by some visual reference. Instructors usually develop different methods of teaching attitude references.

### Overcontrolling

Two factors contribute to overcontrolling the helicopter, the most common difficulty for the beginning student. First, the student fails to notice attitude deviation until it becomes rather large. Second, in the attempt to recover to level attitude, too much control is applied because a student is not prepared for the helicopter's quick response to control inputs. Generally, the beginning student does not know when to remove a control input and usually holds it until after the required attitude is passed. This results in an overshoot, followed by another large control application, another overshoot, and so on. Explain that controls are operated by pressure rather than movement, and it is not necessary to return immediately to the level attitude. As soon as the student understands these two items and loses the sense of urgency, overcontrolling diminishes. It is also helpful to remind a student that when a deviation from the desired attitude is noticed, the proper technique is to first stop the deviation, then make a smooth correction to return to the original attitude.

### Trim

To reduce control pressures in helicopters with electric trim, it is imperative to have the helicopter properly trimmed before the student takes control, otherwise the term control pressure is meaningless. When the helicopter is out of trim, some control pressure must be held just to maintain the desired attitude and any instruction to relax control pressure can only lead to confusion. If the helicopter has an electric trim (especially the "coolie hat" type), the student should be shown how it functions before starting the engine and turning the blades. If installed, have the student use the trim and remind him or her to always trim the pressures off. The on/off type (force trim) is often of less value and very often just left off by experienced pilots as a personal preference. When trim is mentioned in this handbook, it is in reference to the antitorque trim of the helicopter unless otherwise indicated.

### Coordination

Most beginning students have difficulty relating the effect one control has on another. The most obvious of these is the change in torque as power is changed, requiring the use of antitorque pedal pressure as power is varied. Less obvious is the effect of a power change on pitch attitude in forward flight due to gyroscopic precession and differential lift on the advancing and retreating blades. As power is increased with the collective, the nose tends to pitch up; as power is decreased, the nose pitches down. As speed is increased, the nose of the helicopter tends to rise and begin a roll towards the retreating blade. These effects can be most disconcerting to the student unless the instructor thoroughly explains and demonstrates them.

### Scan

To correct a deviation, it must first be recognized. Most beginning students tend to devote all of their attention to a specific problem. For example, full attention may be devoted to an altitude problem while the helicopter drifts off heading or the airspeed changes. Students may also fail to see other aircraft or obstacles in the vicinity if their attention is fixed on a single item. Some instructors find it helpful to call out or point to the items that should be included in the scan pattern. This helps the student build a good habitual scan. It is important that the student be taught to include the engine instruments in the scan, so an impending engine problem does not go undetected.

### Kinesthesia

The sense of motion and pressure changes through nerve endings and muscular sensations is scientifically named kinesthesia, but is commonly called seat-of-the-pants flying. This sense can be developed more rapidly if the instructor calls attention to the sensations as they occur. Development of this sense enables a student to become aware of changes in the helicopter's attitude more quickly. Point out to the student the importance of the other senses as well. The sounds of the engine, rotor, and transmission give information of rpm and possible mechanical problems. Seeing (vision) allows us to fly safely, maintaining level flight with the horizon and/or the instruments. Unusual smells while in flight may be indications of something getting hot or burning within the helicopter. Touch may be the pressure-counterpressure that we exert on the antitorque pedals or the amount of cyclic input to maintain flight profiles.

## Normal Climb

The objectives in practicing climbs are to achieve proficiency in establishing a climb attitude and airspeed, setting climb power while maintaining a specified rpm, and coordinating the use of flight controls. Proficiency is also gained by understanding the techniques used in leveling off at a designated altitude and establishing level cruise flight.

### Instructional Points

For both climbs and descents, focus on:

1. Transitioning from one power setting to another (cruise power to climb power, then back to cruise power; for the descent, cruise power to approach power, then back to cruise power, if desired).

2. Coordinating the controls as a result of a power change.

3. Clearing the helicopter above or below prior to initiating a climb or descent.

To enter a constant airspeed climb from cruise airspeed, simultaneously increase the collective and throttle rpm to obtain climb power. Adjust antitorque pedals to maintain helicopter in yaw trim.

Note: In a counterclockwise rotor system, a left pedal input is required for an increase in torque (right pedal for a torque decrease).

An increase in power causes the helicopter to start climbing, and only slight back pressure on the cyclic is required to change from a level to a climb attitude.

NOTE: Discuss with the student how helicopter design affects climb and descent attitudes. For example, some helicopters are designed to increase the usable CG range of the helicopter at higher airspeeds. Nose-high or nose-low attitudes are also based on load conditions or aircraft designs. Therefore, climb attitude is slightly different from one helicopter design to the other. Point out to the student that simply pulling aft cyclic (while in cruise flight) initiates a climb and, in a short period of time, airspeed lowers and a descent begins. If the available lift/thrust is completely converted to vertical lift by using aft cyclic, the helicopter will begin descending when it slows enough to lose translational lift. This is not the coordinated climb that you are attempting to achieve. While in cruise flight, forward cyclic initiates a descent and, in a short period of time, your airspeed increases. However, depending on altitude, this could result in retreating blade stall. These types of maneuvers are discouraged during training so the student can learn to control the helicopter within stated parameters. This results in a better trained pilot with better skills.

As the airspeed approaches normal climb airspeed, adjust the cyclic to hold this airspeed. Throughout the maneuver, maintain climb attitude, heading, and airspeed with the cyclic; climb power and rpm with the collective and throttle; and yaw trim with the antitorque pedals. To level off from a climb, start adjusting the attitude to the level flight attitude a few feet prior to reaching the desired altitude. The amount of lead depends on the rate of climb at the time of level off (the higher the rate of climb, the more the lead). Generally, the lead is 10 percent of the climb rate. For example, if your climb rate is 500 feet per minute, you should lead the level-off by 50 feet. To begin the level-off, adjust cyclic to adjust and maintain a cruise flight attitude. You should maintain climb power until the airspeed approaches the desired cruising airspeed, then lower the collective to obtain cruising power and adjust the throttle to obtain and maintain cruising rpm. Throughout the level-off, maintain yaw trim and heading with the antitorque pedals. The instructor should remind the student of the effects of inertia, which require some lead time and efforts. Just as one applies the brakes in a car before a stop sign, a pilot should apply control inputs prior to the desired point, be that an altitude or heading.

### Common Student Difficulties
#### Attitude

As in straight-and-level flight, students frequently have difficulty visualizing and establishing the proper attitude for the climb. Use whatever references are available in the helicopter, such as the tip-path plane, canopy crossbars, or any other structural reference. Depending on the helicopter, horizontal stabilizer, and rate of change in altitude, the climb attitude may be the same as the cruise attitude, or slightly higher.

#### Overcontrolling

The difficulty in establishing the correct climb airspeed may be the result of overcontrolling. Since establishing the correct airspeed is usually accomplished by a series of pitch attitude adjustments, students may not hold the attitude long enough for the airspeed to stabilize. This leads to excessive maneuvering while chasing the airspeed. At this point, frustration and tension begin to build. When it becomes apparent that the student is getting frustrated, the instructor can try one of the following three things; have the student return to straight-and-level flight, take over for a brief demonstration while the student relaxes, or allow the student to continue climbing until they achieve a stabilized climb and perceive the sight picture and control pressures. If climbing to an altitude is too complicated, just strive for a coordinated climb at first. Instructors should never be afraid to break any maneuver down to it component parts and allow the student to practice those individual skills until they are ready to assemble those skills into a complete maneuver.

## Coordination

In the process of beginning a climb, all controls are utilized. Each control input causes something else to change, and a beginning student may have difficulty, not only in accomplishing the actions in the proper sequence, but also in compensating for control inputs. During the level-off, some students have a tendency to decrease power before adjusting attitude (cyclic) for cruise flight. Talk the student through the maneuver to remove any doubt about what is to be accomplished, as well as how and when it is done.

## Scan

The scan pattern mentioned in straight-and-level flight becomes more important when the flight condition is constantly changing. Several things are happening at once and the task becomes more difficult unless the student has rehearsed the actions and reactions of the helicopter.

## Normal Descent

A normal descent is a maneuver in which the helicopter loses altitude at a controlled rate in a controlled attitude. The objective in practicing descents is to gain proficiency in establishing the attitude necessary to maintain the desired airspeed, setting power as required to maintain the desired rate of descent while maintaining a constant rotor rpm and correcting for changing torque.

### Instructional Points

To establish a normal descent from straight-and-level flight at cruising airspeed, lower the collective to obtain proper power, adjust the throttle to maintain rpm (a slight amount of cyclic adjustment is normally necessary to maintain desired airspeed), and adjust antitorque pedals to maintain heading. Throughout the maneuver, maintain descending attitude and airspeed with the cyclic, descending power and rpm with the collective and throttle, and yaw trim with the antitorque pedals. To level off from the descent, lead the desired altitude by approximately 10 percent of the rate of descent. For example, a descent rate of 500 feet per minute would require a 50-foot lead. At this point, increase the collective to obtain cruising power, adjust the throttle to maintain rpm, adjust antitorque pedals to maintain yaw trim, and adjust the cyclic to obtain cruising airspeed and a level flight attitude.

## Common Student Difficulties
### Attitude

Visualization of pitch attitude may be difficult for the student in the initial stages of flight training. Make use of any available reference points on the helicopter in order to maintain some sort of visual reference. Helicopter attitude is primary for acceleration and deceleration control. The airspeed indicator is going to be the primary indicator for airspeed control. If the airspeed is slow, then the nose must be lowered to accelerate the helicopter until the airspeed increases. As the airspeed increases, the pilot must plan on indicator delays and allow the helicopter attitude to stabilize and neutralize equilibrium at the desired airspeed. The descent attitude may be the same as the level attitude was once the helicopter is stabilized in the descent.

### Coordination

The student may have difficulty adjusting throttle and antitorque pedals while simultaneously adjusting the collective to set descent power. Emphasize that power is to be changed slowly and smoothly to minimize coordination problems.

### Scan

It is common for a student to concentrate on one factor to the exclusion of others. Students have difficulty with two areas in this maneuver: maintaining a constant angle of descent and leading the level-off sufficiently. These two problems often result in recovery below the desired altitude.

## Turns

Turns are practiced to develop skill in establishing and maintaining a desired angle of bank, while holding the pitch attitude that is appropriate to the desired maneuver. Level turns are practiced first, using bank angles of approximately 15–20°. *[Figure 10-2]* As the student is developing his or

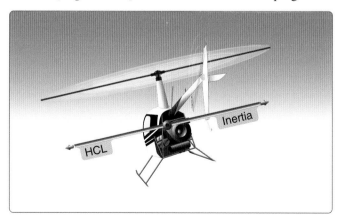

**Figure 10-2.** *During a level, coordinated turn, the rate of turn is commensurate with the angle of bank used, and inertia and horizontal component of lift (HCL) are equal.*

her skills, turns should be practiced at different airspeeds. As the student progresses, turns at $V_H$ or $V_{NE}$ should be added to the program as well as turns below effective translational lift. Care should be exercised to avoid LTE at low altitudes. If possible, a demonstration of LTE at higher altitudes is a good teaching point for the new student.

## Instructional Points

Before making any turns, make sure the student clears the area in the direction of the turn, as well as above and below the helicopter.

To enter a turn from straight-and-level flight, apply sideward pressure on the cyclic in the direction the turn is to be made. This is the only control movement needed to start the turn. Antitorque pedals are not used to assist the turn. Airplane pilots transitioning to helicopters attempt to use the antitorque pedals as they would a rudder pedal. Use the pedals only to compensate for torque to keep the helicopter in yaw trim.

How fast the helicopter banks depends on how much lateral cyclic pressure is applied. How far the helicopter banks (the steepness of the bank) depends on how long the cyclic is displaced. After establishing the proper bank angle, return the cyclic toward the neutral position. Explain to the student that the cyclic tilts the rotor disk relative to the horizon. The amount of tilt or bank depends on how much and how long the cyclic is displaced from perpendicular to the horizon. The rotor disk always follows the cyclic. As the pilot places the cyclic in a neutral position in relation to the helicopter, the cyclic is simply maintaining the rotor disk tilt as referenced to the horizon and the helicopter follows the rotor. Returning the cyclic to the neutral position simply stops the bank (rotor tilt) from increasing or decreasing.

Use the collective and throttle to maintain altitude and rpm. As the torque increases, apply more pressure to the proper antitorque pedal to maintain longitudinal trim. Depending on the degree of bank, additional forward cyclic pressure may be required to maintain airspeed.

When rolling out of a turn, the cyclic is moved back to perpendicular to the horizon, which brings the rotor back to level with the horizon. Lead or lag on the rollout is necessary to complete the maneuver on the desired heading.

## Common Student Difficulties
### Attitude

Visualization of the bank angle is one of the most common problems for students. The angle between the tip-path plane and the horizon should always remain stable and consistent when banking or turning the helicopter. As the bank angle is established and the perspective changes, there is a tendency

to use the center of the canopy as the pitch reference. It must be emphasized that the correct pitch reference is directly in front of the student. The pitch reference point should remain stationary as the helicopter is rolled into the bank, with a helicopter appearing to pivot around the pitch reference. The correct pitch attitude is confirmed by reference to the altimeter and a level turn. If the student attempts to maintain altitude solely by reference to the altimeter, overcontrolling usually results and the student begins chasing the altitude.

### Leaning Away From a Turn

There is a natural tendency to keep the body, or at least the head, level. When the student leans away from the turn, perspective changes, making it even more difficult to maintain the correct attitude.

### Failure to Clear the Area

The student is frequently so occupied with the problems associated with maintaining altitude, airspeed, bank angle, etc., that the responsibility of seeing and avoiding other aircraft is neglected. Clearing the area in the direction of the turn must be included in the items the instructor calls out while talking the student through the maneuver.

Stress to the student that clearing of the helicopter is continuous. Other traffic occupies the same airspace such as traffic helicopters, crop dusters, rescue, police, power-line patrols and many others. If operating near military training areas, remind the student of low level VFR and IFR routes and the increase in flight activity. Birds or anything above ground level (towers, power lines, etc.) present flight hazards as well.

### Rolling Out of a Turn

Difficulties associated with rolling out of a turn are usually related to scan problems. The student who is preoccupied with other factors often loses track of heading. Select a prominent landmark and instruct a student to anticipate the rollout by an amount equal to about half the bank angle.

## Climbing and Descending Turns

Climbing and descending turns are used to further develop control and coordination. They also provide the practice required for departures and landing approaches.

### Instructional Points

As always, before making any turns, clear the helicopter in the direction of the turn, as well as above and below the helicopter.

The turn and climb/descent are usually initiated simultaneously. Until the student gains proficiency, it may be easier first to establish each maneuver separately. For example, to enter a climbing/descending turn, establish the turn first then adjust the controls for the climb/descent. You may reverse the order as well, with enough practice the student will learn to simultaneously perform the required control inputs to accomplish a climbing/descending turn.

## Common Student Difficulties
### *Attitude*
Combining turns with climbs/descents introduces new helicopter attitudes and the initial perception of these attitudes may be difficult for the student to comprehend. A thorough briefing and demonstration minimizes this problem. The first climbing/descending turns should be established by beginning the climb/descent and then rolling into the desired bank angle in order to reduce the number of simultaneous control movements required.

### *Scan*
As with the previous maneuvers, the scan pattern is easily interrupted by concentrating on a specific aspect of the maneuver. During early practice of climbing/descending turns, the instructor should call out all the items that require attention, even if no correction is required. As proficiency improves, the instructor should call attention only to the items that require corrective action.

## Coordination Exercises
Once level flight, turns, climbs, and descents have been introduced, coordination exercises should be practiced to assist a student in developing subconscious coordinated control and proficiency. A good exercise to teach compensation for power changes is to make airspeed changes while maintaining straight-and-level flight. At a safe altitude, and while maintaining a constant rpm, altitude, and heading, have the student reduce airspeed to 40 knots by simultaneously applying aft cyclic and reducing power. Now, instruct the student to accelerate to approximately 80 knots by increasing forward cyclic and power. The maneuver may be repeated, as necessary, for proficiency. During these maneuvers, allowing a stable flight for a few moments and pointing out the helicopter's attitude at the different airspeeds will help the student become familiar with that particular helicopter's attitudes set by the horizontal stabilizer for those airspeeds and loads. There may be an 11 percent change in the weight of some helicopters when the instructor deboards for the student's first solo flight, and a considerable sight picture change with that much weight out of one side of the helicopter.

Another exercise that develops smoothness and coordination is rolling from a medium bank to the left into a medium bank to the right, then back to the left and continuing the series while maintaining a constant base heading and altitude. Each of these exercises helps develop smoothness, coordination, and an active scan pattern.

If a student is having trouble coordinating rpm and manifold pressure, an exercise in throttle/collective coordination can be used. Have the student maintain a constant attitude while disregarding altitude. Now, instruct the student to change the manifold pressure with the collective while holding a constant rpm with the throttle. Then, reverse the procedure by having the student change the rpm with the throttle while maintaining a constant manifold pressure with the collective. This exercise allows the student to concentrate on throttle/collective coordination without devoting attention to other factors.

## Approaches
An approach is defined as the transition from traffic pattern altitude to either a hover or to the surface. In day-to-day operations, approaches in a helicopter may be dictated more by existing conditions than by formal patterns. For training, however, a formal pattern is used to give the student a basis upon which to build the modified patterns a particular situation may require. Downwind, base, and final approach legs should be flown in accordance with the patterns the instructor outlines. A normal approach uses a descent profile of 8°–12° starting at approximately 300 feet AGL. *[Figure 10-3]* The rectangular pattern explained on page 10-20 serves as a good basis for helicopter traffic pattern with the downwind at 500 feet AGL or higher as needed for noise, traffic and aircraft characteristics.

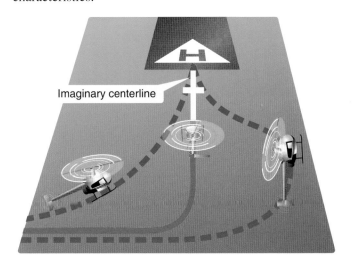

**Figure 10-3.** *Plan the final so the helicopter rolls out on an imaginary extension of the centerline for the final approach path. This path should neither angle to the landing area, as shown by the helicopter on the left, nor require an S-turn, as shown by the helicopter on the right.*

Emphasize to the student that aligning with the landing direction may allow the pilot to detect winds sooner and detect obstructions. Flying any path that is less than straight into an area on the approach azimuth decreases the time available for hazard detection and low reconnaissance.

## Instructional Points

For the beginning student, each approach should be started at approximately the same position and at the same airspeed and altitude. This allows a consistent basis for the student and instructor to evaluate each approach. To accomplish this, concentrate on each leg of the traffic pattern so the helicopter arrives at the point the approach is started and at the correct position, speed, and altitude.

As the approach angle comes into view, begin the approach by lowering the collective sufficiently to get the helicopter descending down the approach angle. The approach angle should be an imaginary angle from the landing gear to the landing point, and not from the pilot's eyes to the landing point. With the decrease in the collective, the nose tends to pitch down, requiring aft cyclic to maintain the recommended approach airspeed attitude. Adjust antitorque pedal as necessary to maintain yaw trim.

Maintain entry airspeed until the apparent groundspeed and rate of closure appear to be increasing. At this point, slowly began decelerating with slight aft cyclic, and smoothly lower the collective to maintain approach angle. Use the cyclic to maintain an apparent rate of closure equivalent to a brisk walk.

Explain to the student that a helicopter pilot should plan an approach to keep the skids/landing gear at a constant angle to a 3-foot hover over the intended landing area. Keeping the landing area in one spot in the windshield or "bubble" does not result in a good approach for a helicopter pilot. It is best for the pilot to visualize and fly the skids/landing gear down the approach angle to the hover point.

At approximately 25–40 feet AGL, depending on wind, the helicopter begins to lose effective translational lift. To compensate for this loss, increase the collective to maintain the approach angle, while maintaining the proper rpm. The increase of collective pitch tends to make the nose rise, requiring forward cyclic to maintain the proper rate of closure. On short final, this is also when the airflow in the aft portion of the rotor disk is disturbed, so the increased pitch in the forward portion of the disk is not balanced by the same lift in the aft portion. Depending on the amount of deceleration used, forward cyclic may be needed to maintain the hovering position.

As the helicopter approaches the recommended hover altitude, increase the collective sufficiently to maintain the hover. At the same time, apply aft cyclic to stop any forward movement, while controlling the heading with antitorque pedals.

## Common Student Difficulties
### *Ground Track*

There can be no basis upon which to build unless the approach path is consistent. Therefore, the student must start the pattern from the same indicated airspeed, altitude, and distance from the landing spot. Thereafter, the student should be encouraged to maintain the correct pattern so each approach does not present a new set of circumstances. During initial training, explain to the student how different wind conditions can affect the helicopter and teach them how to adjust the flight controls so that they are always flying the helicopter rather than letting the helicopter take control of them. As the student gains experience, the instructor should have the student brief the effects of the winds on the expected flight maneuvers and what actions must be taken by the student to counteract those effects.

### *Altitude*

The same comments concerning ground track are applicable to altitudes on downwind, base, and the turn to final approach. Changing altitude requires modifications in some other parameter, resulting in a different approach pattern. Therefore, turns to each leg of the approach should be made from the same spot and at the same altitude during a single training period.

### *Airspeed*

Airspeed control is important if the student is to establish and maintain a consistent approach. Thus, it is important for the student to be aware of, and adhere to, recommended approach airspeeds.

### *Approach Angle*

The student must understand the reason for utilizing a standard approach path. It is to establish the final approach leg at a distance and altitude that requires the same angle to the landing spot on each approach. In this manner, the student learns to visualize the correct approach angle, making it easier to learn the techniques for making corrections.

If there is a visual approach slope indicator (VASI) of some type near, it is good to have the student look up the VASI's glide slope angle and then fly to the VASI. It is often helpful for the instructor to fly while the student observes how the stated angle appears. Because VASI's are often near 3°, about three times that angel (9°) can be a normal helicopter approach angle. The instructor will then announce when the VASI angle is doubled and finally begin the descent when triple the VASI angle. This allows the student to have a gauge or standard for the normal approach angle.

### Traffic

With all the other factors requiring the student's attention, it is very easy to relax vigilance for other aircraft. Before turning to base leg, the student should be required to check for approaching traffic and state whether the pattern is clear of conflicting traffic. Then on final approach, the area should be checked in all directions to make sure there are no other aircraft on, or about to turn onto, the final approach leg.

### Power Adjustments

During the approach, the power setting is usually quite low. In a hover, it is quite high. Most beginning students wait until they are very close to the ground before adding power. This can easily lead to overcontrolling. This is usually done while transitioning through translational lift. The instructor should remind the student that, as power is added by increasing the collective, the cyclic must be used to ensure the extra power is all directed to replace the lift lost as translational lift is lost. This is done by adjusting the cyclic aft in most conditions. As this is happening, the student will also correct for yaw from the loss of translational thrust and place the helicopter in a slip to align the landing gear with the ground track. This habit is important when a landing to the surface begins.

When power is added, the attitude must be changed to continue moving forward and down to the intended landing spot. Forward cycle may be needed because too much deceleration will excessively slow the rate of closure. While paying attention to airspeed and the height/velocity diagram, it may be necessary to slow the helicopter more than usual in the beginning phases of training so the student understands, and is comfortable with, the transition from the approach to the hover.

## Go-Around

Before solo, a student must be taught the procedures and techniques used in a go-around. Encourage the student to use the go-around procedure as a safety precaution at any time he or she is uncomfortable with continuing the approach. Go-arounds should be taught early and often. Every student should know that go-arounds are good maneuvers for the best pilots. If in doubt, go-around! Students should learn to abort landings when the circumstances feel uncomfortable.

### Instructional Points

A go-around is initiated by adding power to the climb power setting and accelerating to climb speed. When power is added, two common errors may occur:

1. With the initial power change, the rate of descent may stop, and the student may not add enough power to continue a climb (they level off).

2. When power is added, the nose of the helicopter begins to rise, giving the impression that the helicopter is climbing. This results in a loss of airspeed if no forward cyclic is added. If allowed to continue, the helicopter may begin to settle.

When the decision to initiate a go-around is made, carry it out without hesitation.

### Common Student Difficulties
#### Initiating the Go-Around

Even experienced pilots may be hesitant to initiate a go-around, either from failure to recognize the need for one or as a matter of pride. Teach the student to recognize the need for a go-around early in the approach instead of waiting until the last moment. The safety of the aircraft and its occupants is the first consideration, and a go-around should be executed at the first indication of an unsatisfactory approach or any unsafe conditions on the intended landing point. Also discuss with the student the difference in helicopter reactions (power requirements) while performing go-arounds above and below ETL.

#### Coordination

Many things must be accomplished simultaneously as a go-around is initiated. Collective is increased, rpm is adjusted as necessary, antitorque pedal corrections are made, and the attitude is adjusted to first accelerate to climb speed and then to maintain it. In the process, the student might overlook one or more of the required adjustments. It may help to practice the first few go-arounds at higher altitudes so the proximity to the ground is not a distracting factor.

## Normal and Crosswind Takeoff From a Hover

The normal takeoff from a hover is the transition from hovering flight into a climb over a specified ground track. During the climb, airspeed and altitude should be such that the crosshatched or shaded areas of the height/velocity diagram are avoided. Other types of takeoff may be performed; however, the student needs to learn early how and why he or she is performing a specific takeoff. The pilot should be making a risk assessment to determine which type of takeoff is the safest.

### Instructional Points

Discuss with the student during the preflight, what control inputs are required during hovering flight and takeoffs. There is more to a normal takeoff than just adding forward cyclic. Bring the helicopter to a hover and make a performance check, which includes power, balance, and flight controls. The power check should include an evaluation of the amount of excess power available. The balance condition of the helicopter is indicated by the position of the cyclic when maintaining a stationary hover. Wind may necessitate some cyclic deflection, but there should not be an extreme deviation from neutral. Flight controls must move freely, and a helicopter should respond normally.

Visually clear the area all around and above. Start the helicopter moving by smoothly and slowly easing the cyclic forward. As the helicopter starts to move forward, increase the collective as necessary to prevent the helicopter from sinking and adjust the throttle to maintain rpm. The sink is caused by diverting lift into forward thrust. Discuss with the student the aerodynamic effects of the rotor system during hovering flight and during normal takeoff. The increase in power requires an increase in the proper antitorque pedal to maintain heading. Emphasis should be placed on aligning landing gear precisely with the direction of travel to avoid dynamic rollover should the student allow the helicopter to touch the surface during takeoff.

Select ground reference points to maintain a straight takeoff path throughout the takeoff. Ensure the student chooses several ground reference points during the maneuver to maintain a ground track. As the forward portion of the rotor system gains undisturbed air, the lift on the forward portion of the rotor tends to lift the front of the disk and stops the acceleration. A little forward cyclic to maintain the accelerating attitude is necessary to continue the takeoff.

As the aft portion of the rotor system gains undisturbed air, the nose tends to tilt forward, causing an excessively fast acceleration. At about the same time, translational lift becomes apparent and some forward cyclic motion is required to gain sufficient airspeed to avoid the H/V chart shaded areas, while continuing a safe climb avoiding the shaded areas.

As airspeed increases, the streamlining of the fuselage reduces engine torque effect, requiring a gradual reduction of antitorque pedal pressure. Just as translational lift occurs, translational thrust follows. This gives two effects that must be countered by adjusting the pedals as the climb begins. As the helicopter continues to climb and accelerate to best rate of climb, apply aft cyclic pressure to raise the nose smoothly to the normal attitude.

Ensure that the student understands that he or she must make constant corrections during all phases of flight (this is to compensate for actions and reactions).

## Crosswind Considerations During Takeoffs

When the takeoff is made during crosswind conditions, the helicopter is flown in a slip during early stages of the maneuver. In this case, the cyclic is held into the wind a sufficient amount to maintain the desired ground track for the takeoff. The heading is maintained with the use of the antitorque pedals. In other words, the rotor is tilted into the wind so the sideward movement of the helicopter is just enough to counteract the crosswind effect. To prevent the nose from turning in the direction of the rotor tilt, it is necessary to increase the antitorque pedal pressure on the side opposite the rotor tilt.

After approximately 50 feet of altitude is gained, crab the helicopter into the wind as necessary to maintain coordinated flight over the desired ground track. The stronger the crosswind, the more the helicopter has to be turned into the wind.

## Common Student Difficulties
### Attitude Control

As in most other maneuvers, smooth, positive attitude control is the key to success in the takeoff. If the student is properly briefed and understands the changing forces during takeoff, it is possible to anticipate and correct deviations promptly.

At about five knots, ground effect diminishes and a helicopter begins to sink. Depending on available power and instructor technique, power should be added to prevent this sink. Any power change requires an antitorque pedal adjustment, which in turn requires a cyclic adjustment to accommodate the increased translating tendency. If the student does not make these adjustments, the ground track will not be straight. Shortly after forward movement is initiated, translational lift is encountered and the nose pitches. This requires forward cyclic to keep the helicopter accelerating.

### Heading Control

Pedal control requirements also change during the transition into a climb. From the hover, additional power is added in helicopters with counterclockwise main rotor blade rotation, the left pedal requirement increases. As speed increases, directional stability increases, so the need for left pedal decreases. The tail rotor achieves translational thrust due to clean airflow and begins making more antitorque thrust than is required, especially if the helicopter has an effective vertical fin that also helps with the antitorque task. These actions require a decrease in antitorque pedal to maintain heading. A throttle change may be necessary since the power demand from the tail rotor decreases. A governor hides this control change.

### Crosswind Corrections

If the takeoff is made in a crosswind condition, the student may not be aware of the corrections required during the climb unless briefed in advance. In the hover and during the initial portion of the climb, cyclic must be applied toward the wind, and downwind, pedal applied to keep the helicopter heading straight along the ground track. As speed and altitude are gained, the cyclic is used to establish a crab. As the helicopter is placed into a crab, the antitorque pedal pressure must be decreased from the slip into coordinated or "trimmed" flight. Failure to correct for the crosswind results in a downwind drift from the specified ground track. After the helicopter has transitioned through effective translational lift (ETL), the student should begin to crab the helicopter (trim), improving the climb performance. Once the desired or required climb airspeed is attained, the cyclic should be adjusted to maintain the stabilized airspeed attitude. Recommend using slip to align the fuselage with the ground track below approximately 50 feet, and trim the helicopter above 50 feet.

## Traffic

The student may concentrate so completely on achieving the stated objectives that conflicting traffic and obstructions, such as towers and powerlines, go unnoticed. Instructing students to look well ahead of the helicopter and to scan for traffic not only helps their awareness of other traffic, but also helps improve attitude control.

# Hovering

Learning to hover can be a frustrating experience for some students as it may take a few flights to learn the maneuver. Instructors should emphasize that the student should relax. Hovering is very difficult at first, but it does work and its importance cannot be overemphasized.

## Vertical Takeoff to a Hover and Hovering

Prior to ascending to a hover, teach the student to always check for unreleased tiedowns or restrictions to the freedom of the landing gear. Pilots should always take off to a hover slowly to avoid dynamic rollover accidents from hung landing gear. A slow takeoff also gives the pilot more time to adjust for the translating tendency, making the takeoff smoother and more controlled.

Explain to the student that a vertical takeoff to a hover involves flying the helicopter from the ground vertically to a landing gear height of two to four feet, while maintaining a constant heading. Demonstrate various hovering heights and allow the student to see how a hover at each height appears. *[Figure 10-4]* Once the desired landing gear height is achieved, the helicopter should remain nearly motionless over a reference point at a constant altitude and on a constant heading. The maneuver requires a high degree of concentration and coordination.

## Instructional Points

Prior to any takeoff or maneuver, have the student ensure the area is clear of other traffic, persons, equipment or obstacles. This can be accomplished by instilling the habit of clearing the helicopter to the left, right, and overhead prior to performing any maneuver.

One additional instructional point to make is to teach the student to assess the helicopter control response prior to each flight. Have the student get the helicopter light on the skids/gear and ensure the helicopter ascends to a hover in a nearly level attitude. Ensure that you have enough cyclic control to continue. Once at a three-foot hover, ensure the helicopter remains nearly level. Point out to the student what a normal hover attitude looks like. If things do not feel or look right, slowly lower the collective and land the helicopter. Attempt

**Figure 10-4.** *Vertical cues. Note the relative difference vertically between the top and bottom edges of the sign and things behind it, or the vertical distance between the sign and the edge of the runway.*

to determine why the helicopter is responding in such a way. Adjustment or reduction of load may be necessary.

To begin the maneuver, head the helicopter into the wind, if possible. Place the cyclic in the neutral position with the collective in the full down position. Increase the throttle smoothly to obtain and maintain proper revolutions per minute (rpm), then slowly raise the collective. Emphasize a smooth, continuous movement, coordinating the throttle to maintain proper rpm.

As the collective is increased, the helicopter becomes light on the landing gear, and torque tends to cause the nose to swing or yaw to the right unless sufficient left antitorque pedal is used to maintain heading. (On helicopters with a clockwise main rotor system, the yaw is to the left and right pedal must be applied.)

As the helicopter becomes light on the landing gear, cyclic pitch control adjustments are necessary to maintain a level attitude. Unless the helicopter design compensates, translating tendency requires constant left cyclic in helicopters with counterclockwise rotating main rotor blades. Many helicopters are designed to hover in a left-landing-gear-low attitude to correct for right drift.

NOTE: When the term "level attitude" is used during hovering flight, it is in reference to the helicopter remaining stationary without excessive tilt of the fuselage. Each helicopter tends to hover at some attitude that may not be exactly level or parallel with the surface. Translating tendency, winds, and weight and balance all contribute to the fuselage hanging at some off angle to the surface. When the tailrotor is under the plane of the main rotor disk, the fuselage is titled to stop the travel induced by the lateral tail rotor thrust counteracting the rotor torque.

When the manufacturer designs the helicopter with an elevated tailrotor, which places the antitorque in the same plane as the main rotor torque, the fuselage tilt is much less in calm wind conditions. If enough wind is blowing from the opposite direction, the helicopter deck or cabin floor may be level or parallel to the surface, as the cross wind pushes on the side of the fuselage. However, if the wind is blowing from the other side, the deck tilt angle is increased because the rotor must develop sufficient thrust to counteract the tail rotor thrust and the crosswind to maintain position over the surface.

If the helicopter only has a single, lightweight pilot aboard, the fore and aft deck or cabin floor angle is probably pitched nose-up compared to having two heavy pilots up front with no other loading. Depending on the helicopter, unloading one person can result in a 12 percent change in gross weight and

a corresponding change in pitch. Helicopters usually hover in a normal attitude for that helicopter in that wind condition under that load condition. The student should be taught what the normal attitude for hovering is for the particular helicopter that they are flying.

When airborne, at a hover, antitorque pedals are used to maintain heading and directional control while application of collective ensures continuous vertical ascent to the normal hovering altitude. When hovering altitude is reached, use the throttle and collective to control altitude, the cyclic to maintain a stationary hover, and the antitorque pedals to maintain heading.

Initially, the student will probably overcontrol the helicopter. Excessive movement of any flight control requires a change in the other flight controls. For example, if the helicopter drifts to one side while hovering, a student naturally moves the cyclic in the opposite direction. When this is done, part of the vertical thrust is diverted, resulting in a loss of altitude. To maintain altitude, the student must increase the collective. This increases drag on the blades and tends to slow them. To counteract the drag and maintain rpm, the throttle needs to be increased. Because torque increases, the student must add more pedal pressure to maintain the heading. This can easily lead to overcontrolling the helicopter. However, as the student's level of proficiency increases, problems associated with overcontrolling decrease.

To maintain a hover over a point, have the student look for minute changes in the helicopter's attitude and altitude by checking the rotor disk changes against the horizon. When these changes are noticed, the student should make the necessary control inputs before the helicopter starts to move from the point. To detect small variations in altitude or position, the student's main area of visual attention needs to be some distance from the aircraft, using various points on the helicopter or the tip-path plane as a reference. Looking too close or looking down leads to overcontrolling. Obviously, in order to remain over a certain point, the student should know where the point is, but his or her attention should not be focused there.

Note: Helicopter pilots tend to use their peripheral vision more than most pilots. At some distance out in front of the helicopter, the horizon is used for attitude control. Peripheral vision is mainly used close in, near the helicopter, enabling the helicopter pilot to discern the clues of movement from a point (stationary reference). At some point during pilot training, many pilots learn to view the horizon with their peripheral vision and view the landing area with their center vision. Inexperienced students that attempt this at the beginning of their training tend to concentrate only on their intended landing area (tunnel vision). They pay little or no attention to the clues of movement close to the helicopter or their attitude on the horizon. This can lead to loss of control.

After a student gains experience, he or she develops a certain "feel" for the helicopter. The student feels and sees small deviations, so corrections can be made before the helicopter actually moves. A certain relaxed looseness develops, and controlling the helicopter becomes second nature, rather than a mechanical response. [Figure 10-5]

## Common Student Difficulties

### Failure To Position Controls Properly

The beginning student rarely knows how to position the controls so the helicopter lifts off the ground in a level attitude with no tendency to turn. Have the student check the tip-path plane of the rotor before raising the collective, looking forward and to each side to see that it is level.

### Visualizing Attitude

The problems of visualizing attitude in the early stages of training can be compounded in hovering flight by looking at a point that is too close to the helicopter. This is a natural tendency when trying to stay over a spot. The student should pick a point well in front of the helicopter, so the horizon is within normal peripheral vision. This makes it easier to perceive the helicopter's attitude while keeping the exact position in view.

### Overcontrolling

The natural tendency to overcontrol is accentuated by the responsiveness of the helicopter and the student's eagerness to get back over the takeoff spot immediately. While the ultimate objective in hovering is to stay exactly over a spot, the problem of overcontrolling can be alleviated by simply having the student stay within a general area, with the stated objective of gently stopping any drift that develops.

Note: Remind the student of Newton's first law of motion. An object in motion stays in motion with the same speed and in the same direction unless acted upon by an unbalanced force.

This is often referred to as "the law of inertia." The student can get ahead of the helicopter and make control inputs before the helicopter has had time to completely respond to any inputs. This is the normal progression of the student gaining experience and getting the "feel" of the helicopter.

### RPM Control

In the initial attempts at hovering, the student usually does not check rpm and make the necessary corrections. On helicopters equipped with a governor or correlator, rpm control is considerably easier. Periodically call attention to the rpm, and point out that changes in engine rpm can also be heard.

### Coordination

It is not uncommon for a student, particularly a student transitioning from airplanes, to attempt to gain altitude by applying rearward cyclic pressure, or attempt to turn by using lateral cyclic. To correct this tendency, many instructors operate one or two of the controls while allowing the student to concentrate on the reaction produced by the remaining control. For example, the instructor can operate the pedals and collective while the student experiments with the reactions produced by cyclic inputs.

### Tension

Tension is the natural result of a student's efforts to perform satisfactorily. The initial stage of training requires a great deal of patience on the part of the instructor and a lot of positive reinforcement and encouragement for the student. When tension builds to a point where the student is incapable of performing with an acceptable degree of proficiency, the instructor should take over and allow the student to relax for several moments. Usually, it is best to land and talk over the problems and to ensure neither the student nor instructor compares the student's performance with the instructor's performance. After encouragement and constructive criticism, another takeoff can be performed.

### Hovering Height

Many beginning students have a tendency to hover too high or too low. Hovering too high can create a hazardous flight condition, while hovering too low creates a risk of touching the ground with lateral movement and possible dynamic

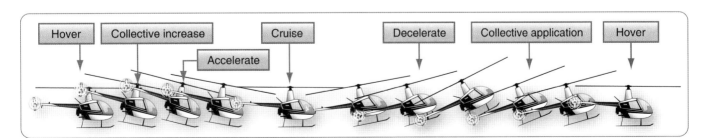

**Figure 10-5.** *Sequence of attitudes from hover to hover.*

rollover. Abrupt aft cyclic movement can also lead to the tail guard/stinger striking the ground and, in worse cases, the tail rotor. To help avoid this problem, continually reinforce what the correct height should look like and continuously remind the student that a good scan helps prevent unwanted altitude changes.

## Hovering Turn

Demonstrate how a hovering turn is accomplished by manipulating the antitorque pedals while the helicopter remains over a designated spot at a constant altitude. *[Figure 10-6]* The turn should be made at a low, constant rate through varying degrees of heading.

### Instructional Points

A hovering turn is initiated in either direction by applying antitorque pedal pressure toward the desired direction. Explain that during a turn to the left more power is needed because application of left pedal increases the pitch angle of the tail rotor which, in turn, requires additional power from the engine. A turn to the right requires less power. (On helicopters with a clockwise rotating main rotor, right pedal increases the pitch angle and, therefore, requires more power.)

As the turn begins, use the cyclic as necessary (usually into the wind) to keep the helicopter over the desired spot. *[Figure 10-7]* To continue the turn, additional pedal pressure is required as the helicopter turns to the crosswind position.

This is because the wind is striking the tail surface and the tail rotor area, making it more difficult for the tail to turn into the wind. As pedal pressures increase due to crosswind forces, additional cyclic pressure into the wind is required to maintain position. Use the collective with the throttle to maintain a constant altitude and rpm.

After the 90° portion of turn, pedal pressure is decreased slightly to maintain the same rate of turn. Approaching the 180° (downwind) portion, opposite pedal pressure must be anticipated due to tail movement from an upwind position to a downwind position. At this point, the rate of turn has a tendency to increase at a rapid rate due to the weathervaning tendencies of the tail surfaces. Because of the tailwind condition, hold rearward cyclic pressure to keep the helicopter over the same spot.

Note: The horizontal stabilizer can increase the difficulty of hovering with a strong tailwind. Depending on the specific design and mounting of the stabilizer, a tailwind may tend to lift the tail, requiring more aft cyclic, or lower the tail boom, decreasing the amount of cyclic needed into the wind to hold your position. The student needs to understand the stabilizer's response to tailwind conditions so as to anticipate control movements required.

Because of the helicopter's tendency to weathervane, maintaining the same rate of turn from the 180° position

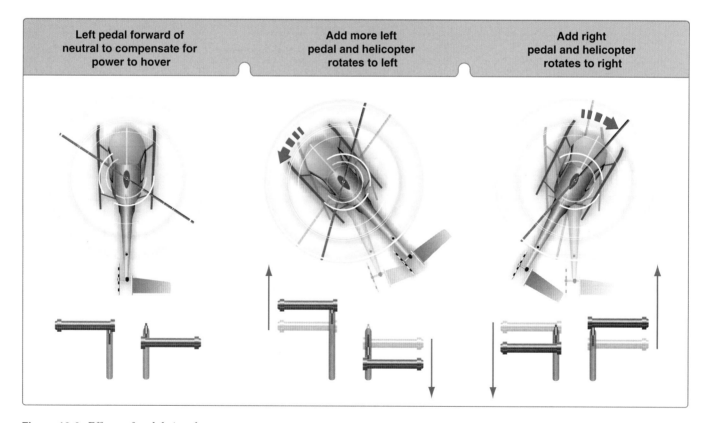

| Left pedal forward of neutral to compensate for power to hover | Add more left pedal and helicopter rotates to left | Add right pedal and helicopter rotates to right |

**Figure 10-6.** *Effects of pedals in a hover.*

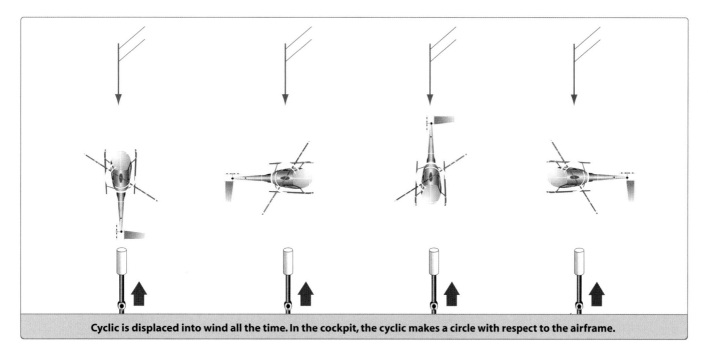

Cyclic is displaced into wind all the time. In the cockpit, the cyclic makes a circle with respect to the airframe.

**Figure 10-7.** *Hovering turns with winds.*

actually requires some pedal pressure opposite the direction of turn. If opposite pedal pressure is not applied, the helicopter tends to turn at a faster rate. The amount of pedal pressure and cyclic deflection throughout the turn depends on the wind velocity. As the turn finishes on the upwind heading, apply opposite pedal pressure to stop the turn. Gradually apply forward cyclic pressure to keep the helicopter from drifting.

Control pressures and direction of application change continuously throughout the turn. The most dramatic change is the pedal pressure (and corresponding power requirement) necessary to control the rate of turn as the helicopter moves through the downwind portion of the maneuver.

The instructor can have the student make turns in either directions; however, in a high wind condition the tail rotor may not be able to produce enough thrust, which means the student will not be able to control a turn to the right in a counterclockwise rotor system. Therefore, if control is ever questionable, have the student first attempt to make a 90 degree turn to the left. If there is sufficient tail rotor thrust to turn the helicopter crosswind in a left turn, a right turn can be successfully controlled. The opposite applies to helicopters with clockwise rotor systems. Hovering turns should be avoided in winds strong enough to preclude sufficient aft cyclic control to maintain the helicopter on the selected surface reference point when headed downwind.

### Common Student Difficulties

In addition to the difficulties already discussed in the *Takeoff to a Hover* section, there are some difficulties associated specifically with the hovering turn.

#### Improper Rate of Turn

Until the student has gained some experience in hovering turns, the amount of pedal required for the desired rate of turn is not known. The result is a turn that is either too slow or too fast, often varying rapidly between the two. The first hovering turns should be practiced in calm or light winds, so a certain pedal input results in a specific rate of turn.

#### Compensating for Crosswind

Students usually fail to anticipate the effect of the wind as the helicopter turns. The student must understand that, throughout the turn, the cyclic is displaced into the wind, and is independent of the direction of the turn. Also, pedal input must be increased as the turn approaches the crosswind position, then decreased as the downwind position is approached. Passing the downwind position, the student should anticipate an increase in the rate of turn as a result of the wind force.

#### Coordination

Before attempting hovering turns, the instructor should explain and demonstrate the effects of pedal input. For example, explain how a left pedal input causes a right drifting tendency, which must be compensated for by using left side cyclic. Even more noticeable is the effect on engine rpm. Left pedal input causes a decrease in rpm and right pedal input causes an increase. As the student gains an understanding of these effects, the tendency to overcontrol the antitorque pedals should diminish.

## Hovering Forward

Forward hovering should be accomplished at hovering altitude and at a speed no faster than a brisk walk with heading remaining constant. The forward track should be defined by markings on the ground or by the alignment of two reference points. *[Figure 10-8]*

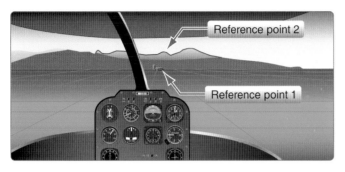

**Figure 10-8.** *To maintain a straight ground track, use two reference points in line and at some distance in front of the helicopter.*

### *Instructional Points*

Instruct students on the importance of maintaining a landing gear height high enough to allow adequate ground clearance before hovering in any direction. Stress to the student that the risk of dynamic rollover is greatest during any hovering maneuver. This also stresses the importance of keeping the landing gear aligned with the direction of travel.

Apply forward cyclic to start the forward motion, then release some cyclic pressure to prevent the helicopter from accelerating. Hold enough forward cyclic pressure to keep forward motion no faster than a brisk walk. Any speed higher than this requires a higher landing gear height to allow adequate ground clearance for the tail landing gear when bringing a helicopter to a stop using rearward cyclic. As the helicopter begins to move forward and lift is diverted, add a little power to compensate for the loss of lift.

Throughout the maneuver, maintain a constant ground speed and path over the ground with the cyclic, a constant heading with the antitorque pedals, altitude with the collective, and the proper rpm with the throttle.

To stop the forward movement, apply rearward cyclic pressure so that the helicopter glides to a halt at a hover, taking care not to lower the tail rotor into the ground. As forward motion stops, return the cyclic to the neutral position to prevent rearward movement. Forward movement can also be stopped by simply applying rearward pressure to level the helicopter and let it drift to a stop.

## *Common Student Difficulties*

### *Altitude Control*

The student may not understand that an airspeed of about 5 knots requires the most power to maintain altitude as ground effect diminishes and translational lift has not begun to help. As the helicopter begins to move forward in a calm wind, it also tends to sink. The student may think this is caused by too much forward cyclic, and the resulting correction causes a helicopter to stop. Point out that a slight amount of increased collective is required as forward motion starts. This usually alleviates the problem.

Note: Moving forward requires forward thrust, whereas at a stationary hover, only enough thrust is needed to overcome the wind. Begin a forward hover by diverting some lift to thrust, if that lift is not restored, the hovering altitude decreases. The more lift diverted to thrust, the lower the hover altitude, or the more collective must be increased, with the power increase sufficient to maintain rpm with more antitorque needed as well.

### *Sideward Drift*

Drift to the side of the planned ground track can be the result of concentration on trying to maintain the heading and altitude. If the ground track is being maintained by reference to a line on the ground, the student may be looking too close to the helicopter and may not notice changes in the altitude. If the student has trouble maintaining the specified ground track, refocus attention to reference points that are farther away from the helicopter.

## Hovering Sideward

Sideward flight begins in a hover and is performed at a constant heading, altitude, and airspeed.

### *Instructional Points*

Explain to the student that the risk of dynamic rollover is highest during sideward hovering maneuvers. Maintain adequate landing gear height. Also ensure that sufficient clearance exists for the expected and possible path of the tail rotor for sideward hovering.

Before starting sideward flight, make sure the student clears the area. This may require some clearing turns. Then have the student pick two points of reference in a line in the direction of sideward flight to help maintain the proper ground track. These reference points should be kept in line throughout the maneuver. *[Figure 10-9]*

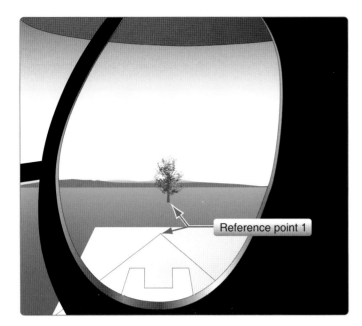

**Figure 10-9.** *The key to hovering sideward is establishing at least two reference points that help maintain a straight track over the ground while keeping a constant heading.*

The maneuver is begun at a normal hovering altitude by applying cyclic toward the side in which the movement is desired. As the movement begins, return the cyclic toward the neutral position to keep the groundspeed low. Throughout the maneuver, maintain a constant groundspeed and ground track with cyclic. Maintain heading, perpendicular to the ground track in this maneuver, with the antitorque pedals, and a constant altitude with collective. Use the throttle to maintain the proper operating rpm. As with all maneuvers, instructors should emphasize the importance of scanning. Viewing objects or obstacles while flying sideways can be deceptive. From a distance, trees and wires may look smaller or higher than they actually are. Terrain may look flat until you hover closer to it and quickly realize that it slopes up and you could possibly contact it with either the landing gear or tail rotor.

To stop the sideward movement, apply cyclic pressure in the direction opposite to that of movement and hold it until the helicopter stops. As motion stops, return the cyclic to the neutral position to prevent movement in the opposite direction. Applying sufficient opposite cyclic pressure to level the helicopter may also stop sideward movement. The helicopter then drifts to a stop.

## Common Student Difficulties

*Speed Control*

In sideward flight, lateral cyclic input controls speed. If the student is looking primarily to the side in an attempt to maintain the track, roll attitude can be difficult to maintain. Scan must be continuous if the correct attitude is to be maintained. The student must continuously check to the side, then look in front to check attitude. This is followed by a check of the rpm, then a look back to the side.

*Drift*

Drift can also be an attitude problem. If the student concentrates too much to the side, pitch attitude can deviate from level, resulting in drift from the desired track.

*Heading*

As the helicopter begins to move sideward, the nose tends to weathercock into the direction of flight. Again, this may not be noticed if the student is concentrating his or her attention in the direction of flight.

## Hovering Rearward

Rearward hovering is conducted using reference points ahead of the helicopter to maintain track. Altitude and heading should remain constant, and groundspeed should be no faster than a brisk walk.

### *Instructional Points*

Before beginning the maneuver, make sure the area behind the helicopter is clear. This is accomplished by making a 90° clearing turn. Choose two reference points in front of, and in line with, the helicopter as if hovering forward. The movement of the helicopter should be such that these points remain in line.

Begin the maneuver from a normal hovering altitude by applying rearward pressure on the cyclic. Once the movement has begun, position the cyclic to maintain a low groundspeed (no faster than a brisk walk). Throughout the maneuver, maintain constant groundspeed and ground track with the cyclic, a constant heading with the antitorque petals, constant altitude with the collective, and the proper rpm with the throttle.

When hovering backwards, the helicopter is tilted so the tail is low to the ground. Therefore, maintain a slightly higher than normal hovering altitude.

To stop the rearward movement, apply forward cyclic and hold it until the helicopter stops. As the motion stops, return the cyclic to the neutral position. Also, as in the case of forward flight and sideward flight, use opposite cyclic to level the helicopter and let it drift to a stop.

### *Common Student Difficulties*

*Speed Control*

The student may not realize that it takes a steeper pitch attitude to start the helicopter moving than it does to continue motion at a steady speed. If the nose is not moved down slightly as the desired rearward speed is attained, the helicopter continues to accelerate.

Note: Acceleration and force are vectors. In Newton's second law of motion, the direction of the force vector is the same as the direction of the acceleration vector. In other words, an object with a certain velocity maintains that velocity unless a force acts on it to cause an acceleration (that is, a change in the velocity). If the pitch attitude is not returned to a neutral (non-accelerating or decelerating) attitude, stabilization of the speed and velocity cannot occur.

*Heading*

The faster the helicopter travels rearward, the greater the tendency for the nose of the helicopter to swing around toward the direction of flight. With the tail directly into the relative wind, there is little tendency for it to weathervane, but if the relative wind is a little bit on one side, the tail tends to continue to the downwind side. The resulting heading correction requires a fairly large pedal input, which may cause an overshoot to the other side, and the process must be repeated with opposite pedal input. Speed must be reduced to regain control. During preflight, show the student which surface areas of the helicopter are affected by relative wind from different directions (what is causing the weathervane). Discuss how you need to overcome the effects of the wind during different directions of hover (forward, rearward, sideward).

## Landing From a Hover

The helicopter is stabilized in a hover directly over the intended landing spot, then gently lowered onto the ground. It should not drift in any direction at the point of touchdown. The instructor should remind the student pilot of the flight control changes that must occur during this seemingly simple task. As the helicopter begins to settle onto the surface, all of the flight controls must be manipulated simultaneously and in coordination to achieve a smooth landing.

When choosing a landing area for the student to practice landing from a hover, instructors should keep in mind that certain conditions usually dictate a landing directly to the ground with little or no hover. For example, dust, sand, or snow landings are very difficult and should not be attempted until the student has shown considerable proficiency with takeoffs and landings. These types of landings are discussed in Chapter 12, Helicopter Emergencies. Landing in a grassy field or in a spot with puddles of water can also cause problems for flight students. The grass or water motion presents a false picture of helicopter movement to the new pilot and causes them to incorrectly respond when attempting to land.

## *Instructional Points*

The student should be instructed to look outside and ahead of the helicopter. Focusing on the ground through the chin bubble leads to overcontrolling and makes it difficult to land on the desired spot.

Always keep the rpm within limits. This allows for quick transition back to hover if the landing is not suitable. Never allow the helicopter to settle on the ground, which might occur if the throttle is reduced below the rpm limits. To prevent overspeed, the correct technique requires simultaneously lowering the collective and reducing the throttle.

Do not abruptly lower the collective once ground contact is made. First, ensure the ground is sufficiently stable to support the helicopter. This requires a slow and deliberate lowering of the collective. The cyclic may be moved in a small circular motion to determine that the helicopter is firmly on the ground before lowering the collective fully. Once the helicopter is firmly on the ground, the collective should be lowered completely.

Landing a helicopter requires the same attention as takeoffs to a hover. Point out which skid/gear will contact the surface first and why. What control inputs will be applied to stop sliding (moving forward)? Forward or aft cyclic may be required during the landing to maintain the position. How is excessive slope recognized before the tilt is too much to overcome?

## *Common Student Difficulties*

### *Attitude Control*

The closer the helicopter comes to the ground, the more likely it is for the student to focus on a point almost directly beneath the helicopter. As the helicopter descends, ground effect tends to increase, thereby creating a greater workload for the pilot to maintain coordination of all the flight controls until the helicopter is on the surface. Without proper attitude technique, the student may overcontrol when the helicopter begins to drift, and the situation may go from bad to worse. The student must be taught to look far in front of the helicopter and then gently lower it until the touchdown is felt, not seen. Remind the student that most likely one side of the landing gear contacts the surface before the other due to winds or translating tendency. Due to loading and winds, the front or the rear of the landing gear may touchdown first depending on the hovering attitude of that helicopter. In any event, as the landing gear first touches down, cyclic and pedal corrections are continually necessary to maintain heading and position until the remainder of the landing gear is firmly on the surface.

### *RPM Control*

During the landing, rotor rpm tends to increase due to the effects of increased ground effect and decreased collective pitch. During touchdown, for those helicopters not equipped with a governor, the throttle may need to be reduced to avoid an overspeed.

# Taxi

Once a student has learned the basic skills required to hover the helicopter, those skills translate into the practical application of taxiing the helicopter. In order to accomplish this task the student should understand the terms used and the limitations placed on each action and the basic information and safety involved in selecting hover altitudes and speeds derived from the H/V chart.

## Hover Taxi

The term hover taxi is used to describe a helicopter movement conducted above the surface and in ground effect at airspeeds less than approximately 20 knots. [Figure 10-10] The actual height may vary, and some helicopters may hover taxi above 25 feet above ground level (AGL) to reduce ground effect turbulence or provide clearance for cargo slingloads.

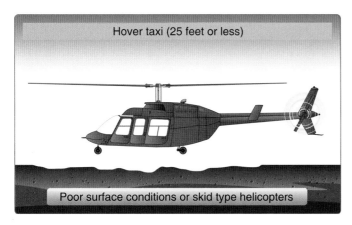

**Figure 10-10.** *Hover taxi.*

## Air Taxi

The term air taxi describes helicopter movement conducted above the surface, but normally not above 100 feet AGL. [Figure 10-11] The helicopter may proceed either via hover taxi or flight at speeds more than 20 knots. The pilot is solely responsible for selecting a safe airspeed/altitude for the operation being conducted.

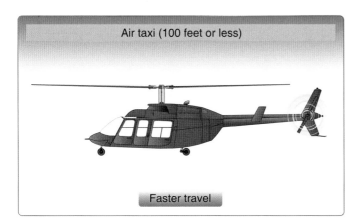

**Figure 10-11.** *Air taxi.*

## Surface/Ground Taxi

Taxi, surface taxi, or ground taxi is the movement of an aircraft under its own power actually in physical contact with the surface of an airport. It also describes the surface movement of helicopters equipped with wheels. [Figure 10-12]

**Figure 10-12.** *Surface taxi.*

Ground taxi can form the basis for a running takeoff used by older, underpowered helicopters. Ground taxi creates less downwash, since less thrust is required to slide the skids than to support the entire weight of the helicopter. Wheeled taxi is very efficient. Water or ski taxi is between those two extremes in terms of power required.

Surface taxi includes ground contact taxiing with wheels, floats, skis, or skids. Ground contact requires less power than hover taxiing and produces less rotor wash, depending on the surface friction. If skis are stuck to the surface, it may take some power to break them loose. Surface taxi also provides an alternative method of taking off in white or brownout conditions by blowing the obstructing material behind the helicopter as some airspeed is reached and certainly by translational lift speed. Caution should be observed to prevent the landing gear from being stuck to the surface or striking something on the takeoff surface, leading to dynamic rollover. Surface taxi is not a common or preferred maneuver, but it has been used in certain situations, generally in much older, underpowered helicopters.

## Instructional Points

For hover taxi, air taxi, and taxi, have the student review the Aeronautical Information Manual (AIM), paragraphs 4-3-17 and 7-5-13c. Prior to air taxi demonstration and practice, the instructor should review the H/V chart and have the student assist in planning or developing a safe air taxi profile of airspeeds versus altitudes. Since the AIM gives 100' AGL as the upper limit, the instructor should find 100'AGL on the chart and determine minimum and maximum airspeeds for the maneuver, remembering the helicopter requires power to decelerate and come to hover without translational lift and thrust. Discuss with the student the H/V chart to determine the safest flight profile that should be used during any taxi operation.

The advantages of ground or surface taxi for skid-equipped helicopters should be discussed so the student will understand some criteria for deciding on their own when ground taxi is safer and the recommended maneuver for that occasion.

Discuss with the student the H/V chart to determine the safest flight profile that should be used during any taxi operation.

## Ground Reference Maneuvers

Ground reference maneuvers are training exercises that can be used to develop coordination, division of attention, and situational awareness. The maneuvers themselves are not evaluated during the practical test, but the skills developed by accomplishing them are evaluated during the conduct of other tasks. By performing ground reference maneuvers, a student develops a better understanding of the effects of

wind drift and how to compensate for winds from different directions during flight.

## Rectangular Course

The rectangular course helps the student develop recognition of drift toward or away from a line parallel to the intended ground track. *[Figure 10-13]* It is important that he or she understand the effects of the wind and how to compensate for it. The rectangular course also simulates an airport traffic pattern, as well as many of the maneuvers a helicopter is tasked with performing, such as tracking an event for photographic purposes, aerial surveys, and observation duties. This is an opportunity to point out to the student what each segment of a traffic pattern represents (upwind, crosswind, downwind, base, and final leg).

For this maneuver, pick a square or rectangular field, or an area bounded on four sides by section lines or roads, where the sides are approximately a mile in length. The area selected should be well away from other air traffic. Fly the maneuver between 500' and 800' feet AGL, which is the altitude normally required for an airport traffic pattern to avoid the flow of fixed wing traffic. If the student finds

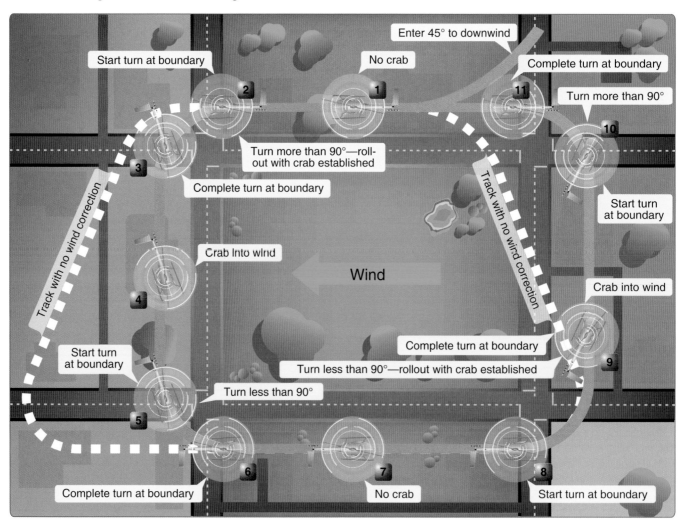

**Figure 10-13.** *A rectangular course.*

it difficult to maintain a proper ground track at that higher altitude, adjust the altitude for better ground reference until the student feels more comfortable and is able to grasp the concept better. The helicopter should be flown parallel to and at a uniform distance from the field boundaries, not above the boundaries. Demonstrate that by flying directly above the edges of the field, there are no usable reference points to start and complete the turns. In addition, the closer the track of the helicopter is to the field boundaries, the steeper the bank necessary at the turning points.

The student should understand that when trying to fly a straight line and maintain a specific heading, the helicopter should always be kept in trim with the antitorque pedals and the pilot should crab into the wind (with the cyclic) to stay on the proper ground track. The concept of "crabbing" can be difficult for the student to understand when first learning how to fly a traffic pattern. The instructor should show the student what happens when you try to fly a straight line with a crosswind, and point out how far off course the wind can take you. Keeping the helicopter in trim and keeping the helicopter straight are done with two different flight controls, and the instructor should ensure the student understands the effects of crabbing to allow for wind drift. Since the helicopter is not headed exactly parallel to the rectangle course, turns at the corners of the rectangle may be more or less than 90 degrees with a shallower or steeper bank angle to hold the correct distance from the rectangle. The rectangular course requires the student to adjust for winds from each quadrant. Also ensure that the cyclic trim (if installed) is properly trimmed in order to decrease the pressures on the pilot.

Prior to takeoff, the instructor should discuss with the student how the wind velocity aloft are greater than those reported on the ground and may possibly shift. If the winds are known, the instructor can have the student calculate the amount of wind crab necessary to track the boundaries. This allows the student more insight to the use of crab angles to track courses. Not enough emphasis can be placed on always knowing which way the winds are coming from and how to be vigilant in seeking ground cues for hints of a wind change.

## S-Turn

The S-turn is another training maneuver that requires a student to compensate for winds *[Figure 10-14]* This maneuver requires turns to the left and right. Choose a road, a fence, or a railroad for a reference line. Regardless of what is used, it should be straight for a considerable distance and should extend as close to perpendicular to the wind as possible.

An S-turn is a pattern of two half circles of equal size on opposite sides of a reference line. A standard radius for S-turns cannot be specified, since the radius depends on the airspeed of the helicopter, the velocity of the wind, and

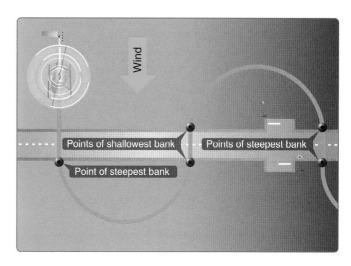

**Figure 10-14.** *S-turns across a road.*

the initial bank chosen for entry. The maneuver should be performed at a constant altitude. While S-turns may be started at any point, it may be beneficial during early training to start the maneuver with the helicopter flying into the wind.

## Turns Around a Point

This training maneuver requires the student to fly constant-radius turns around a preselected point on the ground, using a bank of approximately 30° while maintaining both a constant altitude and the same distance from the point throughout the maneuver. *[Figure 10-15]* The objective, as in other ground reference maneuvers, is to develop the ability to control the helicopter subconsciously while dividing attention between the flightpath, how the winds are affecting the turn, and ground references, while still watching for other air traffic in the vicinity.

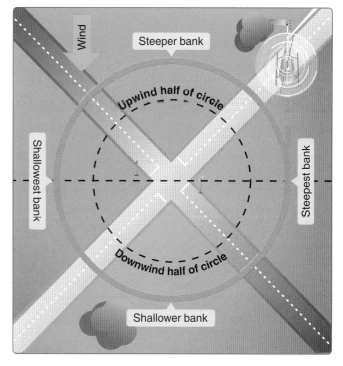

**Figure 10-15.** *Turns around a point.*

A reason to perform turns around a point, other than for proficiency, is high reconnaissance, photography, survey, and search and rescue. Great care should be exercised to remain clear of obstructions and other low-level traffic during ground reference maneuvering.

The factors and principles of drift correction that are involved in S-turns are also applicable in this maneuver. As in other ground track maneuvers, a constant radius around a point requires the student constantly to change the angle of bank and make numerous control changes to compensate for the wind. The point selected for turns around a point should be prominent and easily distinguishable, yet small enough to present a precise reference.

## Common Student Difficulties

### Failure To Plan Properly

Ensure the student plans properly for the ground reference maneuver to be flown, that should include checking the current and forecasted weather. Students often fail to take the wind into consideration when setting up to perform a maneuver. Point out that winds can be determined by observing ground cues, such as flags, ripples on a pond, or smoke. The student should be able to explain, prior to takeoff, how to maintain desired ground track while in flight.

### Coordination

Remind the student that each control input causes something else to change. Remember, a beginning student may have difficulty not only in accomplishing the actions in the proper sequence, but also in compensating for control inputs. Talk the student through each maneuver to remove any doubt about what is to be accomplished, as well as how and when it is done.

### Division of Attention

When performing a ground reference maneuver, students have a tendency to focus attention outside the helicopter, excluding cockpit or instrument checks or focus entirely on the interior checks to the exclusion of outside references. If the student focuses all attention outside the helicopter, a good way to refocus attention inside is to pose a question, such as "What is the helicopter's altitude?" or "What is the airspeed?" If the student's attention is focused inside the helicopter, have the student clear the helicopter or describe what he or she is using as an outside reference.

### Attitude

Remember that the forward seating position and the excellent visibility in most helicopters may make it difficult for a student to visualize the attitude of the helicopter. It is important for the instructor to provide all the assistance possible to ensure the student can determine an attitude by some visual reference such as a cross-member or point on the wind screen. A water soluble marker can be used to place dots on the bubble representing the sight picture of the horizon at a given airspeed. Instructors usually develop different methods of teaching attitude references.

### Scan

It is common for a student to concentrate on one factor to the exclusion of others. Due to poor scanning technique, the student may select a ground reference that fails to offer a suitable emergency landing area within gliding distance.

## Instructor Tips

- A student will attempt to imitate instructor actions. Do not take shortcuts. Instill safety from the first day. Insist the student clear the helicopter in all directions prior to performing any maneuver.

- Stay close to the controls at all times, but especially during the hover. Be prepared for both the expected and the unexpected.

- Students transitioning to helicopters from airplanes may have "air sense" but, remember, they are still students. Students do the unexpected, especially students who are transitioning from airplanes to helicopters, and should be closely supervised. There are many negative transfers of training from airplanes to helicopters. In instances of stress, the airplane pilot can be expected to revert to "first learned" behaviors, which can have deadly consequences in helicopters. The differences should be well explained and briefed before each flight.

- When a student encounters difficulty in mastering an objective, find a means of allowing some degree of success. For example, when practicing climbing and descending turns, rather than have the student attempt the entire maneuver, try having him or her practice climbing and descending. When no difficulty is experienced, add the turn, continue until the entire maneuver is completed. Should difficulty still occur, back up a step and work on climb and level-off and descent and level-off rather than cause too much frustration. Sometimes instructors make the mistake of continuing to have students attempt a maneuver when performance is deteriorating. It is better to quit at that point and go back to something the student can do well to rebuild his or her confidence. Remember, 3 or 4 iterations of each task is sufficient training. After that point, the instructor is probably wasting the student's time and money. Fatigue occurs, and training ends. More preflight briefing and ground school may be needed.

- Beginning with the first flight, students learn to preplan the performance parameters for the entire flight.

  1. Power required to hover (OGE, IGE)

  2. Total power available

  3. Power available at the highest altitude and temperature for the day.

  4. Hover requirements at the highest altitude and temperature for that day.

It is essential to discuss the meeting of these parameter values and to ensure that the student understands the limitations imposed by these values, even if the RFM does not state them as limits. *[Figure 10-16]*

---

## Basic Flight Maneuvers

### Objective

Tho purpooc of thio lcoaon io for the student to learn how to turn at a hover. The student will demonstrate a basic ability to turn the helicopter at a hover.

### Content

1. Preflight Discussion
   a. Discuss lesson objective and completion standards.
   b. Review normal checklist procedures coupled with introductory material.
   c. Review weather analysis
   d. Explain to the student that maneuvering close to the ground and obstacles is a major component of the helicopter's operational environment, particularly in confined areas and when clearing a parking area. This is an important exercise that must be mastered completely.
   e. Review basic helicopter aerodynamics.

2. Instructor Actions
   a. Preflight used as introductory tool
   b. Describe the techniques for making hovering turns and stress the following points:
      1) Discuss Loss of Tail Rotor Effectiveness (LTE).
      2) There can be problems with yaw control and a need for increased power when the helicopter is downwind or crosswind in strong wind conditions.
      3) Clearing the helicopter is important during all hovering maneuvers and, in particular, for low obstacles that are hard to see and that can snag the landing gear or tail rotor.
      4) In strong or gusty wind conditions, a turn away from into the wind should be in the opposite direction to the torque reaction (i.e., to the left in a helicopter with a counterclockwise turning rotor). In this way it is possible to ensure there is sufficient tail rotor control available. If control limits are reached at this stage, a safe return to into-wind is easily accomplished.
      5) No turns or any movements from the hover should be initiated until the helicopter is settled in an accurate hover at the required rpm and power setting.
      6) The continuous use of high power in this exercise means that a careful watch should be kept on engine temperatures and pressures. Prolonged hovering out of the wind should be avoided on some types of helicopter because of the dangers from carbon monoxide in the cockpit.
      7) In some helicopters at certain center of gravity (CG) configurations (i.e., high cabin loading), it is possible to reach the aft cyclic limits when hovering downwind. Warn the student of this possibility and describe the safe recovery actions when:
         - Turning into the wind
         - Landing straight ahead

3. Student Actions
   a. Student practices turning the helicopter at a hover.

### Postflight Discussion

Review the flight, preview the next lesson, and assign *Helicopter Flying Handbook,* Chapter 10, Advanced Flight Maneuvers.

**Figure 10-16.** *Sample lesson plan.*

## Chapter Summary

This chapter presented some training techniques and instructional points an instructor can use to teach hovering flight and basic flight maneuvers. Common difficulties students encounter when attempting to perform hovering flight and basic maneuvers were also discussed.

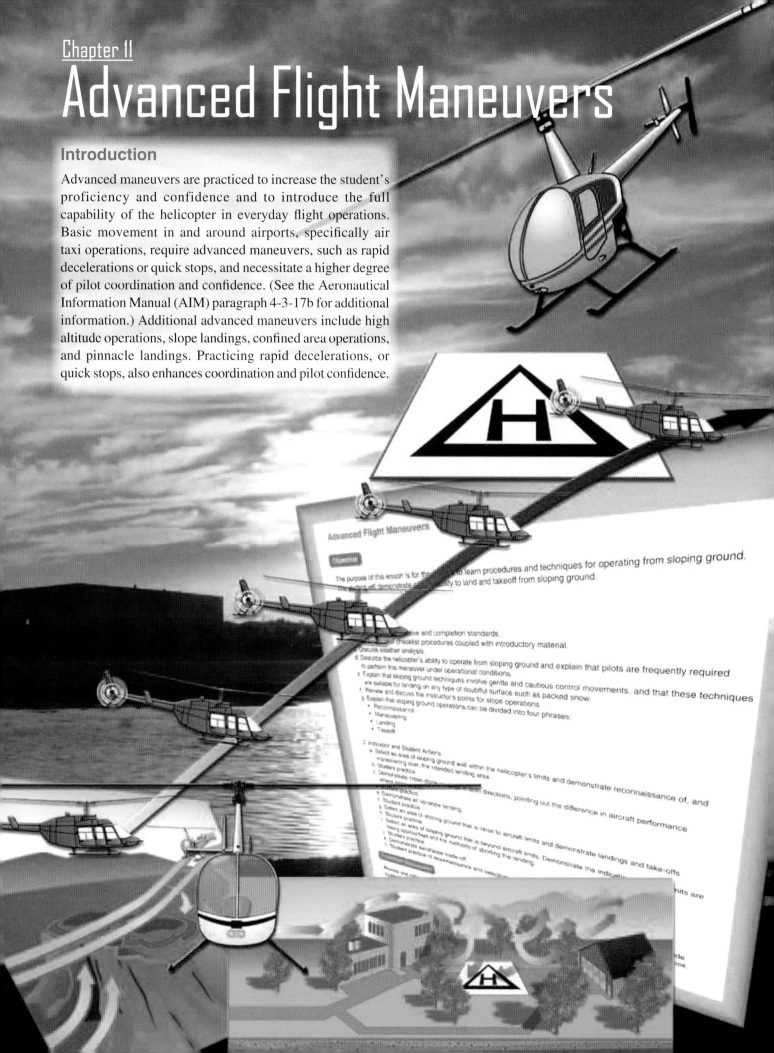

# Chapter 11
# Advanced Flight Maneuvers

## Introduction

Advanced maneuvers are practiced to increase the student's proficiency and confidence and to introduce the full capability of the helicopter in everyday flight operations. Basic movement in and around airports, specifically air taxi operations, require advanced maneuvers, such as rapid decelerations or quick stops, and necessitate a higher degree of pilot coordination and confidence. (See the Aeronautical Information Manual (AIM) paragraph 4-3-17b for additional information.) Additional advanced maneuvers include high altitude operations, slope landings, confined area operations, and pinnacle landings. Practicing rapid decelerations, or quick stops, also enhances coordination and pilot confidence.

## Instructor's Approach

The Commercial Pilot Practical Test Standards (PTS) provides a list of advanced helicopter maneuvers and standards for instruction at this level. The helicopter instructor should develop a plan to teach these maneuvers. Beginning with the simple types and ending with the most complex, the instructor must always keep under consideration the helicopter type, terrain, and ambient conditions prevalent at the time of training.

### Scenario-Based Training

In order to take full advantage of the student's interest in these maneuvers, the instructor should discuss the students intended career path within aviation. By doing this, the instructor can tailor a syllabus to the student's desire with application of specific maneuvers to related job scenarios.

As an example, a student desires to be an Emergency Medical Service (EMS) pilot in a large metropolitan area in the northeastern United States. Rapid decelerations, slope landings, confined areas, and pinnacle landings take on greater meaning to the student when portrayed as events typically encountered during EMS operations. The instructor can demonstrate the need for proficient and safe conduct of these maneuvers while replicating probable profiles the student will encounter. Scenarios involving patient pickup or drop-off in varying conditions and locations, such as sloping confined areas, pinnacle approaches to rooftop helipads, etc., maximizes the student's interest in these maneuvers, as well as, correlates directly to future situations likely to be encountered.

Another example would be a student desiring to become a long line pilot in the mountainous areas of western Canada. Scenarios designed to elevate the student's awareness of high altitude, out of ground effect (OGE) training will probably engage and energize the student.

Regardless of the student's desired career path, all scenarios should replicate environmental and situational conditions. Using the building block theory, begin training these tasks at normal helicopter configurations. Once the student demonstrates the understanding and coordination necessary for these tasks, progress to more demanding scenarios. For instance, add additional weight to simulate a max gross weight simulated condition. If environmental conditions allow, conduct brownout/whiteout training. Always stress the need to accomplish and review performance planning before each flight.

The end state goal is for the student to not only demonstrate proficiency of a maneuver, but to conduct the maneuver with a firm understanding of the environmental and situational impacts that will assist the student in avoiding potentially life threatening circumstances.

### Identification, Prevention, and Recovery

A very important instructional point is to ensure students are thoroughly familiar with identification, prevention, and recovery from the hazards discussed in the Helicopter Flying Handbook, Chapter 13, Helicopter Emergencies.

The student must understand that for some hazards, prevention is the only action, because there may be no recovery once the hazard is fully developed. The old adage of "I'd rather be on the ground wishing I could fly, than being in the sky wishing I were on the ground" holds a lot of weight.

Identification, prevention, and recovery can be illustrated using the following example.

Takeoffs from a muddy or tall grass location entail forethought. Identify the hazard before action is taken. A skid may stick in the mud or a log or other obstacle may elude detection from the pilot on takeoff. Identifying the potential hazards and taking the appropriate steps could prevent the potential for dynamic rollover or other unrecoverable mishaps. Understanding this, the best approach is to slowly lift off, while being mindful of the necessary recovery steps should you feel the aircraft begin to roll.

### Reinforce Fundamentals

The importance of positive habit transfer and the role the instructor plays in demonstrating the proper procedures cannot be overstated. A less experienced pilot will do what he or she has seen the instructor do. Therefore, it is imperative that the instructor continue to demonstrate good basic maneuver skills throughout the advance maneuver training.

The helicopter instructor should continue to emphasize appropriate use of checklists and instruction on how to accomplish the checklist while maintaining control of the helicopter and situational awareness. Clarify abort procedures with the student and stress that it is the pilot's responsibility to brief a passenger on abort options.

Discuss the various methods for determining aircraft takeoff, cruise, and departure performance. Integrate the use of performance charts in the preflight planning for every flight.

## Reconnaissance Procedures

An important component of advanced flight maneuvers is the ability to obtain information by first conducting a reconnaissance. Quite simply put, reconnaissance is gathering information. The student should not only understand the types of reconnaissance, but also what type of information is obtained by each reconnaissance. The student must understand the correlation and value of this information.

By correlation, it is implied that a lone building near a field that the pilot intends to land, should in most cases, have electric power. Not seeing the wires does not mean they are not there. Thorough reconnaissance may reveal an unseen set of wires. Certain expectations (wires) are correlated to certain situations (buildings).

Remember, reconnaissance begins at the planning table. Conduct thorough performance planning to determine if OGE power is available. Once arriving in the area is not the time to find out if it is needed. Use available map data to determine the terrain in the area to be flown. Identify valleys, canyons, and mountains that can produce unforgiving hazards.

## High Reconnaissance

There are three recommended flight patterns flown to conduct high reconnaissance: circular, racetrack, and when terrain dictates, the figure eight. Regardless of the type flown, the flight pattern should be conducted at an altitude that maintains clearance from any and all obstructions and at an altitude that allows safe egress from any possible environmental phenomenon, such as downdrafts, updrafts, turbulence, and varying wind velocities. Always be vigilant of the terrain and its effects on wind.

When beginning the high reconnaissance, ensure the student maintains visual awareness with the terrain and maintains aircraft airspeed limitations. Point out that the flight pattern should be maintained relatively close to the landing area and viewed from the pilot's side of the helicopter. Stress, however, that aircraft maneuvers should be limited to bank angles of 30° or less. The airspeed chosen should allow comfortable control of the aircraft, definitely above effective translational lift (ETL).

During the high reconnaissance, the following elements should be assessed:

- Wind—determine direction, speed, and location of the demarcation line and any other variables of wind flow.

- Obstacles—identify all obstacles to flight, physical structures, wires, towers, trees, etc.

- Approach path—pick an approach path over the shortest obstacles or an area void of obstacles. Account for winds and plan on an approach into the wind if obstacles allow. (Demonstrate crosswind approaches that utilize an approach path over shorter obstacles while avoiding tail winds).

- Landing area—the suitability of the landing area must be evaluated by the pilot.
  - o  Is the area large enough for landing and takeoff?
  - o  Determine slope of the intended landing area. Perhaps the scenario calls for loading or unloading of cargo or passengers, is that feasible with the perceived slope?
  - o  Obstacles in and around the area. How do the obstacles affect the wind inside the area?
  - o  Any surface debris that could damage the aircraft. Is there tall grass? (Bushes or saplings may cause tail rotor/under carriage damage.)
  - o  Dust or snow may cause whiteout/brownout.
  - o  Is there uneven terrain, or slopes?

- Takeoff route— locate the takeoff direction (into the wind) and lowest obstacles, and identify potential forced landing areas. Does the confined area permit repositioning to allow more room for departure?

Remember, when teaching a student, stress that options are always available. The primary location may not always be acceptable. Do not force the approach or landing. If the landing area is found to be unsuitable during any part of the reconnaissance, abort the maneuver and attempt to identify alternate locations nearby. Forcing the situation may lead to catastrophic results.

## Low Reconnaissance

Discuss with the student how a low reconnaissance is performed to verify information gathered by high reconnaissance. If the information from high reconnaissance was sufficient, then low reconnaissance can be combined with the approach. Emphasize that the availability of power for approach and landing is determined during the performance planning. Stress to the student that if at any time during low reconnaissance it is determined that conditions around the landing area are unsafe, reconnaissance and/or the approach are discontinued.

Ensure the student understands the following specific conditions, which are evaluated during low reconnaissance:

- Pinpoint wind direction and effects of the wind on surrounding terrain.

- Evaluate the touchdown point, size of the landing area, slope, type of surface, and any obstructions.

- Determine whether the approach should be terminated to the ground or to a hover.

- Evaluate the approach and departure path.

## Ground Reconnaissance

Explain to the student that, once the helicopter is in the area, ground reconnaissance is used to determine a landing point, the takeoff point, and the takeoff direction. Ensure the student understands the following specific items, which should be evaluated during the ground reconnaissance:

- Determine wind direction and effects within the area.

- Determine the location of the lowest obstacles and their relation to the wind direction.

- Evaluate the area to determine the most advantageous takeoff location and direction.

## Maximum Performance Takeoff

Maximum performance takeoff is practiced to simulate a takeoff from a confined area with a climb over an obstacle. Normally, it is begun from the ground with the collective raised to obtain maximum power while the pitch attitude is adjusted to establish a near vertical climb to clear an obstacle. Height of obstacle permitting, at an altitude of about 50 feet, the nose is lowered gently to accelerate to normal climb speed, attaining efficiency of ETL. The pilot should always be watching the rotor disk in the climb and the distance between the rotor disk and the obstructions should not narrow. If the rotor disk is not clearing the obstruction, then the helicopter, which is suspended under the rotor disk, will not clear the obstructions as well. This is an early indication to abort the takeoff and to possibly try another position. Some penetration of the crosshatched or shaded areas of the height/velocity diagram may be unavoidable during this maneuver.

Stress to the student that the lowest climb angle possible should be used, both to improve climb performance and to minimize the time in the restricted area of the high/velocity diagram.

### Instructional Points

Before attempting a maximum performance takeoff, bring the helicopter to hover and determine the excess power available by noting the difference between the power available and the power required to hover. Under certain conditions, there may not be sufficient power available to complete the maneuver. Also perform a balance and flight control check and note the position of the cyclic.

Explain to the student the aerodynamic advantages of initiating this maneuver from the ground. That is, an in-ground-effect (IGE) hover requires less power due to:

- The reduced induce flow through the rotor disk (resulting from the displacement of induced airflow by the ground).

- The lift vector increases (using a lesser blade angle for the same amount of lift).

- Because the surface area disperses the airflow outward, the blade tip vortices are reduced resulting in larger portions of the blade producing lift.

This discussion helps the student understand the need to use the maximum power available, while IGE, to establish a climb over the obstacle.

Explain the practice of selecting go/no-go criteria, both from a performance planning aspect and as a reference point along the flightpath. If, upon reaching this predetermined point, a climb has not been established that allows clearance over the obstacle within power/rotor limits, abort the maneuver.

Have the student establish a proposed flightpath that maintains rotor and skid/wheel clearance along the entire flightpath. It is crucial that the student understand that at any time it appears that the maneuver cannot be completed (due to lack of available power or other limitations) or that the obstacle cannot be cleared, the student can abort the maneuver. Descending rearward along the same flightpath is a viable option. It cannot be stressed enough that pulling more power may cause rotor RPM droop with horrendous consequences.

Have the student position the helicopter into the wind and return the helicopter to the surface. Normally, this maneuver is initiated from the surface. *[Figure 11-1]* After checking the area for obstacles and other aircraft, select reference points along the takeoff path to maintain ground track. Consider alternate routes in case the maneuver cannot be completed. Begin the takeoff by getting the helicopter light on the skids/landing gear. At this time, have the student pause and neutralize all aircraft movement. The student should then slowly increase the collective (to allow the engine to

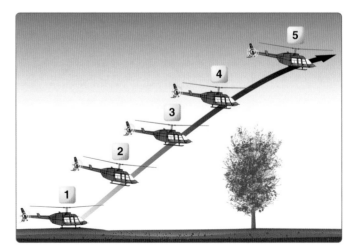

**Figure 11-1.** *Maximum performance takeoff.*

achieve and maintain full power) and position the cyclic so the helicopter leaves the ground in a 40-knot attitude. This is approximately the same attitude as when the helicopter is light on the skids/landing gear. Continue to increase the collective slowly until reaching the maximum power available. Ensure the student is aware that the large collective movement requires a substantial corresponding antitorque pedal input to maintain heading. During the maneuver, use the cyclic as necessary to control movement toward the desired flightpath and climb angle.

An alternate method is a vertical climb with allowing a vertical abort back down whereas the 40 knot attitude lift off requires aborting with backwards flight. If the 40 knot attitude takeoff is chosen, some landmark should be selected as a reference in the case of an aborted takeoff to ensure the tailrotor is not backed into an obstruction.

Maintaining rotor revolutions per minute (rpm) at its maximum is vital. A rotor droop and/or decrease in power causes the aircraft to descend. Maintain these inputs until the helicopter clears the obstacle or until reaching 50 feet for demonstration purposes. Then, establish a normal climb attitude and reduce power.

Reinforce to the student that smooth, coordinated inputs coupled with precise control allow the helicopter to attain its maximum performance.

Maximum performance takeoff in most light helicopters require operation within the crosshatched or shaded areas of the height/velocity curve. An engine failure while operating within the shaded area may not allow enough time for the critical transition from powered flight to autorotation. Check engine condition by monitoring the engine instruments and apply maximum power smoothly and slowly in order to prevent exceeding the engine limitations.

## Common Student Difficulties
### Coordination
Power, pitch attitude, and directional control are all essential when performing a smooth transition from the surface to a maximum performance climb. The student must set power smoothly, yet promptly, to the maximum allowable manifold pressure while maintaining maximum rpm. Proper pitch attitude must be established to ensure the helicopter accelerates to the desired climb speed as it gains altitude. The student should avoid abrupt or uncoordinated control application.

### Airspeed
A common problem for students when doing a maximum performance takeoff is finding the proper airspeed for the maneuver. A steep climb coupled with a low airspeed could

result in insufficient power to take off. This may result in low rpm followed by sinking and a possible hard landing or landing in the trees. Neither option is desirable. The other extreme is accelerating to an airspeed higher than necessary, resulting in a takeoff profile that is not steep enough. Revisit the positive effects of ETL on power requirements while stressing the importance of maintaining clearance along the departure path. The increased rotor efficiency of passing through ETL should be achieved as soon as possible to reduce power requirements.

## Running/Rolling Takeoff
A running takeoff, while not normally used, is practiced to simulate conditions that could exist as a result of high density altitude and/or a high gross weight. The student must be aware of the performance characteristics of the helicopter and the techniques to be used if sufficient power is not available to permit hovering in ground effect.

While conducting performance planning, have the student calculate power for a series of weights at 3 feet and check loading versus performance for that particular helicopter. Use this power-to-weight ratio, if feasible, in scenario-driven lesson plans utilizing increased gross weights. During the flight, validate the calculations and demonstrate to the student that power required to hover can be used to conduct running/rolling takeoffs. From a hover, accelerate the helicopter forward along the ground, until ETL permits a takeoff. Instructors should point out that, during this maneuver, altitude should remain constant until ETL is reached and the student should not lift up and down on the collective, which could cause the skids to touch the ground.

### Instructional Points
Demonstrate the following to the student:

- For helicopters equipped with skids, it may be better to practice running takeoffs on a hard surface runway instead of a grassy field. There is less probability of catching a skid, which could lead to dynamic rollover. In addition, check the condition of the skid shoes before and after practicing running takeoffs and landing.

- To begin the maneuver, align the helicopter to the takeoff path. *[Figure 11-2]* Next, increase the throttle

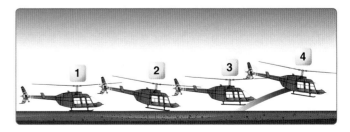

**Figure 11-2.** *Running/rolling takeoff.*

to obtain takeoff rpm (if applicable) and smoothly increase the collective until the helicopter becomes light on the skids for landing gear.

- Move the cyclic slightly forward of the neutral hovering position and apply additional collective to start the forward movement. To simulate a reduced power condition during practice, use one to two inches/pounds of torque less power than required to hover. Do not apply any forward cyclic to start the forward motion until simulated maximum power has been applied. Applying forward cyclic before maximum available power is obtained may cause the helicopter to dig into the ground, requiring even more power to break free.

- Maintain a straight ground track with lateral cyclic and heading with antitorque pedals until a climb is established. As ETL is gained, the helicopter becomes airborne in a fairly level attitude with little or no pitching.

- Maintain an altitude to take advantage of ground effect and allow the airspeed to increase toward normal climb speed. Then, follow a climb profile that takes the helicopter through the clear area of the height/velocity diagram. During practice maneuvers, after climbing to an altitude of 50 feet, establish the normal climb power setting and attitude.

### Common Student Difficulties
### *RPM*

It takes self-discipline to keep from raising the collective when the helicopter is about to become airborne. Emphasize to the student that increasing collective above the maximum available power may only result in a loss of rpm, producing a loss of lift, rather than an increase. This can be demonstrated in a hover by raising the collective while reducing throttle to maintain manifold pressure or torque pressure.

### *Attitude Control*

It initially requires more forward cyclic control to accelerate the helicopter on the ground, against the resistance of the skids/landing gear, than it does to maintain the slightly nose-down attitude required for acceleration in the air. As effective translational lift is gained, the tendency may be to hold the attitude and climb too rapidly. However, lowering the nose too much after becoming airborne may result in the helicopter settling back to the surface. Once airborne, the helicopter should be held in ground effect until climb speed is reached.

### *Wind*

Explain to the student that the greater the headwind component, the easier it is to get the helicopter off the ground when power is limited. All available means should be exercised to determine wind direction accurately before attempting the takeoff. In a crosswind, cyclic must be applied into the wind to keep the ground track parallel. Maintain this attitude even after leaving the ground. Only after climb speed is reached and the climb has begun, should the helicopter be crabbed into the wind.

## Rapid Deceleration or Quick Stop

A rapid deceleration, or quick stop, is used to decelerate from forward flight to a hover. The objective of a rapid deceleration or quick stop is to lose airspeed rapidly while maintaining a constant heading, ensuring adequate tail rotor to ground clearance at all times. Quick stops are practiced to improve coordination and to increase proficiency in maneuvering a helicopter. As the student gains coordination and proficiency in the maneuver, conduct the maneuver with a crosswind. From previous discussion of aerodynamics, the student should realize that downwind decelerations or quick stops are not recommended and that every effort should be made to avoid them.

### Instructional Points

During initial training, always perform this maneuver into the wind. As discussed above, once the student has demonstrated sound coordination and proficiency, conduct training with crosswinds. It is essential that the student begin crosswind training with light wind velocities. *[Figure 11-3]* For the instructor, it is important to know that the maneuver is conducted in IGE and just above ETL, which facilitate recovery.

After leveling off at an altitude between 25 and 40 feet, depending on the manufacturer's recommendations, accelerate to the desired entry speed, which is approximately 45 knots for most training helicopters. The altitude chosen should be high enough to avoid danger to the tail rotor during the flare, but low enough to stay out of the crosshatched or shaded areas of the height/velocity diagram throughout the maneuver. In addition, this altitude should be low enough that the helicopter can be brought to hover during the recovery.

Initiate the deceleration by applying aft cyclic to reduce forward speed and simultaneously lowering the collective, as necessary, to counteract any climbing tendency. Emphasize to the student that the timing must be exact. If too little down collective is applied for the amount of aft cyclic applied, a climb results. If too much down collective is applied, a descent results. A rapid application of aft cyclic requires an equally rapid application of down collective. As collective pitch is lowered, apply proper antitorque pedal pressure to maintain heading and adjust the throttle to maintain rpm.

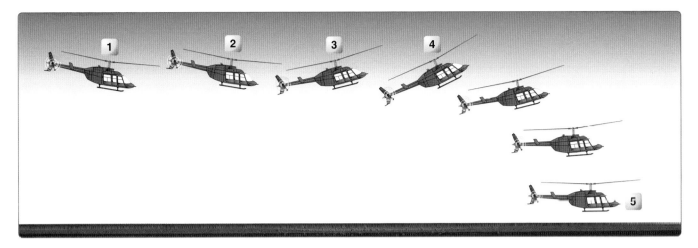

**Figure 11-3.** *Rapid deceleration or quick stop.*

As the speed dissipates, transition to a hover by lowering the nose and allowing the helicopter to descend to a normal hovering altitude in level flight and at zero groundspeed.

During the recovery, increase collective pitch as necessary to stop the helicopter at normal hovering altitude, adjust the throttle to maintain rpm, and apply proper antitorque pedal pressure to maintain heading.

Ensuring the student understands at all times where the tail rotor is relative to the ground is the key to success for this maneuver. As a teaching point prior to take off and, if the helicopter rpm allows (or consult maintenance), have the student sit in the helicopter and pull the tail boom down until the tail stinger or guard almost touches the surface. The student then gains the visual picture of the most nose-high attitude in which the helicopter needs to be in most situations.

## Common Student Difficulties
### Coordination

Because the quick stop demands a high degree of coordination, the student may encounter difficulties during the initial attempts. All flight controls are used: the cyclic to establish the pitch attitude for the desired rate of deceleration, collective to control altitude, throttle to maintain rpm (if applicable), and antitorque pedals to control heading. Initial quick stops should be practiced with a gentle deceleration rate to reduce the amount of control required. As the student gains proficiency, steepness of the initial flare can be increased until full down collective is required to prevent an excessive gain in altitude.

### Recovery

During the recovery, the helicopter should settle gently toward the hovering altitude. However, some students fail to recognize the need for recovery action and allow the helicopter to settle too rapidly as airspeed diminishes. Late application of collective requires an abrupt input to stop the rate of descent. As translational lift is lost and collective is increased, forward cyclic should be applied to return to a level attitude. In addition, as translational thrust is lost, even more antitorque pedal must be applied and more power produced to provide the antitorque. If stopping downwind, LTE could occur.

## Steep Approach to a Hover

Steep approaches are practiced to simulate approaches to confined areas where obstacles make a normal approach impossible. Steeper approach angles are normally used to maintain obstacle clearance along the flightpath. After completing a high reconnaissance, the pilot begins the final leg approach at an altitude and angle (approximately 15°) that permits a continuous stabilized approach to the upper portion of the landing area to keep the tail rotor out of contact with other objects. This continuous approach allows the pilot to establish and maintain that angle throughout the approach. The pilot should be cautious of rapid rates of descent and should continually check the closure rate. Using this technique allows much more time to complete the low reconnaissance, thereby allowing more time and altitude to conduct a go-around if needed. It could terminate either in a stabilized hover about a designated spot or in a landing. *[Figure 11-4]*

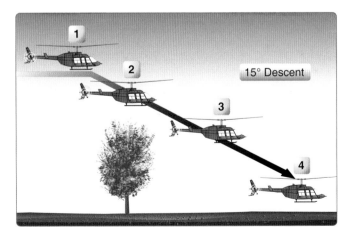

**Figure 11-4.** *Steep approach to a hover.*

## Instructional Points

On final approach, head the helicopter into the wind or approach path and align it with the intended touchdown point at the recommended approach airspeed.

Begin the approach by intercepting an approach angle of approximately 15°, lowering the collective sufficiently to start the helicopter descending along the approach path, and decelerating. Any approach angle of greater than 15° increases the likelihood of encountering settling with power; therefore, make certain the pilot understands the purpose of a steep-approach angle is to ensure tail rotor clearance, and does not enter at an excessive angle. Use the proper antitorque pedal for trim. Since this angle is steeper than a normal approach angle, reduce the collective more than that required for normal approach and control the rate of descent.

Continue to decelerate with slight aft cyclic, and smoothly lower the collective to maintain the approach angle. Remind the student to visualize the approach angle as an imaginary line from the landing gear to three feet above the landing point and attempt to maintain that angle throughout the maneuver. *[Figure 11-5]* Aft cyclic is required to decelerate sooner than on a normal approach, and the rate of closure becomes apparent at higher altitude.

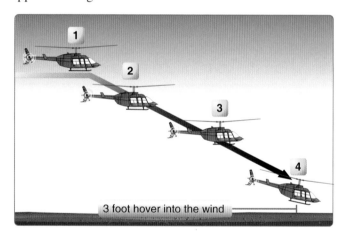

**Figure 11-5.** *Approach to a three foot hover over landing spot.*

Maintain the approach angle and rate of descent with the collective, rate of closure with the cyclic, and trim with antitorque pedals. Initially, use a crab above 50 feet and above the obstacles to compensate for wind. Use a slip below 50 feet for any crosswind that might be present.

The loss of ETL occurs higher in a steep approach, requiring an increase in the collective to prevent settling, and more forward cyclic to achieve the proper rate of closure.

For training, terminate the approach at hovering altitude above the intended landing point with zero groundspeed. If power has been properly applied during the final portion of the approach, very little additional power is required in the hover. The instructor should remind the student of ground effect as the helicopter gets closer to the surface. This effect on a well executed approach increases lift and allows the helicopter to glide to a halt over the landing point.

## Common Student Difficulties

Students frequently attempt to dive toward the selected touchdown point when they seem to be overshooting it. This only builds speed and moves the helicopter farther forward. Frequently, the next move is to lower the collective, which does accomplish a loss of altitude, but also results in a high rate of descent. Conversely, when undershooting, it is a natural tendency to raise the nose in an attempt to maintain altitude. This leads to collective corrections, which could lead to settling with power, or in the case of large collective increases, leads to rotor droop, engine overspeeds, or loss of tail rotor thrust. The student should be taught to regulate the steepness of the approach path with collective and use the cyclic to control the airspeed. Near the landing area, the collective should be increased to slow the rate of descent so the transition to a hover is accomplished as smoothly as possible. Constant reminders to use the imaginary approach path from the landing gear to 3 feet above the intended landing point helps the student to not chase the landing point.

It must be stressed to the student pilot that, if the approach is unstable, it is best to conduct a go-around prior to descending below the obstacles. Once below the obstacles, the required power to conduct a climb out may not be available. If after several attempts, the angle is too steep or environmental conditions dictate, have the student select another landing.

## Shallow Approach and Running/Roll-On Landing

The objective in a running landing is to maintain a sufficient forward speed to take advantage of translational lift until touchdown. As in the running takeoff, power should be limited to less than that required to hover. A shallow final approach should be used to maintain a low rate of descent. Since the helicopter is sliding or rolling to a stop during this maneuver, the landing area must be smooth and long enough to accomplish this task. *[Figure 11-6]*

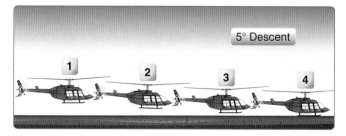

**Figure 11-6.** *Shallow approach and running landing.*

It may become useful to the student to practice these approaches for the purpose of relating conditions where minimal powered approaches are required, such as fixed or loss of tail rotor control or mechanically induced engine underspeeds.

Additionally, under certain whiteout or brownout conditions, shallow approaches can be made to keep the aircraft ahead of the obscurants, allowing the pilot to see the landing area. Extreme caution should be used when conducting this technique. The landing area must be smooth, and precise heading control must be maintained to avoid the possibility of dynamic rollover.

In any case, do not rapidly lower the collective after touchdown. Smooth reduction of collective prevents a rapid stop that could result in damage to the aircraft or injury to the crew.

## Instructional Points

For helicopters equipped with skids, it may be better to practice running landings on a hard surface runway instead of on a grassy field because there is less probability of catching a skid, which can lead to dynamic rollover. In addition, check the condition of the skid shoes before and after practicing running takeoffs and landings.

A shallow approach is initiated in the same manner as the normal approach except that the student intercepts the final approach path at a shallower angle (approximately 5° or less). The required power to maintain this shallow angle is more than that required for a normal approach but less than that needed to land due to increased ground effect and partial translational lift, depending on the wind.

As the collective is lowered, maintain heading with proper antitorque pressure and rpm with the throttle. Maintain approach airspeed until the apparent rate of closure appears to be increasing, then began to slow the helicopter with aft cyclic.

As in normal and steep approaches, the primary control for the angle and rate of descent is the collective, while the cyclic primarily controls the groundspeed. However, there must be a coordination of all the controls for the maneuver to be accomplished successfully. The instructor should remind the student that with any aft cyclic movement, any collective control increase is also increasing the aft cyclic adding more negative thrust, which can slow the helicopter too quickly. The student should always remember the cyclic's effects are affected by the amount of total thrust selected by the collective. The helicopter should arrive at the point of touchdown at, or slightly above, ETL. Since ETL diminishes rapidly at low airspeed, the deceleration must be coordinated smoothly with not too much aft cyclic applied, while keeping enough lift to prevent the helicopter from abruptly ballooning

upward, or worse, allow the rotor blades to contact the tail boom if the landing is too harsh.

Just prior to touchdown, place the helicopter in a level attitude with the cyclic, and maintain heading with the antitorque pedals. Use the cyclic to keep the heading and ground track identical. Allow the helicopter to descend gently to the surface in a straight-and-level attitude, cushioning the landing with the collective.

After surface contact, move the cyclic slightly forward to ensure clearance between tail boom and the rotor disk. Use the cyclic to maintain the surface track. In most cases, hold the collective stationery until the helicopter stops. However, if more braking action is required, lower the collective slightly. Keep in mind that due to the increased ground friction when the collective is lowered, the helicopter's nose might pitch forward. Exercise caution not to correct this pitching movement with aft cyclic, this movement could result in main rotor contact with the tail boom. During the landing, maintain normal rpm with the throttle and directional control with the antitorque pedals.

As a teaching point, the instructor can locate a nearby visual approach slope indicator (VASI) and help the student look up the information on the VASI to determine the correct angle for running or roll-on landings. Most VASI glide slopes are set in the 3° range, so that would be a good sight picture for the beginning student when learning the approach angle.

## Common Student Difficulties
### Approach Angle

The desired approach angle for a running landing is recommended to be approximately 2°–5°, somewhat shallower than a normal approach. The student may have difficulty visualizing and maintaining this approach angle and achieving the correct attitude, airspeed, and rate of descent on touchdown. Also, remember that sight pictures for students vary, as do aircraft design. It is possible the student has had previous instruction in another airframe or even a fixed wing aircraft.

### Attitude Control

In order to achieve the desired airspeed at the proper altitude above the touchdown point, correct pitch attitude is vital throughout the approach and touchdown in a running landing. Initially, the helicopter must have a slightly decelerating attitude (nose up slightly); in the middle of the approach, more of a gliding neutral stabilized attitude may be required. Finally, as power must be increased to replace lift lost due to loss of ETL, the cyclic must be adjusted to obtain the landing attitude necessary to the landing gear to touchdown.

As a teaching point, the instructor should have the student fly a restricted power go-around so the student gains the experience of the slow response under those conditions.

### Collective Control

Collective should not be fully lowered until the helicopter has stopped. Students tend to lower the collective immediately after landing, as in a normal landing from a hover. Rapidly lowering the collective while the helicopter is still sliding results in a deceleration rate that can impose undue stress on the landing gear support, rotor mast, transmission, or supporting structure.

### Touchdown

Common problems on touchdown include improper use of the collective and cyclic controls. Just prior to touchdown, there is a tendency to apply aft cyclic to cushion the landing. This causes the tail of the skids to touch first, followed by a forward pitching moment. A student may attempt to correct this action with more aft. In extreme cases, this can cause the main rotor blades to contact the tail boom. The correct technique is to level the helicopter, use collective to cushion the touchdown, and then apply a small amount of forward cyclic if needed after touchdown. Collective should be held to maintain a low rate of deceleration.

These general procedures are for landings to a smooth hard surface. Procedures vary for other types of surfaces, which means that the amount of collective applied varies depending on the surface. The instructor should never hesitate to increase power and collective and terminate the maneuver if any parameter is being exceeded.

## Slope Operations

### Slope Landings

While landing on a gentle slope is similar to landing in a crosswind, landing on a slope approaching the maximum capability of the helicopter requires smooth, yet positive, control. *[Figure 11-7]* Prior to flight, revisit the basics with the student prior to performing maximum angle slopes. Putting

the right skid or wheel upslope allows greater lateral control due to translating tendency (counterclockwise rotating rotor systems). Conversely, a left skid/wheel upslope restricts the amount of lateral cyclic available. Depending on winds, discuss with the student that it is easier to land on the slope where the uphill side is to the right and the wind is coming from the downslope side.

The student should also understand it is preferable to land the helicopter uphill from people so they can approach the helicopter from the downhill side. Approaching from the downhill side provides greater main rotor clearance between the rotor system and the downward sloping ground.

Slope limitations vary from manufacturer to manufacturer. It is important to emphasize that personal, physical, or mechanical limits may be reached before manufacturer's limits. Personal limits are reached when the pilot simply feels uncomfortable with continuing the approach. A pilot does not need to reach any of the other limits to get to this point. If it does not feel right to the student, abort the maneuver and try again.

In the case of physical limits, it is possible the student will find the cyclic restricted by either a leg or a kneeboard. Perhaps in situations where dual controls are installed the pilot may make cyclic contact with the passenger's leg or his/her own leg or kneeboard. In that case, abort the maneuver and find another location and/or reposition objects within the flight control area.

The surface of the slope can also be a limiting factor. If the surface is gravel, skid landing gear may not rest on the slope without sliding. If the surface is damp grass, it may be too slick for the helicopter to stay on the slope. An icy slope may only be suitable for offloading or loading but not for a complete shutdown. A thorough visual assessment of the surface should always be done when possible before attempting to land on a slope.

Mechanical limiting may occur if the aircraft rigging limit is preventing continuation of the maneuver. Mechanical resistance against the cyclic is felt. In this case, abort the

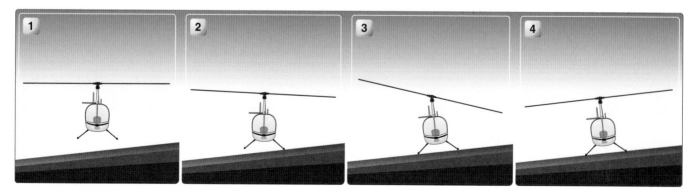

**Figure 11-7.** *Slope landing.*

maneuver, analyze the contributing factors, and reposition to another location, selecting a shallower slope.

If, at any point during the slope landing the student feels uncomfortable, abort the landing and reposition to a location where the limiting factors can be eliminated.

The limits of the helicopter's capability are discernible before the helicopter is committed to landing, as long as the student proceeds slowly and remains alert for sliding or for the cyclic control approaching the lateral travel limits. Proper technique on a slope landing eliminates the risk of dynamic rollover or sliding downslope.

The performance and techniques involved with different types of helicopter on sloping ground vary. This exercise should be tailored to meet the performance of the training helicopter. Students tend to be very tense when introduced to sloping ground operations, are likely to overcontrol, and tire quickly. Before introducing this exercise, ensure the student demonstrates proficiency at hovering and standard takeoffs and landings.

### Instructional Points

Prior to performing this maneuver, several briefings and discussions should take place. The first point to discuss is the student's depth of understanding of dynamic rollover. As previously stated, take time to discuss the reasons and events that lead up to dynamic rollover. Let the student lead the discussion and determine his or her level of understanding. Does the student simply apply rote memorization in the discussion or does there appear to be a level of understanding that leads to application and correlation?

The second vital discussion centers around the decision and procedure of aborting the maneuver. As already stated, the student will be tense during the maneuver. Discussion on the proper flight control inputs required to recover the aircraft is essential to safe execution of the maneuver.

The third discussion should be about where to go after the maneuver was terminated and then to analyze why the maneuver was aborted. Remember to provide positive, meaningful feedback to the student.

Make the initial approach to the slope at a 45° angle to check the suitability of the landing site. At the termination of the approach, move the helicopter slowly toward the slope, being careful not to turn the tail upslope. Position the helicopter across the slope at a stabilized hover headed into the wind over the spot of intended landing.

Maintain heading with proper antitorque pressure, using caution to not overcontrol with the pedals. Ensure the student relates left yaw inputs as a critical condition for dynamic rollover in helicopters with counterclockwise rotating rotor systems.

Downward pressure on the collective starts the helicopter descending. As the upslope skid touches the ground, hesitate momentarily in a level attitude, then apply lateral cyclic in the direction of the slope. This holds the upslope skid against the slope while the downslope skid is lowered with the collective. Ideally, the student understands that by keeping the rotor disk level, the lift vector is kept in a near vertical state. As the collective is lowered, continue to move the cyclic toward the slope to maintain a fixed position.

The slope must be shallow enough so the helicopter can be held against it with the cyclic during the entire landing. A slope of 5° is considered maximum for normal operation of most helicopters.

Be aware of any abnormal vibration or mast bumping, which signals maximum cyclic deflection. If this occurs, abandon the landing because the slope is too steep. In most helicopters with a counterclockwise rotor system, landings can be made on steeper slopes when the cyclic is being held to the right. When landing on slopes using left cyclic, some cyclic input must be used to overcome the translating tendency. If the wind is not a factor, consider the drifting tendency when determining landing direction.

After the downslope skid is on the surface, reduce the collective to full down, and neutralize the cyclic and antitorque pedals. In helicopters requiring manual throttle manipulation, normal operating rpm should be maintained until the full weight of the helicopter is on the landing gear. This ensures adequate rpm for immediate takeoff in case the helicopter starts sliding down the slope. Use antitorque pedals as necessary throughout the landing for heading control. Before reducing the rpm, move the cyclic control as necessary to make sure the helicopter is firmly on the ground.

Once proficiency has been demonstrated, let the student make the decision as to where to land. This allows the instructor to judge the student's ability to evaluate slopes.

Start the student on a shallow slope and gradually increase the difficulty level as proficiency improves.

Ensure the student is shown some slopes which are a mix of cross slope and up/down slope, so that the helicopter has to be landed diagonally on the slope.

## Common Student Difficulties

### Overcontrolling

The student's concern about the possibility of sliding or overturning usually leads to more uphill cyclic input than is required. The effects of excessive cyclic are not noticeable as the collective is lowered, since the helicopter has no tendency to slide uphill, but the limits of cyclic travel may be reached earlier than if the student were using only enough uphill cyclic to prevent sliding.

### Loss of Reference

Students tend to look at the ground close to the aircraft. Overcontrolling frequently results and it is often necessary to remind the student to raise his or her eyes and use the horizon as a reference.

## Slope Takeoff

The takeoff from a slope is much easier than the landing because the helicopter is already resting on the slope. *[Figure 11-8]* The student must be briefed not to turn away from the slope while still at a hover because this moves the tail toward the slope. Have the student slide away from the slope and conduct a turn about the tail to ensure the tail is clear of the sloping terrain. Begin by initiating the climb on the takeoff heading or, if obstacles dictate taking off downslope, move sideward before beginning the turn.

### Instructional Points

Begin the takeoff by increasing rpm to the normal range with the collective full down, then move the cyclic toward the slope. Position the cyclic to level the rotor disk and apply enough lateral cyclic to hold the landing gear on the slope in position as the collective is slowly increased.

As the skid comes up, move the cyclic toward the neutral position. If properly coordinated, the helicopter should attain a normal attitude as the cyclic reaches the neutral position. At the same time, use antitorque pedal pressure to maintain heading and throttle to maintain rpm. With the helicopter at a normal hovering attitude and the cyclic centered, pause momentarily to verify everything is correct, and then gradually raise the collective to complete the lift off.

After reaching a hover, take care to avoid hitting the ground with the tail rotor. If an upslope wind exists, execute a crosswind takeoff and then make a turn into the wind after clearing the ground with the tail rotor.

## Common Student Difficulties

Eager to get the helicopter back into a level attitude, the student may use excessive cyclic into the slope or apply up collective too rapidly. Emphasize that smoothness is essential to a safe, comfortable takeoff. Cyclic should be positioned into the slope before the collective is raised. The rotor disk should be checked visually to see that it is level with the natural horizon or inclined slightly toward the slope. Increasing the collective slowly enables the pilot to remove the lateral cyclic input so the helicopter is level when the upslope skid leaves the ground. Antitorque pedal input is gradually increased as the cyclic is raised, preventing a turning tendency. Caution the student to stabilize the hover before beginning the climb.

## Confined Area Operations

A confined area is an area in which the flight of the helicopter is limited in some direction by terrain or the presence of obstructions. For example, a clearing in the woods, a city street, road, parking lot, or a building roof or medical landing pad can be regarded as a confined area. *[Figure 11-9]* Generally, takeoffs and landings should be made into the wind to obtain maximum airspeed with minimum groundspeed. This is a comprehensive exercise that may require more than one session of preparatory ground instruction.

Another critical aspect of confined area operations involves the student's ability to estimate the size of the confined area as it relates to the helicopter. Tips on how to visualize the placement of the helicopter in the confined area are very helpful. A good exercise for obstruction judgment is to hold a broom around the edges of the rotor disk, not just the front.

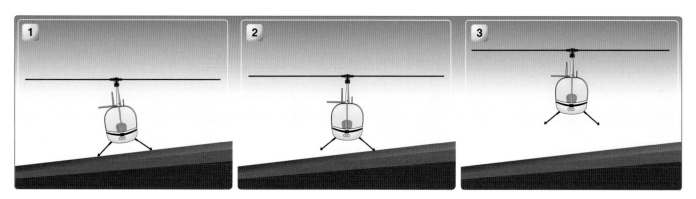

**Figure 11-8.** *Slope takeoff.*

**Figure 11-9.** *If the wind velocity is 10 knots or greater, expect updrafts on the windward side and downdrafts on the lee side of obstacles. Plan the approach with these factors in mind, but be ready to alter plans if the wind speed or direction changes.*

Blade strikes usually occur on the side of the helicopter. Ask the student if the blades would miss the broom to help build judgment. As training progresses, the instructor should point out wires and dead tree limbs hiding in plain sight, invisible under most circumstances when airborne.

If there is a shortage of suitable confined areas in the local training area, consider planning cross-country navigation exercises to locations that may have more suitable training areas. When introduced to this exercise, most students require more than one orbit of the area to obtain all the information they require. Encourage cutting down this number as proficiency increases, until experience reduces it to a practical minimum.

## Instructional Points
### *Approach and Landing*

There are several things to consider when operating in confined areas. One of the most important is maintaining a clearance between the rotors and obstacles forming the confined area. The tail rotor deserves special consideration because in some helicopters it cannot always be seen from the cabin. This not only applies while making the approach, but while hovering as well.

Exercises in distance and size estimation can assist the student in creating a reference base for the dimensions needed to land in a confined area. Simple cues such as the known dimensions of a football field (50 × 120 yards) can give the student a large-scale reference. An automobile parking space (9 × 18 feet) can help the student judge smaller distances as they correlate to the footprint of the helicopter. Ask the student for the size of the confined area while he or she is conducting the high reconnaissance. This provides information about the student's ability to capture this information.

Keep in mind that wires are especially difficult to see, but their supporting devices, such as poles or towers, serve as an indication of their presence and approximate height. If any wind is present, expect some turbulence.

Consider the availability of forced landing areas during the planned approach. Think about the possibility of flying from one alternate landing area to another throughout the approach, while avoiding unfavorable areas. Always leave a way out in case the landing cannot be completed or a go-around is necessary.

A high reconnaissance should be completed before initiating a confined area approach. Start the approach phase using the wind to the best possible advantage. Keep in mind areas suitable for a forced landing. It may be necessary to choose between an approach that is crosswind but over an open area and one directly into the wind but over heavily wooded or extremely rough terrain where a safe forced landing would be impossible. If these conditions exist, consider the possibility of making the initial phase of the approach crosswind of the open area and then turning into the wind for the final portion of the approach.

Always operate the helicopter as close to its normal capabilities as possible considering the situation at hand. In all confined area operations, with the exception of the pinnacle operation, the angle of descent should be no steeper than necessary to clear any barrier in the approach path and still land on the selected spot.

Always make the landing to a specific point and not to some general area. This point should be located well forward, away from the approach end of the area. The more confined the area, the more essential it is that the helicopter be landed precisely at a definite point. Keep this point in sight during the entire final approach. Use forward and lateral cues near the termination point to assist in maintaining the desired glidepath.

When flying a helicopter in confined areas, always consider the tail rotor. A safe angle of descent over barriers must be established to ensure tail rotor clearance of all obstacles. After coming to a hover, take care to avoid turning the tail into obstructions.

Alert the student to the problem of losing the wind when the helicopter descends below the height of the obstacle, normally on the final stages of the approach. Have the student plan and think ahead of the aircraft. Discuss the consequences of less wind upon descending into the confined area. Does the student comprehend that losing favorable wind velocity may have the same effect as losing ETL? Help the student anticipate the potential increased rate of descent and corresponding need for power.

### Takeoff

A confined area takeoff is considered an "altitude over airspeed" maneuver. This means it is more important to gain altitude than airspeed. Make the best takeoff possible to attain a normal attitude and climb airspeed to avoid or exit the crosshatched or shaded areas of the height/velocity diagram as quickly as possible. Because this is primarily an altitude acquiring maneuver, it usually requires lowering the nose to gain airspeed after the obstructions are cleared.

Before takeoff, make a ground reconnaissance to determine the type of takeoff to be performed. Doing so helps determine the point in which the takeoff should be initiated to ensure the maximum amount of available area, and how to best maneuver the helicopter from the landing point to the proposed takeoff position.

If wind conditions and available area permit, the helicopter should be brought to a hover, turned around, and hovered forward from the landing position to the takeoff position. Under certain conditions, sideward flight to the takeoff position may be necessary. If rearward flight is required to reach the takeoff position, the pilot must attempt to see what is behind the helicopter prior to hovering backwards. Hovering backward in a confined area is not recommended unless the pilot is absolutely sure that the area is clear of trees, high sloping terrain, or any other obstacle that the tail rotor could strike. If there is no other option and there are two pilots available, it would be beneficial to get out of the helicopter and visually inspect the area first.

When planning a takeoff, consider power available, wind direction, obstructions, and forced landing areas. The angle

of climb on takeoff should be normal, or at least no steeper than necessary to clear any barrier. It is better to clear a barrier by a few feet while maintaining normal operating rpm with perhaps a reserve of power, than to clear a barrier by a wide margin but with a dangerously low rpm and no power reserve.

As an aid in helping to fly up and over an obstacle, have the student form an imaginary line from a point on the leading edge of the helicopter to the highest obstacle to be cleared. Fly this line of ascent with enough power to clear the obstacle. As a rule of thumb, during the climbout, if there is an observed distance between the rotor tip-path plane and the obstacle, the obstacle will most likely be cleared. After clearing the obstacle, maintain the power setting and accelerate to the normal climb speed. Then, reduce power to the normal climb power setting. The instructor should also explain that the leading edge of the landing gear needs to clear the obstructions as well. Often, the student begins to get in a hurry and nose the helicopter over to speed up the process, but this lack of patience could cause the landing gear to contact leaves and limbs. The student must observe where both points are relative to the obstruction(s) during takeoff.

In the event that it appears the helicopter will not clear the obstacle (i.e., the tip-path plane is below the obstacle), take the appropriate action to abort the takeoff. This decision should be made earlier in the approach rather than waiting until the student finds the helicopter at maximum power and with fewer options available. Demonstrate that aircraft control can be maintained by flying the aircraft back down the flightpath if necessary.

If the student is going to pursue flight at higher elevations and small areas, training in the vertical takeoff method with aborts back into the confined area would be advisable to demonstrate the different performance of the helicopter. Proper heading control is very important to maintain tail rotor clearances. Aborted confined area takeoffs should terminate close to the forward barriers to ensure tail rotor clearance.

All instructors should have designated confined areas established to conduct various proficiency based levels of training. A new student will probably need a large open area. As coordination and proficiency levels increase, the student should be introduced to smaller, more challenging confined areas. Those same smaller confined areas are suitable for training of more advanced students, such as a commercial pilot applicants.

Several areas should be used to give the student every opportunity to consider all of the factors that influence operation into and out of confined areas. Have the student fly over a projected landing area and then describe the factors involved in a landing to that specific area. Items such as wind direction, favorable approach path, suitable forced landing areas, obstacles to be cleared, where turbulence might be encountered, and the helicopters expected performance should be discussed relative to each approach. The student must also consider the performance capability of the helicopter when planning a departure from a confined area.

As an instructor, maintain awareness of the disadvantages of seasonal training. A student that learns these maneuvers in the fall or winter months will not be prepared for the environmental consequences of warmer weather. Discuss this with the student, increasing general awareness of the potential impact of different seasonal conditions. A good discussion on performance would include having the student compare the charted and experienced performance values against those from a typical hot summer day. If the charts furnish sufficient information, the instructor might be able to adjust the power to simulate that of a hot summer day and how the helicopter performance is different.

## Common Student Difficulties

Safety of the approach depends mainly on the thoroughness of the planning that precedes it. Students may not recognize and prepare for hard-to-see obstacles, such as power lines. Explain that supporting structures indicate the presence of lines.

In a desire to make an approach directly into the wind, the student might fly over areas unsuitable for a safe landing. Advise the student that it is often preferable to make the approach with a slight crosswind, remaining downwind of potential landing areas, if possible. Should there be a need to conduct an emergency landing, turning toward the landing area would allow the approach to be conducted into the wind.

Upon completion of the maneuver, discuss with the student any points that may need to be addressed. It is easier for the student to talk this through on the ground then trying to fly and listen.

Before the takeoff is begun, ask the student to explain the factors being considered and the procedures being planned. This provides an opportunity to introduce factors the student may not have considered and gives the instructor a chance to evaluate the student's judgment.

## Pinnacle and Ridgeline Operations

### Pinnacle Landings

A discussion with the student on mountain weather and flying techniques, regardless of the geographical location of training, should be conducted prior to flight. Particular care must be taken to be alert to the impact of updrafts and, most importantly, downdrafts around the pinnacle.

Before attempting a pinnacle landing, the student must demonstrate proficiency in precision approaches to a spot along a constant approach path. This ability is the essence of a good pinnacle landing. Have the student fly over the selected pinnacle so he or she can observe obstacles and decide on a suitable approach path, as well as determine a plan of action if the approach does not go as planned. The final approach should be started at a sufficient distance from the pinnacle to enable the student to establish an approach angle appropriate to the approach path and wind conditions. If at all possible, a normal approach angle should be used.

Depth perception may be difficult because the surrounding terrain is lower and the approach angle is the only means of judging altitude. If the landing spot stays in the same angular relationship to the helicopter, the approach angle is constant. If at any time during the approach it appears to be unsafe, the approach should be abandoned according to the alternate plan. An approach that requires excessive maneuvering near the landing spot is unsatisfactory. Closure rate and approach angle should be carefully monitored during the approach because the visual cues, normally used when performing a normal approach, may not be available, and ETL must be maintained until the helicopter is nearly over the landing spot.

### Instructional Points

If a climb is needed to reach the pinnacle or ridgeline, do it on the upwind side, when practicable, to take advantage of any updrafts. Avoid the areas where downdrafts are present, especially when excess power is limited. [Figure 11-10] The

**Figure 11-10.** *When flying an approach to a pinnacle or ridgeline, avoid the areas where downdrafts are present, especially when excess power is limited. If downdrafts are encountered, it may become necessary to make an immediate turn away from the pinnacle to avoid being forced into the rising terrain.*

approach flightpath should be parallel to the ridgeline and into the wind as much as possible.

Load, altitude, wind conditions, and terrain features determine the angle used in the final part of an approach. As a general rule, the greater the wind, the steeper the approach needs to be to avoid turbulent air and downdrafts. Groundspeed during the approach is more difficult to judge because the visual references are farther away than during approaches over trees or flat terrain.

If a crosswind exists, remain clear of downdrafts on the leeward or downward side of the ridgeline. If the wind velocity makes the crosswind landing hazardous, a low coordinated turn into the wind just prior to terminating the approach might be an option. When making an approach to a pinnacle, avoid leeward turbulence and keep the helicopter within reach of a forced landing area as long as possible.

It is good practice to have the student plan the approach to a 3-foot hover. Most pinnacle sites do not provide suitable landing areas. In situations where the surface is known to be flat, such as a roof top helipad, approaches can be made to the surface.

Discuss with the student that actual pinnacles may not provide a cushion of IGE. Because the pinnacle surface may be uneven, rocky and sloping away in all directions, the student should anticipate the need for OGE power all the way to the surface.

Upon landing, take advantage of the long axis of the area if wind conditions permit. Touchdown should be made in the forward portion of the area. Always perform a stability check prior to reducing rpm to ensure the landing gear is on firm terrain that can safely support the weight of the helicopter.

### Common Student Difficulties
#### Planning
As in the approach to a confined area, there are many factors to consider. The approach should be planned to fly over the most favorable areas and with an approach angle only as steep as conditions warrant.

#### Approach Angle
It may be difficult for the student to maintain a constant approach angle due to the different visual cues, as compared with a normal approach in flat terrain. The cyclic is used to control the closure rate and the collective is used to regulate the rate of descent. If the landing spot seems to be moving away, a little forward cyclic corrects it. If the landing spot seems to be moving under the helicopter, a slight rearward cyclic correction is appropriate.

#### Airspeed
Explain the importance of controlling airspeed during the final stages of the approach. An airspeed that is too low can cause a loss of ETL before the landing spot is reached, requiring an OGE hover. In some cases, the entire approach may require OGE power. An excessively high airspeed may require excessive maneuvering to avoid overflying the landing spot. The pitch attitude and airspeed control procedures become more critical on a pinnacle approach.

## Pinnacle Takeoff and Climb

The terrain features affecting a pinnacle takeoff and a plan to cope with the situation should be formulated on the reconnaissance that precedes the landing. A change in wind, temperature, or takeoff weight may make it necessary to consider obstacles that were not a factor on the previously planned departure. To build the habit and to be certain that all items are being considered, the student should be required to review and describe the factors affecting the takeoff.

### Instructional Points
A pinnacle takeoff is an "airspeed over altitude" takeoff made from the ground or from a hover. Since pinnacles and ridgelines are generally higher than the immediate surrounding terrain, gaining airspeed on the takeoff is more important than gaining altitude. Quickly gaining airspeeds above ETL also allows greater separation from hazardous terrain. In addition to covering unfavorable terrain rapidly, a higher airspeed affords a more favorable glide angle, and thus contributes to the chances of reaching a safe area in the event of a forced landing. If a suitable forced landing area is not available, a higher airspeed also permits a more effective deceleration prior to making an autorotative landing.

As the helicopter moves out of ground effect on takeoff, maintain altitude and accelerate to normal climb speed. When normal climb speed is attained, establish a normal climb altitude. Never dive the helicopter down the slope after clearing the pinnacle.

### Common Student Difficulties
#### Planning
Failure to consider all factors involved in the takeoff and climb, or failure to take advantage of wind, the lowest obstacle, and favorable terrain are items that may need to be discussed frequently. Anything the instructor notices that could enhance the margin of safety of this or any other operation helps build the safety consciousness of the student.

## RPM

As the helicopter leaves the pinnacle, ground effect is lost almost immediately. The student may increase collective beyond the capability of the engine to maintain rpm, especially at high density altitudes. This possibility should be discussed prior to takeoff and watched carefully during the takeoff.

### Airspeed

The helicopter will be at altitude as soon as departing the pinnacle. The student should be instructed to gain airspeed rather than try to climb away from the pinnacle.

## Night Flying

Night flying introduces a new environment to the student and must be preceded by thorough preparation. Briefing for the first night flight should include, but is not limited to, the following items:

1. Equipment required for night flight
2. Airport and heliport lighting
3. Night flying physiology
4. Physiology of the eye
5. Weather considerations
6. Night flying techniques
7. Light discipline

The student's first night flight can be conducted at dusk so visual impressions are introduced gradually and adaptation to the night environment is accomplished over a period of time rather than instantaneously. *[Figure 11-11]* The regulations now require a night cross-country aeronautical experience for both private and commercial pilot applicants.

**Figure 11-11.** *The student's first night flight can be conducted at dusk so visual impressions are introduced gradually.*

The instructor needs to teach the student light discipline starting with the first night flight. As dusk wanes, it will be darker on the ground sooner, so night hovering could be the first task. Hovering should be accomplished with and without the landing light. In many cases at a lighted airport, hovering without the landing light can allow more visual clues as the eyes adjust to the ambient lighting conditions.

## Common Student Difficulties
### Takeoff

Attitude control problems during the takeoff and climb to 500 feet above the ground may be caused by several factors. Initially, the student might be tempted to look at reference points that are too close to the helicopter, focusing on things that can still be seen clearly. When airborne, a reference well in front of the helicopter should be used for attitude control. During the initial departure from a lighted area into the darkness beyond, it is necessary to refer to the airspeed indicator and altimeter frequently in order to confirm the desired attitude.

### Airborne

Following the first night takeoff, spend a few minutes away from the traffic pattern in a poorly lighted area. This allows the student to relax and become acclimated to the night environment and gain confidence in the ability to maintain flight with minimal visual references. During this time, other aircraft should be pointed out so the student can relate to the appearance of their lights to their apparent motion.

### Approach

The standard traffic pattern should be used for training in night approaches. Particular attention should be paid to attitude control, to assist in visualizing the correct approach angle. Since in-depth perception is more difficult at night, the approach angle is especially important.

Landings should be practiced with and without the use of the landing light. If the landing light is used, it should be used only on the final leg, preferably during the last 100 feet or so of descent. The student must be cautioned not to concentrate only on the area illuminated by the landing light, but rather to look ahead a bit for better attitude control and depth perception. Becoming fixated on the landing light can cause the student to misjudge the landing point and result in a hard landing.

## Instructional Points

When discussing lighting, explain to the student the effects lighting has on the eye. During the first flight, continually adjust the cockpit lighting to the lowest level that instrumentation can still be seen. This allows the student to better adapt his night vision.

On the subsequent night flight (with the aircraft on the ground in a dark location), again dim the cockpit lighting to the lowest comfortable level instrumentation can be seen. Have the student identify objects outside the aircraft. Next, turn the lights all the way up and have the student attempt to reacquire those same outside objects. Undoubtedly, this demonstrates the importance of dimming cockpit lights. To further the student's understanding, discuss that helicopters frequently are asked to fly at night into unlit locations. The importance of dim cockpit lighting is reinforced with this demonstration.

It is important for the instructor to be very knowledgeable of human factors affecting night flight. By discussing the anatomy of the eye, for example, the instructor can stress the importance of oxygen for night vision. Explain those factors that deprive the eyes of oxygen, such as illness, fever, or smoking. A new student who smokes tobacco may not see what a nonsmoker sees at night. High altitude flights impact night vision due to hypoxia. Because the eye uses more oxygen than any other part of the body per weight, vision degradation can begin at 5,000 feet pressure altitude.

## Cross-Country Operations

Cross-country flight training should include pilotage and dead reckoning, radio navigation, radar services, diversions, and disorientation procedures. These operations require a good working knowledge of the airspace system, chart interpretation, radio navigation, and communication. This is usually too much to teach while in the helicopter. Therefore, a cross-country training flight should be proceeded with one or more ground training sessions. The regulations require night cross-country instruction for both private and commercial applicants. However, this should be taught only after the student is comfortable with both night and cross-country operations.

### Instructional Points

Make sure your student has a good working knowledge of the airspace system. He or she should be able to interpret airspace boundaries from the charts, as well as understand the operational, communication, equipment, and weather requirements to operate within a particular airspace. The instructor should ask the student to read the charts and maps under the red lighting conditions found in most helicopters at night. Using the red lens flashlight in a dark classroom can also be a good simulation as well.

Make sure the student has a good working knowledge of aeronautical charts. The student must clearly understand all symbols and markings. He or she should also be able to read and understand topography and any potential hazards.

Teach dead reckoning and pilotage first. This is a good foundation for cross-country navigation without having to

rely on navigation equipment. It is equally important for the student to learn how to operate all available navigation radios.

It is imperative that the student demonstrate good crew resource management. The student should consider how to organize maps, charts, checklists and be aware of the difficulty in using light sources. The student must be familiarized with the locations and functions of equipment switches and dials to preclude repeatedly looking inside to adjust these items.

Single-pilot resource management (SRM) is an essential element of all flights, but especially important for cross-country operations. Remind the student that at least one hand must be on the flight controls at all times, so navigation information, such as charts and flight logs, must be easily accessible and ready for use. Have the student use a kneeboard for charts, logs, and other visual aids so that everything is accessible and in one place. See Chapter 15, Single-Pilot Resource Management, Aeronautical Decision-Making, and Risk Management, for a more in-depth discussion of SRM.

### Common Student Difficulties
#### Poor Cross-Country Planning

A thorough understanding of the airspace system and a good working knowledge of aviation charts are prerequisites for any cross-country flight. If the student is lacking in any of these areas, the result is poor cross-country planning. When planning a flight, use checkpoints that are easily recognizable, even if they require a little deviation from the most direct route and remind the student that good daytime checkpoints may not make good night time checkpoints. The differences should be discussed with the student during flight planning. For a beginning student, it may be advisable to skirt certain airspace in order to reduce communication workload. As experience increases, flight into busier airspace can be increased.

#### Reliability on Navigation Equipment

More helicopters now have global positioning system (GPS) navigation equipment. This equipment has a host of features, including moving maps, airspace, and airport information. While GPS is very useful, the beginning student must not rely on it. A thorough knowledge and understanding of pilotage and dead reckoning is required. If the training helicopter is equipped with GPS or any other navigational equipment, use it only as a backup, especially during the initial training with the student.

### Instructor Tips

- A student will attempt to imitate instructor actions. Do not take shortcuts. Instill safety from the first day. Insist the student follow established procedures when performing any maneuver.

- Make the training relevant to the student. *[Figure 11-12]* Explain how the student will use the maneuver in future operations.

- During maneuvers that bring the helicopter in close proximity to the ground or obstacles, instructors should stay close to the controls but not "on the controls." Staying on the controls ruins the control feel for the student and makes it more difficult for the instructor to judge exactly what the student is doing, and may hurt the student's confidence to fly. It is important to not let the helicopter depart the instructor's comfort zone. It does not really matter what the student is doing as long as the helicopter stays within safe recoverable parameters. The helicopter's parameters and situation are the instructor's primary concern.

- Remember, when a student encounters difficulty in mastering an objective, find a means of allowing some degree of success. For example, the lesson is steep

## Advanced Flight Maneuvers

### Objective

The purpose of this lesson is for the student to learn procedures and techniques for operating from sloping ground. The student will demonstrate a basic ability to land and takeoff from sloping ground.

### Content

1. Preflight Discussion
   a. Discuss lesson objective and completion standards.
   b. Review normal checklist procedures coupled with introductory material.
   c. Discuss weather analysis.
   d. Describe the helicopter's ability to operate from sloping ground and explain that pilots are frequently required to perform this maneuver under operational conditions.
   e. Explain that sloping ground techniques involve gentle and cautious control movements, and that these techniques are suitable for landing on any type of doubtful surface such as packed snow.
   f. Review and discuss the instructor's points for slope operations.
   g. Explain that sloping ground operations can be divided into four phrases:
      - Reconnaissance
      - Maneuvering
      - Landing
      - Takeoff

2. Instructor and Student Actions
   a. Select an area of sloping ground well within the helicopter's limits and demonstrate reconnaissance of, and maneuvering over, the intended landing area.
   b. Student practice.
   c. Demonstrate cross-slope landings in both directions, pointing out the difference in aircraft performance where appropriate.
   d. Student practice.
   e. Demonstrate an up-slope landing.
   f. Student practice.
   g. Select an area of sloping ground that is close to aircraft limits and demonstrate landings and take-offs.
   h. Student practice.
   i. Select an area of sloping ground that is beyond aircraft limits. Demonstrate the indications that the limits are being approached and the methods of aborting the landing.
   j. Student practice.
   k. Demonstrate wind/slope trade-off.
   l. Student practice of reconnaissance and selection of landing points.

### Postflight Discussion

Review and critique the flight being sure to discuss student strengths and weaknesses. Remember to provide suggestions on how to improve performance. Preview the next lesson and assign *Helicopter Flying Handbook*, Chapter 11, Helicopter Emergencies.

**Figure 11-12.** *Sample lesson plan.*

turns. Rather than have the student attempt the entire maneuver, try having the student practice the entry. When no difficulty is experienced with the entry, add the next stage, then continue until the entire maneuver is completed. If the student continues to have a difficult time, it may be best to revert to the classroom and discuss what the helicopter should be doing, what the student should be seeing, where the student should be looking, and what responses can be expected. If the student cannot explain the maneuver on the ground, they will take hours to figure it out in the air. Then, as the instructor redemonstrates the maneuver, task the student with talking the instructor through the maneuver. Usually by this point, the student learns or recognizes the missing piece, practice, procedure, or control movement to fly the maneuver properly.

- For any topic in the debriefing, where the student is deficient, the instructor should always have the answer for the student's problem. It is never good enough to simply critique the student's performance without explaining how to improve and correct the student's actions. The instructor must determine if the difficulty is the student's basic understanding of the maneuver, an incorrect perception of what should be happening, the lack of remembering to use the proper reference, or the improper perception of which control to use to correct the performance. Often, the student did not remember or learn one basic requirement. Learn the situation through talking with the student and by having them describe what they are trying to do. The problems will go away as the instructor explains how to correct or compensate for the effect.

## Chapter Summary

This chapter presented training techniques and instructional points an instructor can use to teach advanced flight maneuvers. Common difficulties encountered by students when attempting to perform these maneuvers were also discussed, as well as suggestions for instructional techniques.

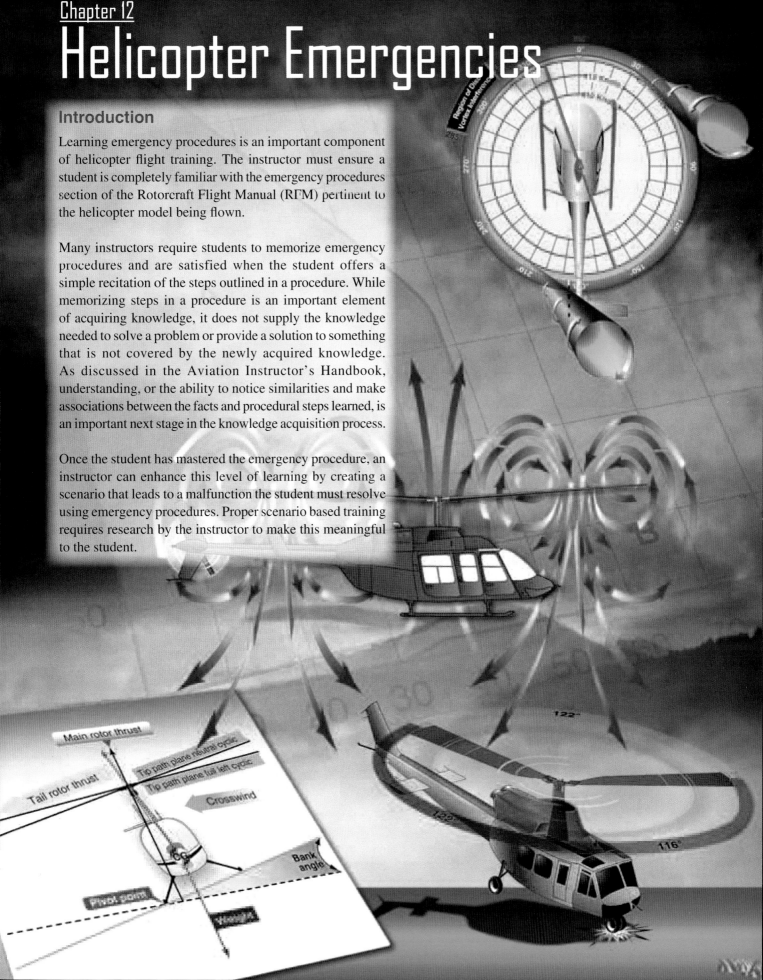

# Chapter 12
# Helicopter Emergencies

## Introduction

Learning emergency procedures is an important component of helicopter flight training. The instructor must ensure a student is completely familiar with the emergency procedures section of the Rotorcraft Flight Manual (RFM) pertinent to the helicopter model being flown.

Many instructors require students to memorize emergency procedures and are satisfied when the student offers a simple recitation of the steps outlined in a procedure. While memorizing steps in a procedure is an important element of acquiring knowledge, it does not supply the knowledge needed to solve a problem or provide a solution to something that is not covered by the newly acquired knowledge. As discussed in the Aviation Instructor's Handbook, understanding, or the ability to notice similarities and make associations between the facts and procedural steps learned, is an important next stage in the knowledge acquisition process.

Once the student has mastered the emergency procedure, an instructor can enhance this level of learning by creating a scenario that leads to a malfunction the student must resolve using emergency procedures. Proper scenario based training requires research by the instructor to make this meaningful to the student.

Take time to research an accident that replicates the training being performed. The National Transportation Safety Board (NTSB) website, www.ntsb.gov, provides several links that may be helpful.

Once research has been completed, sit down with the student and discuss what factors may have caused the accident. By discussing the causes of the accident, a thorough review and check of the student's depth of understanding can occur. As an example, there are many possible causes to an engine failure. Internal engine failure, fuel starvation, contaminated fuel, drive train failure, air inlet interruption, or hardware failure are just some of the possible causes. Discuss in detail, the cockpit indications that the pilot would have seen. What instrumentation differences may have been observed? What is the aerodynamic profile as it applies to this situation? What actions by the pilot were necessary and were they correctly assessed and taken?

Follow the discussion with a training flight that allows the student to view the gauges, indications, and flight profile that permit the student to practically apply the lessons learned from the scenario-based discussion. This type of training requires the student to exhibit knowledge of aircraft systems, as well as higher order thinking skills (HOTS).

The outcome of any emergency that might occur in an aircraft is directly related to the pilot's ability to react instantly and correctly, since there may be limited time to analyze the problem. Although a pilot could spend an entire career in helicopters without actually encountering an emergency situation, the demonstrated ability to cope with any situation is essential to safe operations and pilot confidence. Therefore, it is necessary to introduce emergency procedures early in the training programs and to practice them frequently.

## Autorotative Descents

An autorotative descent is a part of numerous emergency procedures in helicopters. Autorotations should be introduced first with an intentional entry so the student can practice establishing an autorotative glide. When the student develops proficiency in performing autorotations to a selected spot, the instructor should initiate autorotations when the student is not expecting them by announcing "engine failure." *[Figure 12-1]* As the student's ability to do an autorotation progresses, the instructor can use scenarios like decreasing oil pressure or zero oil pressure to stage an engine failure scenario. This will better prepare the student for an actual engine failure as there are usually prior indications that occur and need to be recognized. A student who becomes accustomed to someone announcing "engine failure" may never learn what the prior indications of the actual failure are. Refer the student to the Helicopter Flying Handbook and use it as a guide for a detailed discussion of the techniques for performing autorotative descents.

### Straight-In Autorotation, With Instructional Points

As the student demonstrates the ability to react properly, the instructor should announce "engine failure" as he or she retards the throttle to further develop the student's reactions. Never give the student a simulated engine failure unless an autorotation can be made safely to the surface. Do not assume a power recovery will prevent a landing to an unsuitable area because the engine may hesitate or even completely shut down when rolling off the throttle during the autorotation entry. To prevent an inadvertent engine shutdown during practice autorotations, follow the procedures outlined in the RFM.

Perform practice autorotations to a known area free of obstructions, where a safe landing to the surface can be made at any time. *[Figure 12-2]*

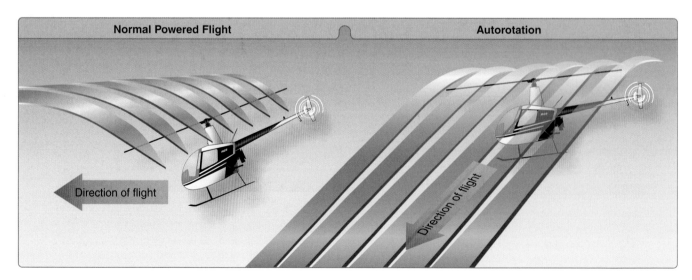

**Figure 12-1.** *During an autorotation, the upward flow of relative wind permits the main rotor blades to rotate at normal speed. In effect, the blades are "gliding" in their rotational plane.*

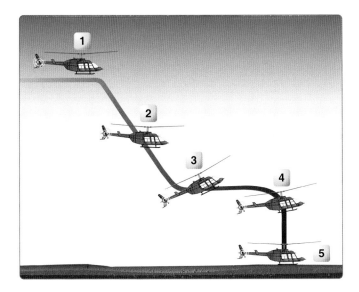

**Figure 12-2.** *Straight-in autorotation.*

Demonstrate the maneuver using the manufacturer's recommended airspeeds for autorotative descent or the best gliding speed. The instructor should require the student to review the performance charts and know the minimum and maximum glide airspeeds for the day's conditions. Especially during the early phases of training, the flights should be conducted at higher altitudes to give the new student more time to understand the different descent angles available, using minimum rate of descent airspeeds and maximum glide airspeeds. Some helicopter RFMs also advise of different rotor rpm values for gliding descents. In those cases, substantial care should be exercised to maintain the proper rotor rpm for the glide and then restoring the proper rpm value for the landing in a safe manner, even if performing a power recovery. Emphasize coordinating the collective movement with proper antitorque pedal for trim to prevent nose tuck. Nose tuck mainly occurs to certain helicopters, usually those with canted tail surfaces. Nose tuck is the rapid nose-down pitching resulting from a yawing motion causing excessive lift in a stabilizer. This only occurs when the stabilizer is canted upward from the horizontal. Instruct the student to apply cyclic control to maintain proper airspeed. After entering the maneuver, have the student verify and call out needle split, rotor revolutions per minute (rpm) in the green arc, helicopter in trim, and airspeed.

During the descent, instruct the student to adjust collective pitch control as necessary to maintain rotor rpm. Emphasize avoiding large collective pitch increases because they result in a rapid decay of rotor rpm. Demonstrate how aft cyclic movements cause a temporary increase in rotor rpm.

At approximately 40–100 feet above the surface, or at the altitude recommended by the manufacturer, demonstrate the deceleration with aft cyclic control. Explain how the deceleration is used to both reduce forward airspeed and decrease the rate of descent. Point out that the cyclic control is not moved rearward so abruptly as to cause the helicopter to climb.

Show how the helicopter is placed in the landing attitude at approximately 8–15 feet, or the altitude recommended by the manufacturer. Explain that extreme caution should be used to avoid an excessive nose high and tail low attitude below 10 feet. Demonstrate the application of collective pitch, as necessary, to slow the descent and cushion the landing. Explain that additional antitorque pedal is required to maintain heading as collective pitch is raised due to the reduction in rotor rpm and the resulting reduced effect of the tail rotor. Emphasize touching down in a level flight attitude.

NOTE: A power recovery can be made during training in lieu of a full touchdown landing. (Refer to the section on power recoveries for the correct technique.)

After touchdown, and after the helicopter has come to a complete stop, lower the collective pitch to the full-down position. Discuss the hazards of using the cyclic or collective to slow the helicopter's ground run.

NOTE: Numerous tail boom strikes have occurred due to improper collective pitch response upon ground contact and completion of the maneuver. Particular attention must be emphasized as to the proper rate and timing of lowering collective to avoid potential damage to aircraft components. Emphasize the minimum requirements for rotor rpm, airspeed, and trim conditions throughout the maneuver.

If the instructor or the student feels uncomfortable with the autorotation, do not hesitate to initiate a go-around. The instructor is responsible and should terminate the maneuver anytime the parameters vary from normal, or if either pilot determines something is not right. These are common comments that are submitted on accident or incident reports. Therefore, the instructor should be especially aware of these occurrences and err on the side of caution by making power recoveries or go-arounds if any doubts exist. The instructor's primary duty is to ensure the student has safe flights and does not get into trouble.

The instructor should highlight the student's understanding of the wind's effects as related to airspeed versus ground speed, and the amount of flare necessary to slow for the landing. The instructor should help the student to cope with the increased airflow through the rotor system in the flare and how to control rpm, both increasing and decreasing.

The importance of landing gear alignment cannot be overemphasized during actually touchdowns. If the helicopter is kept in trim, the rate of descent will be the minimum rate. Out of trim conditions increase the rates of descent considerably. Usually, it is useful to demonstrate how much out-of-trim conditions can increase the rate of descent. Although not usually needed or desired, a steeper descent angle to enable an emergency landing in a very close-in landing area could be useful, but only if the student is trained how to recover and arrest that increased rate of descent.

## Autorotations With Turns, With Instructional Points

When autorotations are first introduced, they are usually conducted straight ahead and into the wind. Discuss with the student that a turn, or a series of turns, may be necessary during an autorotation to land into the wind or avoid obstacles. Emphasize that the turn should be completed as soon as possible during the descent so that the remainder of the autorotation is the same as a straight-in autorotation.

Ensure the student is aware that, during the turn, the rotor rpm must be closely monitored as it has a tendency to build during the turn. Airspeed, on the other hand, may be unreliable until the turn is completed. The importance of maintaining the helicopter in trim cannot be overemphasized.

Instruct the student to adjust the collective as necessary in the turn to maintain the rotor rpm in the green arc. If the collective pitch was increased to control the rpm, it may need to be lowered on rollout to prevent a decay in rpm.

During training, the turn should be completed and the helicopter in position to land in the intended touchdown area prior to passing through 100 feet above ground level (AGL). Initiate an immediate power recovery if the helicopter is not aligned with the touchdown point or if the rotor rpm and/or airspeed are not within proper limits.

## Power Recovery From Practice Autorotation, With Instructional Points

From an instructor's point of view, it must be remembered that a power recovery, if applied too late or too rapidly, can result in hard landings that sever tail rotor drives (resulting from loss of yaw control), overboosting of turbocharged engines, overtemping and/or overtorques of turbine engines. To avoid this from occurring, think through power application using an early, smooth, and measured response. Do not wait until the helicopter is on the verge of being unrecoverable to initiate the power recovery.

Upon initial training, explain to the student that a power recovery is used to terminate practice autorotations or simulated engine failures at a point prior to actual touchdown. The power recovery itself is not an emergency procedure.

In the event of an emergency that requires autorotation, the student must be prepared to complete the autoroation to touchdown.

Emphasize the importance of the throttle detent to avoid inadvertent shutdown. This may occur by unintentionally closing the throttle to the off position. As the instructor, it is your responsibility to verify the power control detent is in the proper position when manipulating the throttle in flight.

Additionally, stress the need to increase throttle and join the engine and rotor tachometer needles with the collective in the down position. The power or throttle should be increased to engage the clutch or have the rotor and engine together at about 100 feet AGL. If the throttle is increased too much or too fast, an engine overspeed can occur; if the throttle is increased too little or too slowly in proportion to the increase in collective pitch, a loss of rotor rpm results.

Coordinate upward collective pitch control with an increase in the throttle to join the needles at operating rpm. Ensure the student understands the increase of throttle and collective pitch must be accomplished properly.

If a power recovery is to be made, slowly increase the engine rpm during the descent to avoid the rapid application of throttle for the recovery.

If the student is at all apprehensive about autorotations, be prepared for him or her to either overcontrol or freeze on the controls. One technique of teaching autorotations is to use power-on glides that replicate the autorotational profile, terminating to a hover or with a slight ground run, if permissible. By using this technique, a new instructor can get the feel for the power required and subsequent yaw that occurs during the flare and landing. An instructor should never let the student exceed aircraft limits or the instructor's ability to recover the helicopter.

As a teaching technique in many helicopters, a minimum of power can be used for the landing to offer a preview of a full autorotative touchdown. Basically, this offers more cushion and allows a quicker but very controlled touchdown. Having the engine engaged prior to the flare allows the engine to accelerate throughout the flare to produce hover power. The instructor should remember the power recovery is actually a very precise maneuver, requiring the most control movement of almost any maneuver in very little time, and the student is required to transition from power-off flight to almost maximum power in a hover. Because so much is happening so quickly, anything the instructor can do to allow more time for the student to perceive all of the actions that must occur in that brief time period gives large returns for the training.

## Power Failure in a Hover, With Instructional Points

Whether instructing or evaluating, performing a power failure at a hover is another high accident rate maneuver. The following are some of the common mistakes made by instructors.

The first is site selection. Ensure the surface area is a flat, smooth, and hard surface. Students initially tend to have a difficult time accounting for translating tendency. This, in turn, most often causes lateral drift. If conducting this maneuver on a sloping surface, one or more of several phenomena may occur, from dynamic rollover to damaged skid gear. The same can be said for soft or uneven surfaces, as the skid gear may stick, causing a pivot point.

Selecting the proper entry height is also imperative. Often, initiating this maneuver from higher than the prescribed entry altitude may cause the student to prematurely apply collective to arrest the rate of descent for landing, leaving the aircraft at little or no available rpm while not yet on the ground.

At higher entry heights, a tendency is to attempt to lower the collective to build available inertia or keep the rotor rpm up. Loss of power at a hover with the normal high power setting always results in a rapid rotor rpm decay. Review of the H/V chart almost always indicates that there is insufficient power to allow a safe, uneventful landing if the hover height is too high. However the student does it, if the hover height is too high, a hard landing results from a power or driveline failure. There simply is not enough potential energy in the rotor system alone to cushion the landing. This should not be shown to students as it requires a great amount of experience and special training to accomplish successfully. Ensure the student understands a power failure can occur at any time. Hovering autorotations are practiced to develop the automatic response and coordination required to maintain heading and cushion the landing following an engine failure at a hover.

Explain to the student that power failures at a hover may occur during either of the two modes of flight: hovering flight in ground effect (IGE) and hovering flight out of ground effect (OGE). Refer to the RFM to generate discussion of the procedures to be used for each instance.

Stress to the student that for power failures at a hover while IGE, the collective is not lowered, but held in place until approximately 1 foot AGL when collective pitch is increased to cushion the landing. Explain that when hovering OGE, operations usually fall in the hashmarked area of the height-velocity diagram, which may require lowering the collective and adding forward cyclic in attempt to land the helicopter.

As previously mentioned, emphasize the importance of minimizing sideward or rearward drift during the maneuver to reduce the possibility of dynamic rollover or other damage to the helicopter.

It is best to debrief the student after each maneuver. Review the maneuvers with the student and point out what was done correctly, and what was not. When discussing errors, point out to the student why they happened and what could be done to correct or prevent them. When a student is continuously having difficulty with a particular maneuver, the instructor must find a way to help the student understand what is happening. Helping the student pinpoint exactly what they are doing wrong, and how to help the student learn to avoid these errors is what the instructor should focus on. This permits the student to immediately take corrective action on the next repetition. Take time to talk about the hazards of excessive nose low takeoffs, exposure to the hazard area of the height/velocity chart, and the dangers of dynamic rollover. Discussing these hazards while performing this maneuver reinforces the need to carefully choose correct flight techniques and profiles.

## Common Student Difficulties With Autorotation

Students have a tendency to look straight down in front of the aircraft while performing an autorotation. By focusing straight down or on the intended touchdown point, the student loses the ability to determine altitudes. Also, students fail to check trim, rotor rpm, and airspeed, and fail to completely close or place the throttle in the ground idle or override position.

Autorotation is overwhelming the first few times students perform it and can almost put them into a "frozen state" of mind. Rather than looking at all the instruments, flight gages, and directing what flight path the helicopter should be going, they become fixated with the ground and focus only on that. Once the autorotation has been initiated, immediately have the student verbally call out what they should be looking at. For example, have the student say "Rotors are at 100 percent, NG is stabilized, airspeed is at 60 knots and the aircraft is in trim." At first, the student is just rehearsing aloud but after doing more and more autorotations and training them to look at the instruments, gauges, and attitude of the helicopter each time with verbal cues help them remain focused on what is important, landing the helicopter safely.

## Emergency Situations for Discussion Only

There are a few situations unique to helicopter operations that must be understood if they are to be avoided. Some of these emergencies may only be discussed, not demonstrated, because demonstration would probably result in damage to the helicopter. They are included here to remind the instructor to make sure the student is able to discuss the problems and how they can be avoided.

## Vortex Ring State (Settling With Power)

NOTE: Vortex ring state (settling with power) is not a discussion only maneuver in a single-rotor helicopter. It can safely be introduced and practiced at altitudes allowing distance to recover. However, settling with power should not be practiced in twin-rotor helicopters due to excessive roll or pitch rates that occur when one rotor is in the vortex ring state and the other is in relatively more stable air. Coaxial and intermeshing main rotor system helicopters should be safe for settling with power practices.

Ensure the student understands that settling with power can occur as a result of attempting to descend at an excessively low airspeed in a downwind condition, or by attempting to hover OGE at a weight and density altitude greater than the helicopter's performance allows. *[Figure 12-3]*

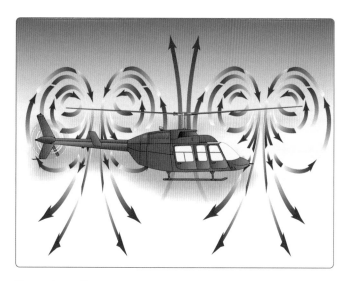

**Figure 12-3.** *Vortex ring state.*

As is always the case, using a scenario that the student finds applicable to his or her future helicopter career is most beneficial. An ideal scenario-type discussion for this can be a confined area approach (in which the winds were incorrectly judged, resulting in a downwind approach) or a job-specific OGE hover-type maneuver (long line, filming for news crews, search-and-rescue hoist operations, external load hookup/release, etc.,) that puts the helicopter in the required settling with power profile.

While common thought is to lower the collective and fly out of the downwash, this may not be the best option due to lower altitudes or obstacles. Through discussion of aerodynamic theory, an option of laterally exiting the 'disturbed air' can be the best option available. In a counterrotating system, lateral flight to the right often results in the quickest exit from the

airflow and affords the pilot the best view of the exit route. The lateral travel is enhanced by the thrust from the tail rotor and introduces turbulent free air sooner to the tail rotor than any other direction.

Adding collective while the helicopter is descending vertically only aggravates the situation. Settling with power can occur only if the rotor is powered. A demonstration of settling with power may be required of the applicant for a commercial helicopter rating. The private pilot applicant may be required only to discuss recognition and avoidance, but should have a demonstration of settling with power to understand its effects better.

The demonstration should begin at an altitude high enough above the ground to allow room for safe recovery. Entry into the condition can be made from the events as described in the scenario, if acceptable. Another set of circumstances causing entry into the vortex ring state is a decelerating airspeed descent, such as that experienced with a high, steep, downwind approach. This can be used to show the student a probable or likely result of poor or rushed planning.

The student should be reminded that settling with power always requires three conditions to occur:

1. Rate of descent greater than 300 fpm

2. Airspeed less than 10 knots in any direction

3. More than 20 percent of engine power applied to the rotor system

By taking away any one of the three conditions, settling with power should not occur. Therefore, the instructor can remind the student to keep the descents slow and less than 300 fpm; keep the helicopter moving into the wind, which normally maintains good airflow at the 10 knot or higher value; and, if altitude allows, reduce the collective to reduce power entering the rotor system. In most instances, a lateral maneuver produces the quickest exit from the disturbed column of air.

When simulating an attempt to hover OGE, the airspeed is gradually slowed and power is added to maintain altitude. Care must be taken to avoid any rearward speed. If the helicopter can hover OGE, it may be necessary to reduce power to begin settling, then add power to increase rotor downwash. As soon as the effects of settling with power are noticeable, recovery should be initiated. The noticeable effects are vibration, reduced control effectiveness, and a high rate of descent.

Recovery is accomplished by either laterally exiting the disturbed wind or, if altitude allows, reducing collective and lowering the nose to increase forward speed. This moves a helicopter out of its downwash and into translational lift.

When the helicopter is clear of the disturbed air, or downwash, confirm a forward speed indication and initiate a climb to regain the lost altitude.

## Retreating Blade Stall

The student must understand the limits of high speed in the helicopter and the reasons for imposing them. While structural and airframe design limit helicopter airspeed, the most frequently referenced aerodynamic limitation to helicopter airspeed is retreating blade stall.

The symptoms of retreating blade stall are main rotor vibrations, nose pitch up, and a rolling tendency, usually to the left in a helicopter with a counterclockwise main rotor blade rotation. High gross weight, maneuvering, and turbulence all tend to aggravate the retreating blade stall condition.

Retreating blade stall normally cannot be demonstrated without exceeding the never-exceed speed ($V_{NE}$) or maneuvering limits of the helicopter. However, the student must be able to explain that the cause of retreating blade stall is excessive forward speed for the existing circumstances. The manufacturer's recommended $V_{NE}$ provides protection for normal situations. If the helicopter is heavily loaded and then flown into turbulence at or near $V_{NE}$, or if it is maneuvered abruptly, retreating blade stall can occur. Ensure the student becomes familiar with the procedure (lower the collective, increase rpm, if possible, and reduce speed) for recovery from retreating blade.

The instructor should advise the student that retreating blade stall is only one of many factors that produce the limiting $V_{NE}$ in coaxial and close tandem rotor systems, such as the Kaman K-Max, $V_{NE}$ may not be characterized as retreating blade stall due to the counterrotating blade system. This design may mask the huge stresses on the rotor system until the stresses overcome the structure. The instructor should ensure the student abides by all limitations.

When corrective action is introduced at the first vibration indicating retreating blade stall, there should be very little effect. If the blade stall is allowed to progress to the point the helicopter pitches up and rolls, trying to stop the pitch and roll with cyclic inputs only aggravates the situation. Allowing a helicopter to pitch up reduces speed and alleviates the blade stall condition. Roll control is then effective. The best way to prevent students from encountering retreating blade stall is to instill in them the practice of flying at air speeds below $V_{NE}$. The margin should be increased if turbulence is encountered.

Emphasize to the student that retreating blade stall can be avoided by adhering to the $V_{NE}$. The decrease in $V_{NE}$ speed with increasing density altitude must also be explained thoroughly.

## Ground Resonance

Ground resonance is a hazard associated with an articulated rotor system. Have the student explain which types of rotor systems are susceptible to ground resonance, the factors that tend to cause it, and the means to avoid it or recover from it if it occurs. The student must understand that if the helicopter is allowed to touch down in a manner that creates a jolt or bounce to the airframe, ground resonance can occur. Improperly serviced landing gear struts in some helicopters can aggravate the tendency to oscillate and contribute to ground resonance. This contact with the ground sends vibrations through the aircraft to the rotor system, causing an imbalance and center of gravity shift in the rotor system. The imbalanced rotor system energy is transferred, in the form of a rhythmic back and forth vibration (normally a lateral, or side to side, motion), from the surface to the rotor system. Although the frequency of this vibration remains constant, the strength of the vibration is amplified until an uncontrollable oscillation develops. *[Figure 12-4]*

**Figure 12-4.** *Ground resonance.*

Since the skids or landing gear wheels contacting the ground cause the rotor to become unbalanced, the obvious means of fixing the situation is to lift the helicopter free of the ground and allow the blades to assume a balanced condition. If rpm is too low for flight, the next best corrective action is to lower the collective to place the blades in low pitch and reduce powerplant to idle or cutoff. The vibration increases in severity only if there is power applied to the system. Simply lowering the collective is not sufficient to stop the destruction; power must be removed.

## Dynamic Rollover

Dynamic rollover is another potentially hazardous situation peculiar to helicopter operation. The student should be able to explain that, for dynamic rollover to occur, some factor must first cause a helicopter to roll and pivot around the skid or landing gear wheel, until its critical rollover angle is reached. Beyond this point, main rotor thrust contributes to the roll and recovery becomes impossible, regardless of any cyclic corrections made. Tell the student that dynamic rollover can occur on a flat surface and most often does due to obstructions such as tiedowns, hot asphalt, and frozen or deep mud.

Once the helicopter reaches an angle such that the cyclic cannot keep the rotor system level with the horizon, it is subject to uncontrollable rolling. It is this horizontal thrust component that makes this effect dynamic. Helicopters tend to be top heavy and easily disturbed in the vertical mode. All of the heavy components, such as engines, transmissions, drive shafts and bearing mounts, tail rotor gearbox, rotor head, swashplates, and blades are mounted high in the airframe in most helicopters. This contributes to a very delicate balance above what can be relatively narrow landing gear. The only possible recovery action is to lower the collective. If the static balance point has not been passed and the roll rate is low, the helicopter might land back roughly on its landing gear.

An easy but dramatic demonstration of dynamic rollover can be illustrated by having the student stand straight with his or her feet shoulder width apart and then gradually lean to the side until his or her body extends past his or her center of gravity. At this point, the student will lose balance and will be unable to straighten up. This exercise demonstrates to the student that once the center of gravity of the human body extends too far in any direction, the person topples. The same principle holds true for the helicopter.

Point out the factors that influence dynamic rollover to include a skid or landing gear wheel in contact with the landing surface, pedal inputs, lateral loading (asymmetrical loading), crosswind condition, and a high roll rate. Explain that smooth and moderate control inputs are most effective in preventing dynamic rollover as it reduces the rate at which lift/thrust is introduced.

Thorough preflights ensure the landing gear is fully free, prudent control movements slowly lift the helicopter off the surface or land without any lateral motion. The rising of one side of the helicopter for compensation for the translating effect should be expected and countered. More or less tilt than normal or usual is reason enough to stop the motion and check for control response. If in doubt, set the helicopter back down and double check for obstructions and skids stuck in hot asphalt, tie down chains or ropes over skids or gear points, etc.

While landing the helicopter, the pilot should feel for any sliding or sinking motions, signaling an unsafe landing surface. If such a surface is encountered, descent must be stopped by increasing collective. If after coming to a level attitude, no further attempts should be made to touch down, the helicopter should be repositioned to a different area for another landing attempt.

## Low-G Conditions and Mast Bumping

A student pilot must understand the potentially hazardous implications of intentionally or inadvertently performing low-G maneuvers. Too often, students study topics but fail to correlate their impact on the aircraft. Low-G maneuvers and the resulting consequences of mast bumping must be thoroughly addressed. Explain to the student that abruptly pushing the cyclic control forward from either straight-and-level flight or after a climb can put the helicopter into a low-G (weightless) flight condition, with catastrophic results.

The student must be reminded that the semirigid rotor system suspends the airframe below the rotor. Therefore, in normal flight, the helicopter is using the pull of gravity for a normal airframe/rotor relationship or normal load on the rotor system. In a low-G maneuver, with the helicopter in an excessively nose-low or tilted attitude, gravity is no longer pulling the airframe down, and the feeling of weightlessness is experienced. With negligible gravitational pull between the rotor and airframe, any cyclic control input tends to fly the rotor disk over. The airframe does not respond, resulting in mast bumping.

Ensure the student understands what is happening and that improper corrective action can lead to mast bumping or may allow a blade to contact the airframe. *[Figure 12-5]* Since a low-G condition could have disastrous results, it cannot be demonstrated, but the student must learn the correct response. The other part of the Low-G effect is the thrust of the tail rotor being above the center of gravity. This is often the initial cause of the fuselage to roll. Stress that abrupt movement of the cyclic and collective should be avoided, but if the student

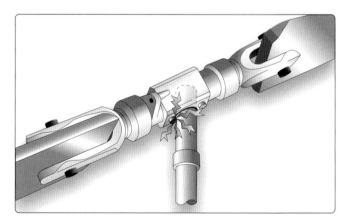

**Figure 12-5.** *Result of improper corrective in a low-G condition.*

gets into a low-G condition, (be sure to describe the feeling of weightlessness and an uncontrolled roll to the right), he or she should immediately and smoothly apply aft cyclic. Warn the student against trying to correct the rolling action with lateral cyclic. Explain that by applying aft cyclic, lift is redirected to counteract weight and gravity restores the balance of forces. This in turn requires antitorque so the tail rotor forces then become balanced as well. Unless the collective is lowered, the thrust never changes. Only the direction of the rotor system thrust and lift is affected. Prevention is the only proper cure for the hazard.

Since the best way to prevent low-G conditions is to avoid the conditions in which they might occur, suggest the student avoid turbulence as much as possible (since this type of weather induces low-G conditions). If the student encounters turbulence, advise him or her to slow forward airspeed and make small control inputs. If turbulence becomes excessive, explain that a precautionary landing should be considered. To help prevent turbulence induced inputs, demonstrate how to properly support the cyclic arm by bracing it against the leg.

## Low Rotor RPM

Although rotor rpm and airspeed are repeatedly emphasized during training, it is important for the student to understand not only that there are limits to both, but also why those limits have been specifically defined. Ensure the student understands and can explain that the limits for powered operation are dictated by the operating limits of the engine. At low rpm, the engine cannot develop full power, and the high limit is imposed by engine structural limits. Also, point out that if the engine and main rotor rpm are allowed to get too low, tail rotor rpm is also greatly reduced. This situation could lead to an inability to keep the helicopter from turning. At the low limit on the rotor tachometer, the rotor may not produce enough lift to sustain level flight. The high limit of rotor rpm is imposed to protect the structural integrity of the rotor and drive components.

### Blade Stall

Impress on the student that low rotor rpm can also lead to blade stall. If the rotor rpm decays to the point at which all the rotor blades stall, the result is usually fatal, especially when it occurs at altitude. In addition to the main rotor blades stalling, if centrifugal force is decreased too much, there is not sufficient force to keep the blades horizontal. Explain that, even at normal operating rpm, the blades cone upward when producing lift. If the required lift remains high coupled with high angles of attack, the blades bend upward further until there is no recovery. The danger of low rotor rpm and blade stall is greatest in small helicopters with low blade inertia. Explain to the student that a pilot can create the situation in a number of ways, such as simply rolling the throttle the wrong way, or pulling more collective pitch than power available, or when operating at high density altitude.

Point out to the student that when the rotor rpm drops, the blades tend to maintain the same amount of lift by increasing pitch. As the pitch increases, drag increases, which requires more power to keep the blades turning at the proper rpm. When power is no longer available to maintain rpm, and therefore lift, the helicopter begins to descend. This changes the relative wind and further increases the angle of attack. At some point, the blades stall unless rpm is restored. If all blades stall, it is almost impossible to get smoother air flowing across the blades.

To make matters worse, the main rotor rpm to tail rotor rpm ratio is one to six. Therefore, a one percent reduction in main rotor rpm can also result in a six percent reduction in tail rotor rpm, which corresponds directly to a loss of antitorque thrust available and loss of yaw control.

Stress that any time the rotor rpm falls below the rpm limits while power is still available, the student should simultaneously add throttle and lower the collective. If in forward flight, gently applying aft cyclic loads up the system and helps increase rotor rpm. If there is no power available, immediately lower the collective and apply aft cyclic.

### Recovery From Low Rotor RPM

Before the student is allowed to solo, the techniques for recovery from low rotor rpm in both a hover and in flight must be practiced. This low rotor training should not exceed any limitations. The instructor should not place undue stress on the dynamic parts of the helicopter, as letting the rpm drop too low can result in excessive blade bending, internal engine stress (including localized overheating not monitored by a gauge or sensor), and blade stop pounding. While in a hover, rotor rpm is reduced until the throttle alone will not increase the rpm. The student should then take the controls and attempt to recover rpm by lowering the collective just enough to allow the helicopter to settle gently toward the ground while increasing the throttle. The objective is to regain rpm without allowing the helicopter to touch down. To prevent touchdown, the collective should be raised slightly to stop the rate of descent. Practicing this maneuver graphically demonstrates to the student that if rotor speed is lost and the helicopter begins to settle, collective pitch alone should not be increased in an attempt to maintain altitude. While in flight, rpm may be regained by lowering the collective slightly and increasing the rpm. Aft cyclic while lowering the collective may also help increase rotor rpm, but is usually not required unless the rpm is critically low.

Under certain conditions of high weight, high temperature, or high density altitude, a situation might exist in which the rpm is low even though maximum throttle is being used. This is usually the result of the main rotor blades having an angle of attack that creates so much drag that engine power is not sufficient to maintain or attain normal operating rpm.

In a low rpm situation, the lifting power of the main rotor blades can be greatly diminished. Therefore, as soon as a low rpm condition is detected, immediately apply additional throttle, if available, while lowering the collective. This reduces main rotor pitch and drag. Under training conditions, make sure the skids or landing gear wheels do not contact the ground. In an actual situation in which the engine does not have sufficient power to accelerate the rotor, smoothly lower the helicopter to the ground, if conditions permit. Once on the ground, the collective can be lowered a little more to regain rpm. Do not try to maintain a hover by raising the collective when the rpm is too low and the throttle is wide open.

As the helicopter begins to settle, smoothly raise the collective to stop the descent. At hovering altitude, this procedure might need to be repeated several times to regain normal operating rpm. This technique is called "milking the collective."

When operating at altitude, the collective may need to be lowered only once to regain rotor speed. The amount collective can be lowered depends on altitude.

Since the tail rotor is geared to the main rotor, low main rotor rpm may prevent the tail rotor from producing enough thrust to maintain directional control. If pedal control is lost and the altitude is low enough that a landing can be accomplished before the turning rate increases dangerously, slowly decrease collective pitch, maintain a level altitude with cyclic control, and land.

As instructors, we are aware that in most modern helicopters low or inadequate rotor rpm is an indicator of probable overloading or engine performance problems. Show the student that increased awareness to higher density altitudes is critical. Have the student determine the density altitude at the time this maneuver is performed and note to the student that recoveries at higher altitudes may not be possible.

## Common Student Difficulties

As the helicopter begins to settle, students may have a tendency to increase collective pitch to stop the descent. Remember, students often fail to correlate the requirement to reduce collective pitch and increase throttle because previous training emphasized reducing throttle when lowering collective.

## Brownout/Whiteout

Brownout and whiteout are two more helicopter hazards that cannot be demonstrated, but should be discussed with the student. Stress that the helicopter's capability of landing in many conditions make it susceptible to visual obscuration when flying over ground material that can be blown up into the rotors during hover flight at low altitude. Differentiate between the brownout caused by dust and sand and the whiteout experienced in snow conditions.

If during normal training sessions, seasonal conditions warrant the need to encounter such conditions, use the time to discuss the proper actions, taking care not to expose the student to an unfamiliar environment. Most often, an approach to the ground will suffice to compensate for these conditions.

It should be noted that brownout conditions occur to varying degrees for a given landing zone depending on the aircraft in use (single versus tandem rotor), its configuration (weight), tactics being employed (rapidity of approach and landing, accompanying aircraft, time on deck, etc.), and environmental conditions (humidity and/or rain, day versus night, and temperature and density altitude). In general, explain that helicopters tend to begin to experience brownout during an angled, no-hover approach to landing at approximately 1–2 rotor diameters above ground level (50–150 feet), with the most serious conditions being experienced at approximately 50 feet and below. As the aircraft slows, the thrust vector of the main rotor disk becomes more vertical (as the aircraft pitches its nose up to decelerate), and the thrust becomes greater as power is added to sustain a hover, or near-hover condition prior to landing. Also, the rotor thrust tends to circulate down, out, then back up and down again through the rotor disk just prior to touchdown. All of these conditions combine just prior to landing, the most critical time for the pilot to eliminate lateral drift.

With the helicopter engulfed in the cloud of a suspended particulate material, the pilot loses outside reference and horizon cues. The instructor should stress that the loss of pilot visual reference to a fixed point inevitably causes some degree of unintended pilot or aircraft-induced drift away from the intended line of approach and landing. Discuss how, given the top-heavy nature of most helicopters and the aerodynamics of their rotor disks, landing with lateral drift of any kind can cause the aircraft to be damaged, or to roll over, particularly if combined with a pivot point (such as a rock, or rut) upon touchdown. Emphasize the importance of avoiding sideways motion, which can lead to dynamic roll.

While the majority of mishaps related to brownout conditions occur during landings, advise the student that a helicopter takeoff may also cause brownout due to rotor wash. Ask the student to explain why it is less problematic during takeoff. The student should explain that the aircraft is accelerating away from the dust, and lateral drift is less of a safety factor, except for obstructions in close proximity to the line of departure. The pilot is able to set a power and nose attitude and fly the aircraft safely out of a zone, and not be overly concerned with induced drift due to restricted visibility. A maximum performance takeoff should be used when taking off from a possible brownout/whiteout condition to ensure the quickest exit possible. If the surface is sand or snow covered, planning for an approach to the surface and being slightly faster than normal should help the pilot avoid a brown or white out. Likewise, the takeoff in a possible white or brown out condition should be from the surface and not from a hover in order to exit the condition before the horizon is completely obscured. Highlight the need to first check the helicopter for clearance from the surface by bringing it light on the skids or skis to ensure power is good, controls are normal, and the landing gear is free, then reducing power to allow the air to clear before the actual takeoff. The standard 40-knot takeoff attitude is very important to heed to stay ahead of the majority of the cloud dust/snow cloud that is generated by the rotor system and maximum power to climb out of the surface obscuration as soon as possible. Instructors should note the experience level of the students being flown. Do not use instrument flight terms with students who have yet to conduct instrument flight training.

## System or Equipment Malfunctions

### Antitorque System Failure

NOTE: Many helicopters have antitorque failure procedures that can be safely practiced and will most likely terminate the landing with some sort of run on landing.

An antitorque failure can occur in several different forms. Impress upon the student the necessity to become familiar with each type of failure, its effects on flight, and the manufacturer's recommended procedure for coping with the malfunction. A discussion of some of the types of malfunctions, as well as probable effects and corrective actions, should be included in any training syllabus.

The instructor should also include discussion on the effects of structural design and components. The vertical stabilator, for example, provides a streamlining or trim effect under certain conditions and airspeeds. At what speeds will the vertical fin be most effective, and does that correlate to touchdown airspeed for the respective RFM procedure? An instructor's role is greater than regurgitating already printed information. Go beyond the print and incorporate aircraft specifics and procedures to probable conditions that will help the student obtain a higher level of understanding.

### Complete Loss of Tail Rotor Thrust

Ensure the student understands that complete loss of tail rotor thrust involves a break in the drive system, and pedal input will have no effect on helicopter trim. The tail rotor is providing no thrust to compensate for torque.

### Fixed Pitch Settings

Explain to the student that a fixed pitch setting is dependent upon the amount of power applied at the time of the malfunction. If the failure occurs at a reduced power setting (low torque), the helicopter's nose will turn when power is applied. When the failure occurs at an increased power setting (high torque), the nose of the helicopter will turn when power is reduced. Emphasize the use of the manufacturer's recommended procedures for coping with either situation.

### Loss of Tail Rotor Components

Make the student aware that if the loss of tail rotor components occurs, he or she will also experience a shift in the center of gravity. The student should understand that the severity of the situation is dependent upon the amount of weight lost. Impress upon the student that if a major failure allows something like the drive to begin whipping around, it could destroy the entire tail boom. Extended flight is not recommended after a failure of any kind. The helicopter can be repaired at a remote landing area much more quickly and at a much lower cost than rebuilding the helicopter after a crash landing 3 miles closer to the base of operations. If something on a helicopter fails, follow the RFM procedures. If there is any suspicion of a component failure, safely land as soon as possible.

### Unanticipated Yaw/Loss of Tail Rotor Effectiveness (LTE)

Loss of tail rotor effectiveness (LTE), or unanticipated yaw, is not related to equipment malfunction, but rather is a result of the tail rotor not providing the adequate thrust required to maintain directional control. LTE may occur at airspeeds less than 30 knots and can be caused by a number of factors, including main rotor disk interference, weathercock stability, and tail rotor vortex ring state. It also depends upon wind direction and speed, altitude, and helicopter design. Thus, LTE offers a number of opportunities for the instructor to discuss the principles of aerodynamics and the physics involved in helicopter flight while reviewing a helicopter hazard.

To help reduce the onset of LTE, make sure the student understands the limitations of the training helicopter and those circumstances under which LTE is most likely to occur. Additionally, discuss recovery procedures (forward flight into the wind is best, but it must be in an obstacle-free direction, not downwind) and the effects of altitude and terrain on the recovery procedure.

## Main Drive Shaft Failure

Failure of the main drive shaft causes an immediate increase in engine rpm and a decrease in rotor rpm. Explain to the student that an autorotation is necessary to maintain rotor rpm. Engine rpm should be kept within normal limits to provide power to the tail rotor for directional control if necessary for that helicopter.

If the drive system is like that of a Robinson-44, then the engine should be shutdown to minimize damage and heat sources. If the main drive shaft of a BH 206 failed, then the engine would be necessary to power the tail rotor for landing from an autorotation. Until the drive shaft separates from a turning drive, it generates a lot of noise and can cause much damage to structures. The pilot may be told to ignore the noise, autorotate with the rotor tachometer in the green range, and execute an autorotative landing. Once safely on the ground, perform an emergency engine shutdown and kill all electrical circuits in accordance with the RFM.

## Hydraulic Failure

Explain to the student that the effect of hydraulic failure on the control system in a helicopter depends on the model. Since the hydraulic system is used to overcome high control forces, point out to the student that the first sign of a hydraulic failure is generally a need for more force to control helicopter movement. In those helicopters in which control forces are so high they cannot be moved without hydraulic assistance, two or more independent hydraulic systems may be installed. Have the student consult the RFM for the helicopter being flown and then discuss the corrective actions required.

Instructors must be most diligent when conducting hydraulic failure training. Some accidents have occurred due to instructors improperly or erroneously following procedures. Discuss with the student the events that will occur before practicing any emergency training. Always have complete understanding, as a crew, of what steps are taken and in what sequence.

In helicopters equipped with dual hydraulic systems, the instructor should ensure the student understands the purpose of the backup system, which is to get the helicopter on the surface and not continue the flight. The student should be taught the common parts of the systems, such as a common reservoir or drive pad, and how one failure can lead to failure of the other system. The student must realize there is complete lack of control if the remaining system fails.

## Governor Failure

Many training helicopters and all turbine-powered helicopters are equipped with engine governors. It is valuable training for the student to learn how to manually control the throttle in the event of a governor failure. In some helicopters, the governor can be safely turned off for training purposes. Consult the RFM for correct procedures and techniques.

## Multiengine Operations With One Engine Inoperative

Helicopters with two engines are rarely found in the training fleet; however, instructors may transition helicopter pilots into these more advanced aircraft. The advantages of having two engines are obvious, not only in the redundancy of the engines but also in the power available when both are operating. Students should understand that, in the event of an engine failure, there is still one engine operating and the helicopter is still capable of flight although performance is diminished. The RFM outlines single-engine operations and capabilities for multiengine helicopters.

While unlikely, dual engine failure is possible and has occurred. Circumstances such as, icing, fuel system problems, contamination, or drive train failure are just a few examples that may lead to loss of powered flight in multiengine helicopters. When conducting multiengine transitions, training one engine inoperative emergency flight consumes the bulk of training hours. However, ensure adequate time is allocated to loss of powered flight in dual engine aircraft.

## Emergency Equipment and Survival Gear

Ensure the student is familiar with the location and operation of all emergency equipment and survival gear installed or carried in the helicopter. *Figure 12-6* shows an example of a typical survival gear and emergency equipment list for a wooded or densely forested location. Students should also be shown sample lists of survival gear for all weather extremes, such as Alaska and Arizona requirements. In a survival situation you can never have too much equipment. Students should receive instruction on and be familiar with the function of emergency releases of all doors, hatches, and windows, as well as operation of all items contained in the list.

Discuss emergency egress and basic survival requirements for the locale and time of year.

While the intent of this chapter is to discuss instructional points for emergency procedures, it cannot begin to address all emergency equipment and survival gear required for the many different geographical locations and aircraft-specific procedures that instructors may be faced with while conducting training. Therefore, it is important that instructors discuss emergency equipment and survival gear unique to the

## EMERGENCY EQUIPMENT AND SURVIVAL GEAR

Food cannot be subject to deterioration due to heat or cold. There should be at least 10,000 calories for each person on board, and it should be stored in a sealed waterproof container. It should have been inspected by the pilot or his representative within the previous 6 months, and bear a label verifying the amount and satisfactory condition of the contents.

A supply of water

Cooking utensils

Matches in a waterproof container

A portable compass

An ax weighing at least 2.5 pounds with a handle not less than 28 inches in length

A flexible saw blade or equivalent cutting tool

30 feet of snare wire and instructions for use

Fishing equipment, including still-fishing bait and gill net with not more than a two-inch mesh

Mosquito nets or netting and insect repellent sufficient to meet the needs of all persons aboard, when operating in areas where insects are likely to be hazardous

A signaling mirror

At least three pyrotechnic distress signals

A sharp, quality jackknife or hunting knife

A suitable survival instruction manual

Flashlight with spare bulbs and batteries

Emergency Position Indicating Radio Beacon (EPIRB)(406 MHz) with spare batteries

Stove with fuel or a self-contained means of providing heat for cooking

Tent(s) to accommodate everyone on board

Additional items for winter operations:
- Winter sleeping bags for all persons when the temperature is expected to be below 7 °C
- Two pairs of snow shoes
- Spare ax handle
- Ice chisel
- Snow knife or saw knife

**Figure 12-6.** *Emergency equipment and survival gear.*

geographical location and installed on the training helicopter being flown.

Also, discuss the requirements of Title 14 of the Code of Federal Regulations (14 CFR) that apply to the equipment carried on board for the type and area of operation. For instance, 14 CFR part 91, section 91.205(b)(12), states, "For VFR flight during the day, the following instruments and equipment are required: If the aircraft is operated for hire over water and beyond power-off gliding distance from shore, approved flotation gear readily available to each occupant and, unless the aircraft is operating under part 121 of this subchapter, at least one pyrotechnic signaling device. As used in this section, 'shore' means that area of the land adjacent to the water which is above the high water mark and excludes land areas which are intermittently under water."

For example, if you are located near a large body of water, expand beyond the required syllabus to discuss what unique characteristics and limitations are present when flotation devices are installed. What impact could they have upon exiting the helicopter?

Talk about the emergency procedure required for ditching in water. Discuss when to remove the doors in situations of power off (autorotations) or power on (ditching). When forced to land in water, taking the doors off is critical; however, if the helicopter is in an autorotational descent, removing the doors in descent may cause greater problems. If under powered flight and forced to ditch (with little fuel remaining), jettisoning the doors and then hovering a safe distance away to enter the water is the best option. It is important to discuss the danger of turning rotor blades to those exiting the helicopter and to discuss the best manner to exit.

Again, tailor this topic to the equipment and gear installed and used in the training helicopter. This discussion can be quite in depth and should be based on local procedures, as well as the knowledge of more experienced pilots in the flight area.

## Scenario-Based Training

Once the student has mastered engine failure in the hover emergency procedures, this emergency can be incorporated into any scenario by the instructor giving a verbal warning at the appropriate time to indicate the engine has failed.

For an instructor, the initial focus of emergency training is to have the student correctly analyze the malfunction and properly perform the corrective action. This is often accomplished with one specific emergency procedure being taught as the center of attention. But how do we know if the student will be able to perform this action if it actually occurs? An instructor should never let the student exceed aircraft limits or the instructor's ability to recover the helicopter.

One way to check the student's proficiency and confidence is to build emergency training into a routine flight scenario. Give the student the task of planning and executing a flight to a nearby airport. The instructor will have a predetermined series of simulated emergencies to be given at different segments of the flight. This allows the instructor to plan ahead to ensure the specific procedure to be performed will be in an adequate or approved emergency procedure training environment.

For example, one task to be evaluated is antitorque failure. Prior to entering the nearby airport's traffic pattern, the instructor gives the student a verbal description of antitorque controllability issues and fixes the pedals. Another example would be to advise the student that the oil pressure is low or the engine temperature is too high. The instructor could announce that the engine is running rough or is cutting out. Simply saying "engine failure" is counterproductive because the student must learn the signs of an engine failure and not be trained to wait for cues from the instructor. The whole intent of the training is for the pilot to be fully trained and self-sufficient in single-pilot helicopters. At this point, allow the student to work through the situation and take the appropriate actions. If appropriate actions are proficiently executed, the student's confidence level rises and the instructor also feels secure in the student's abilities.

## Instructor Tips

By using simulation or flight training programs, the student experiences the virtual reality of various helicopter hazards, especially ones that cannot be demonstrated due to possible structural damage to the aircraft. Refer to *Figure 12-7* for a general lesson plan to be used to help outline training.

## Chapter Summary

This chapter addressed various emergency procedures and provided the instructor with some topics for discussion with a student. Emergencies were discussed in general terms; this chapter is not intended to replace the procedures recommended by the manufacturer in the RFM.

### Objective

The purpose of this lesson is to teach the student how to land safely following an engine failure at the hover or hover taxi. Engines can fail just as easily at the hover or hover taxi as in flight. The helicopter will land very quickly should this happen, and it is vital that a pilot be able to react quickly and prevent an incident from becoming an expensive accident. The student will demonstrate the ability to land safely following an engine failure at the hover or hover taxi.

### Content

1. Preflight Discussion
   a. Discuss lesson objective and completion standards.
   b. Review hovering, take-off and landing procedures.
   c. Review normal checklist procedures.
   d. Weather analysis.
   e. Point out that at normal hover or hover-taxi heights, it is not possible for the student to flare. In fact, lowering the collective following an engine failure may result in a hard landing. The pilot relies on the inertia in the rotor system to land safely. Review height/velocity chart and discuss shaded areas to emphasize profiles at which successful autorotations can be accomplished.
   f. Describe the reaction of the helicopter when the engine fails: yaw, drift, and sink.
   g. Explain that the yaw and drift must be corrected before touchdown. Sink should be controlled by use of the collective, as appropriate to the type of helicopter and the height above ground, to cushion the landing.
   h. Explain that, should engine failure occur at the hover-taxi, the student should avoid any rearward movement of the cyclic and complete a run-on landing. Review dynamic rollover and the importance of minimizing lateral drift. Re-emphasize the concern of collective pitch being rapidly reduced prior to airframe motion coming to a complete stop. Failure to do so may potentially result in aircraft damage.
   i. Have the student explain what is to be done and highlight the major points of the maneuvers.

2. Instructor Actions
   a. Select a suitable area for practice.
   b. Demonstrate into wind as follows:
      1) Give a verbal warning. 2) Close the throttle. 3) Counteract yaw and drift. 4) Cushion the landing.

3. Student Practice
   After being given a verbal warning by the instructor, the student performs the procedures for an engine failure at a hover or hover taxi.

4. Instructional Points
   a. This exercise should be introduced by providing the student with plenty of warning before each practice. The pace of the maneuver can then be increased to flight test standards where the student is given minimal or no warning of the practice engine failure.
   b. Closing the throttle and cushioning the landing with the collective takes a good deal of manual dexterity in most helicopters. Since the aim of this exercise is for the student to react to an engine failure, the instructor should control the throttle in the beginning.
   c. Always ensure that the surface is suitable for this exercise, particularly after precipitation.
   d. This is a good exercise to demonstrate to the student the landing stage of an autorotation. It is a good skill to practice just before starting a full-on autorotation exercise.
   e. Exercise caution because a student may react to a simulated engine failure by rapidly lowering the collective. Be sure to give a verbal warning before closing the throttle.
   f. The demonstration of this exercise is easily split to show the three control movements separately. Perform three separate demonstrations, letting the student focus each time on an individual control movement, then combine all three before student practice.
   • Always ensure that the engine is reduced to the detent so power does not increase as collective is being pulled.
   • Always ensure that there is no lateral travel at touchdown during a hovering autorotation.
   • Always maintain landing gear alignment with the direction of travel.

### Postflight Discussion

Review and critique the flight, being sure to discuss student strengths and weaknesses. Remember to provide suggestions on how to improve performance. Preview and assign the next lesson. Assign *Helicopter Flying Handbook*, Chapter 12, Attitude Instrument Flying.

**Figure 12-7.** *General lesson planning.*

## Chapter 13

# Attitude Instrument Flying

## Introduction

This chapter is intended to assist the helicopter instructor in further explaining attitude instrument flying in helicopters. When appropriate, refer to the Instrument Flying Handbook and the Advanced Avionics Handbook for the definition or further explanation of a system instrument. Many of the current helicopter instrument systems have been included in those handbooks; however, it is not feasible to include every helicopter instrument system that may be installed in any particular helicopter. Instrument systems on different aircraft serve the same purpose; however, the configuration of the instrument panel and the design may be somewhat different.

Ideally, the helicopter instructor will wait to explain this chapter until after a basic understanding of instruments is achieved. Once the student understands the basics, then actual helicopter flights reinforce what was taught in this chapter. This chapter is intended as a building block towards attaining an instrument rating. Students will be taught attitude instrument flying and should apply all of the basic maneuvering flight skills that have already been mastered.

## Instructor's Objective

An appropriately rated flight instructor is responsible for training the instrument rating pilot applicant to acceptable standards in all subject matter areas, procedures, and maneuvers included within the appropriate instrument rating practical test standard. Title 14 of the Code of Federal Regulations (14 CFR) part 61, section 61.195(c), states that basic instrument maneuver training for private pilot students and lower need not have an instrument instructor rating, if the instructor has instrument privileges on his or her pilot certificate. Because their teaching activities affect the development of safe and proficient pilots, flight instructors should exhibit a high level of knowledge, skill, and the ability to impart that knowledge and skill to students.

It is important to find out the student's background during the initial portion of instrument training. New students with only basic maneuvering instruction provide a different set of challenges for the instructor than a more experienced pilot (e.g., a pilot who has flown by instruments in a fixed-wing aircraft). The instructor must know who is being trained and what tendencies or trends may commonly be observed. Just as with any other instructional approach, instructor ability to identify and correct student error is based on the instructor's ability to cull from knowledge and experience. The instructor should reference the specific helicopter sections in the Instrument Flying Handbook, which includes full discussions on helping an airplane pilot transition to helicopters.

Additionally, the flight instructor must certify that the applicant is able to perform safely as an instrument pilot and is competent to pass the required practical test.

Discuss with the student:

- Flying with the instruments is essentially visual flying with the flight instruments substituted for the various reference points on the helicopter and the natural horizon.

- The IFR helicopter pilot cannot reference the rotor tip path to the horizon, but depend instead on the artificial horizon for a reference.

- There is no difference in helicopter control inputs between flying visual flight rules (VFR) and instrument flight rules (IFR).

- There is no outside reference and a pilot must trust what is presented on the helicopter instruments.

- Basic instrument training is intended to be a building block toward attaining an instrument rating.

## Ground Instruction

In a classroom environment, work with the student to begin developing a basic knowledge of the terms associated with attitude instrument flight. One way to capture the student's level of understanding is by requesting that the student identify the location of the instrument located in the helicopter, explain what each instrument is used for during attitude instrument flight, including its indications and limitations. The following instruments should be covered in the lesson plan: airspeed indicator, altimeter, vertical speed indicator (VSI), attitude indicator, heading indicator, and turn indicator. *[Figure 13-1]*

- The student should learn the names and locations of the pitot static instruments (airspeed indicator, altimeter, and VSI), their use in preflight and airborne checks, and common errors.

- The student should learn the name and location of each gyroscopic instrument (attitude indicator, heading indicator, and turn indicator), their use in preflight and airborne checks, and common errors.

- The student should learn about the magnetic compass to include magnetic variation, magnetic dip, and compass deviation, as well as preflight checks, airborne checks, and common errors.

Students have been known to be intimidated by instrument flying. Lack of experience and/or poor training can contribute to this. An instructor's goal is to keep the training and the lesson plan as interesting as possible. Once the student understands what each instrument does, knows how to use it and develops a good cross-check, he or she will overcome the intimidation factor. The amount of time spent on each area is determined by the individual's ability to achieve a satisfactory level of proficiency. A portion of the instrument training may utilize a flight simulator, flight training device, or a personal computer-based aviation training device (PCATD). *[Figure 13-2]*

**Figure 13-1.** *Instruments that should be discussed when teaching instrument flying: airspeed indicator, attitude indicator, altimeter, turn indicator, heading indicator, and vertical speed indicator (VSI).*

**Figure 13-2.** *Helicopter simulator used for training purposes.*

The instructor must help the student form the correct scanning habits from the very first instrument flight, whether it is in the helicopter or in a flight simulator. Students have a tendency to stare at one instrument, which allows the other instruments to exceed tolerances very quickly. A VFR pilot must scan the instruments and gauges, as well as the sky, for traffic and obstructions. An IFR student must scan the gauges and instruments, and maybe the outside if there is anything to view.

One common problem is attempting to stare at an instrument while a correction is being made. A student should be taught to scan, determine the issue or problem, make a control change, and then continue the scan. The result of the control change is checked on the next scan. The student must remember that inertia of the helicopter and all the changes require some finite period of time, so changes neither occur instantly nor would we usually wish to make abrupt changes of large magnitude.

## Flight Instruction

Once the student has an understanding of the instruments and knows the location of each, then the next logical step would be an actual flight. *[Figure 13-3]* Prior to the flight, the student should review all the instruments that will be used for the particular helicopter being flown and learn how to perform an instrument cross-check, instrument interpretation, and aircraft control.

**Figure 13-3.** *Demonstrate to the student that proper instrument interpretation is the basis for aircraft control.*

When teaching the student about flying a helicopter with reference to the flight instruments, the key is getting that student to understand that proper instrument interpretation is the basis for aircraft control. Skill, in part, depends on understanding how a particular instrument or system functions, including its indications and limitations. With this knowledge, the student can quickly interpret an instrument indication and translate that information into a control response. Start with simple tasks, and then progress to the more complex tasks.

The student should be able to demonstrate performance of the following tasks at a satisfactory level:

- Straight-and-level flight

- Straight climbs

- Straight descents

- Turns: predetermined heading, timed, change of airspeed, 30° bank, climbing and descending, and compass

- Unusual attitudes

- Emergencies

- Instrument takeoff

Perform a learning check by asking the student the location and function of each instrument used for attitude instrument flight. During flight, these demonstrations indicate if the student is able to maintain aircraft control during attitude instrument flight using both cross-check and instrument interpretation. Ensure the student knows what to do if the instrument fails and how that failure affects the scan being used. Allocate adequate time to train recovery from unusual attitudes. The student needs an understanding of the errors inherent in each instrument and common errors for the tasks to be performed.

### Instructional Techniques

Instrument flying is simply composed of level, turning, climbing, and descending instrument flight maneuvers. Do not overwhelm the student on the first instrument training flight. Use the building block technique by introducing one flight task at a time. Allow the student to fly and maintain a set altitude. Then, introduce heading control and then airspeed control. The instructor's main duty is to divide the instrument procedures into small enough tasks to enable the student to grasp the concepts, acknowledge the desired outcomes, and understand the methods to use and when to use them to achieve the necessary performance.

The student should understand that the helicopter never really flies straight and level. Only after much practice does it begin to appear to fly in that fashion. Due to the very small tolerances in the control system, each rotor blade flies a very slightly different path every revolution. Therefore, the helicopter pilot must continually make small corrections to achieve what passes for straight-and-level flight. This characteristic is the practical reason that IFR certified helicopters must have a fully functional autopilot or be crewed by two pilots for IFR operations. Helicopters are very controllable, but not necessarily stable. Therefore, cockpit organization and flight planning is very essential. Depending on the helicopter, some sort of crew training and resource management must be incorporated into this training.

When training instrument flight to a transitioning airplane pilot, instructors should explain or reiterate the differences between an airplane and helicopter. For example, an airplane must be pointed up on the artificial horizon in order to climb, whereas a helicopter can climb quite well with its nose down. Conversely, an airplane is pointed down to descend but in a helicopter, the nose is raised and the collective is lowered to descend. Instrument flight can cause a student to become tense or get behind on tasks, which may cause a transitioning pilot to revert back to the first learned airplane habits.

Often, a student may seem to be depressed from seeing so many errors during IFR flight training periods. That is when the instructor should congratulate the student because that is when flying really begins to improve; the student sees the errors and, with practice, learns the proper amount of control movement to correct those errors in a timely but controllable manner.

## Student Tendencies

Some common student tendencies are:

- Inconsistent or no scanning technique
- Staring too long at one flight instrument
- Not analyzing what they see
- Exaggerated flight control inputs
- Failure to correlate control inputs

While discussion of scanning can be done in the classroom, the actual practice does not yield results until in flight. Depending on the instrument panel layout, have the student determine the most useful scan. (According to the Instrument Flying Handbook, no specific method of cross-checking (scanning) is recommended; the pilot must learn to determine which instruments give the most pertinent information for any particular phase of a maneuver.)

Watch the student's head and eyes to see if fixation is occurring. If the student stares too long at one instrument (heading indicator, as an example) then other parameters are usually affected (altitude, airspeed, trim, etc.). This can have a snowball effect as the student will eventually become overwhelmed.

Ensure the student allows time to see and interpret the particular instrument, within the chosen scanning technique, and makes the necessary flight control input. Failure to take action may be a result of not processing the information present or absent in the scan of a particular instrument.

When applying control input corrections, the student should use small inputs and allow time for them to take effect. Too often the student identifies and responds with the correct input but does not allow adequate time for the input to achieve the result. This can lead to overcontrolling.

Initially, failure to correlate corresponding inputs is a common tendency. The student may need to be reminded of the associated control inputs normally used for the various instrument indications. Frequent breaks and discussion may be needed to allow the student time to process the information presented before continuing practical application.

An example of this occurs when a student notes an increasing rate of climb and reacts by placing forward cyclic. Forward cyclic alone arrests the climb rate, but it also produces an increase in airspeed. In this example, however, during repeated attempts, the student repeatedly fails to make a corresponding reduction in power, and airspeed continually increases. The student may not correlate the impact of forward cyclic on airspeed, instead focusing only on rate of climb. Repeating the fault to the student while he or she continues to fly may result in sensory overload.

Have the student transfer the controls, take a moment, then reemphasize the learning point by demonstrating the correct control inputs with the student watching. Remind the student that the position of the flight controls never stays the same when flying a helicopter. Even flying straight and level requires change and adjustment to the flight controls and, as more fuel is used, the helicopter becomes lighter. Instrument flying is really precision flying and students will slowly start noticing small changes without pilot input and need to be reminded of that. Then, transfer aircraft control back to the student and have the student repeat the maneuver.

Stress to the student the need to maintain a consistent scan technique and to maintain situational awareness of all indicators. Over time, the student's scan and response time will improve.

During advanced instrument training, allowing the student to work through some of these issues can be beneficial to the student's confidence. However, the new student can quickly become overwhelmed and will not understand what is happening. Therefore, it is not good practice to allow the new pilot to become overly frustrated when first learning simple instrument tasks.

Reference the Instrument Flying Handbook for discussion on these additional topics:

- Control instruments

- Performance instruments

- Navigation instruments

- Four-step process used to change attitude: establish, trim, cross-check, and adjust

- Apply the four-step process for: pitch control, bank control, and power control

- Primary and supporting method: pitch control, straight-and-level flight, primary pitch, primary bank, primary yaw, and primary power

- Scanning techniques of attitude instrument flying

- Common errors of attitude instrument flying: fixation, omission, and emphasis

More detailed information, as well as additional explanations can be found in the following references:

- Aviation Instructor's Handbook

- Instrument Flying Handbook

- Helicopter Flying Handbook

- Instrument Procedures Handbook

- Aeronautical Information Manual (AIM)

- Advanced Avionics Handbook

## Instructional Objectives

Continue to reinforce the basic standards throughout this training. Proper performance planning can be used to demonstrate the understanding and use of power settings in stabilized instrument flight techniques. Knowing the power settings used for climbs, level flight, and approaches as well as the various instrument flight modes decreases confusion and/or searching by the student pilot. During use of the checklist, help the student understand the importance of checking items, such as flight instruments during hover checks. Also, tie in the importance of altimeter settings and the fluctuations that may occur when hovering IGE. The instructor should be explaining the reasons for the instrument takeoff following the maximum performance takeoff profile in order to gain altitude as quickly as possible to clear obstructions and the specified minimum instrument airspeed for that helicopter. Special care should be taken to fully explain helicopter approaches, required airspeeds, and the underlying reasons for those restrictions.

The instructor's role is to identify, analyze, and make specific corrective suggestions to help the student. Pointing out parameter errors, such as, "Your heading is off, your altitude is off, your airspeed is off…" without providing detailed corrective action does not help the student. Assess the cause and provide methods or techniques to correct the situation.

Whether a ground or flight training session, each training period should end with a thorough debriefing of what transpired and what will be covered in the next training period. Ask for the student's perception of the training. The student should not walk away unaware of what occurred during this training session, or what will be covered in the next session.

## Instructor Tips

- Review with the student how all the flight instruments operate and the actual location of each instrument inside the helicopter.

- Review and practice with the student what instruments are utilized during attitude instrument flight.

- Practice a "good cross-check" with the student.

- Practice with the student how to interpret the instruments during flight.

- Ensure the student uses smooth control inputs at all times during flight.

- Review with the student the common errors with each task and instrument.

- If appropriate, tell the student well ahead of time what will be covered (task(s) to be flown) for the lesson plan and what the student should study or reference. *[Figure 13-4]*

## Chapter Summary

This chapter discussed all the common instrument references and concepts associated with attitude instrument flying and some common errors associated with helicopter flight. The chapter covered which flight tasks are accomplished during attitude instrument flying and how to accomplish those tasks with the instruments.

### Attitude Instrument Flying

**Objective**

The purpose of this lesson plan is to review with the student the common tasks associated with attitude instrument flying. The student needs to know the name and location of each flight instrument and its functions. Instruments that should be discussed are the airspeed indicator, attitude indicator, altimeter, turn indicator, heading indicator, and the vertical speed indicator. The student needs to know how each flight instrument operates and should correlate each flight control input with the aircraft response during attitude instrument flying.

**Content**

**Ground Training**
1. Discuss lesson objective and completion standards.
2. Review terms associated with attitude instrument flight.
3. Verify that the student has an understanding of each instrument and its location.
4. Conduct performance planning and review as necessary.
5. On subsequent lessons, review previous training session, to include lessons learned, actions taken.

**Flight Training**
1. Perform all required tasks according to the checklist, emphasizing those steps applicable to instrument flight.
2. Review pertinent power settings, flight parameters to be used during the flight.
3. Perform actual flight
   - Perform those maneuvers discussed from the previous lesson, reinforcing corrective actions taken.
   - Progress to current flight training session. Begin with simple tasks, then more complex tasks.

**Postflight Discussion**

- Review what was covered during this phase of training.
- If any problems were encountered during the flight tasks, correct or, if necessary, retrain for that particular task(s).
- Preview and assign the next lesson. Assign Helicopter Flying Handbook, Chapter 13, Night Operations.

**Figure 13-4.** *Sample lesson plan.*

# Chapter 14
# Night Operations

## Introduction

By definition, night flight is flying between the hours of sunset (end of evening civil twilight) and sunrise (beginning of morning civil twilight). During day or night, vision is the sense that makes a pilot aware of the position of the helicopter in space. The eyes can rapidly identify and interpret visual cues during daylight. During darkness, however, visual acuity is decreased proportionally as the level of illumination decreases. Night vision devices improve the capability of the human eye to see at night. Night vision goggles (NVGs) are being utilized more and more by police departments, emergency management services, and civilian pilots. This chapter covers only night flight without the use of NVGs.

FOCAL POINTS

X
10°

X 10°      X 10°

10°
X

### Observer

Once an object is detected in the peripheral field of dark-adapted vision, aircrews maintain continual surveillance by using the off-center vision technique. They look 10 degrees above, below, or to either side of an object, viewing it no longer than 2–3 seconds at each position.

## Instructor's Objective

Vision is the most important sense for flying. Vision allows us to perceive our position relative to the horizon, observe our location, see and avoid obstructions, and read aircraft instruments and charts. However, in comparison to most other mammals, humans have poor vision in low light conditions. As an instructor, ensure that the student has a working knowledge of visual deficiencies and the techniques that the student can employ to overcome them.

Night vision is of particular interest to the helicopter pilot because a helicopter often lands at unimproved sites with no lighting other than what the helicopter furnishes. Night flying can be overwhelming to a new student, especially when leaving the comfort of a lighted airport. Students must learn that potential hazards, such as wires, poles, and towers, become even more dangerous at night as they are not always marked or lighted. The flight profile of the helicopter is typically at a low altitude, which is where most obstacles exist. The instructor should familiarize the student with an area during daylight hours and point out all of the hazards that need to be avoided during a night flight. Another challenge for the student is manipulating the searchlight while flying. It is very easy for the student to become fixated on the light or forget to fly the helicopter while trying to place the light in the correct position. Allow the student time to become familiar with the searchlight while on the ground before attempting it in flight.

Another critical issue to address when beginning night flight training is light discipline in the cockpit and how it can affect the ability to see what is around you. As a protection against damaging itself, the eye always adjusts for the brightest light level. If the helicopter's interior lights are brighter than the outside ambient light level, the eye adjusts to see the brighter inside lights, greatly limiting the ability to see outside the helicopter. The correct cockpit lighting is essential to night flying.

It is important that the student has in-depth knowledge of the eye, visual acuity, and the function of rods and cones. Discuss the terms associated with night flight, including the parts and functions of the eye, visual illusions, and night scanning techniques. This information can be found in greater detail in the Pilot's Handbook of Aeronautical Knowledge, the Aviation Instructor's Handbook, the Helicopter Flying Handbook, and Title 14 of the Code of Federal Regulations (14 CFR) part 91. Additionally, pilots can consult updated information for end of evening civil twilight (EECT) and beginning of morning civil twilight (BMCT) in the document Complete Sun and Moon Data for One Day: U.S. Cities and Towns at http://www.usno.navy.mil. Review these handbooks and regulations with the student and, when appropriate, ask the student to explain night scanning, aircraft

lighting, and visual illusions. Achieving these objectives is crucial for the student when performing night flight and applying the night flight techniques.

## Eye Anatomy and Physiology

The eye is similar to a camera. The cornea, lens, and iris gather and control the amount of light allowed to enter the eye. The image is then focused on the retina, which has two types of cells: cones and rods. [Figure 14-1] Review the anatomy of the eye and discuss the night blind spot. [Figure 14-2] This discussion should focus on vision while in flight to include visual acuity and the eye's rods and cones. The more a student knows about the eyes and how they function, the easier it is to use vision effectively and compensate for potential problems.

### Visual Problems

There are several visual problems, or conditions, that affect night vision. Discuss with the student the visual deficiencies, such as myopia, night myopia, hyperopia, astigmatism, and presbyopia. Refer the student to the Helicopter Flying Handbook for in-depth definitions of each visual problem and how each can affect night flying. Instructors should be aware of any specific visual problems that the student may have that could affect his or her ability to fly at night. For example, if the student wears glasses, are the glasses reflecting glare from an unnecessary light? Do the glasses have a coating that interferes with night vision? Does the student wear sunglasses during the day to protect night vision? If the student is older, has he or she been screened for cataracts, which could cause night vision problems? The instructor must teach the student about all possible visual problems but also be aware of any that the student may have.

## Ground Instruction

In a classroom environment, review all terms associated with night operations. Explain to the student how to protect night vision, how to recognize self-imposed stress, the negative effects on night vision from smoking, and the various scanning techniques for night flight. It is also important that visual illusions are explained in detail with information regarding how to detect and react to the illusions. Aircraft instruments are easier to read under higher levels of interior illumination. However, this need must be balanced with the ability to see outside and the hazard of interior lights reflecting off the interior surfaces. Minimize interior lighting, whenever possible, without hindering reading of essential instruments.

### Night Vision Protection

Explain to the student that protecting night vision should always be a priority. Discuss some of the steps to take to protect night vision such as the use of sunglasses and oxygen. Repeated exposure to bright sunlight has an increasingly

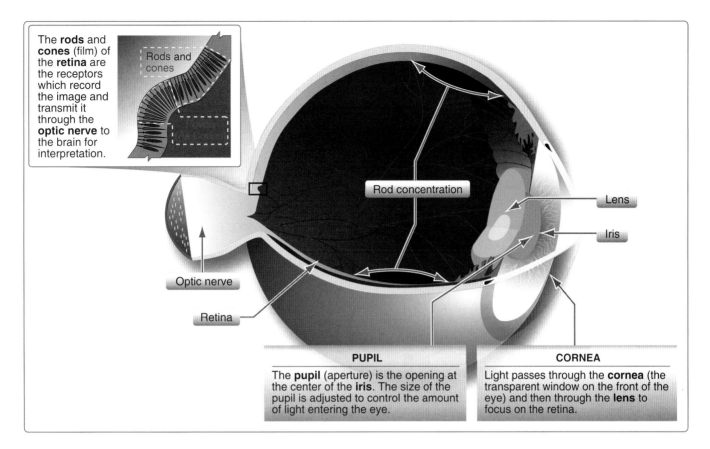

The **rods** and **cones** (film) of the **retina** are the receptors which record the image and transmit it through the **optic nerve** to the brain for interpretation.

Rods and cones

Rod concentration

Lens

Iris

Optic nerve

Retina

**PUPIL**

The **pupil** (aperture) is the opening at the center of the **iris**. The size of the pupil is adjusted to control the amount of light entering the eye.

**CORNEA**

Light passes through the **cornea** (the transparent window on the front of the eye) and then through the **lens** to focus on the retina.

**Figure 14-1.** *The human eye.*

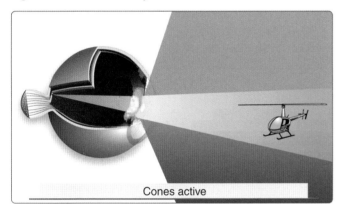

Cones active

Night blind spot

Rods active

**Figure 14-2.** *The night blind spot.*

adverse effect on dark adaptation. Point out that this effect is intensified by reflective surfaces, such as sand and snow. Sunglasses aid in filtering the bright sunlight and increase the rate of dark adaptation at night while improving night visual sensitivity. *[Figure 14-3]*

**Figure 14-3.** *Pilot and passenger wear sunglasses to protect their eyes from bright sunlight.*

Explain to a student that unaided night vision depends on optimum function and sensitivity of the rods of the retina. Lack of oxygen to the rods significantly reduces their sensitivity and increases the time required for dark adaptation, as well as decreases the ability to see at night. Since most helicopters do not carry oxygen, more practical advice to give a student who smokes is either to quit or at least reduce the amount that they smoke and advise that physical conditioning helps not only the heart, but also assists the body's ability to increase oxygen intake. If oxygen is available, pilots should use it when flying above a pressure altitude of 4,000 feet.

Additional precautions to discuss with the student pertain to the airport or heliport lighting. Any light sources that may impair the student's dark adaptation should be eliminated. Tell the student to try to select departure routes that avoid highways and residential areas where artificial light can impair night vision. If bright lights are encountered from a specific direction, turn the aircraft away from the light source when able. If this is not possible, instruct the student to preserve dark adaptation by shutting one eye and using the other to observe. Once the light source is no longer visible, the eye that was closed can provide the required night vision.

## Self-Imposed Stress

Night flight is more fatiguing and stressful than day flight because the brain has to work harder in order to make sense of the limited visual cues or the lack of visual acuity. The pilot must scan more in order to gain sufficient visual cues to maintain the desired flight path. Many helicopters are not well equipped for night flight. Most charts and maps are harder to read and interpret in low levels of light. Inform the student of several self-imposed stressors that limit night vision, such as drugs, illness, fatigue, alcohol, and tobacco. For example, if an individual smokes 3 cigarettes in rapid succession or 20 to 40 cigarettes within a 24-hour period, the physiological effect at ground level is the same as flying at 5,000 feet. More importantly, the smoker has lost about 20 percent of night vision capability at sea level. Review the Pilot's Handbook of Aeronautical Knowledge with the student to ensure a better understanding of these types of stressors and place emphasis on the fact that the student can control this type of stress.

## Scanning Techniques

It is important to teach the student various night vision techniques that enable him or her to overcome many of the physiological limitations of their eyes. These techniques require considerable practice and concerted effort on the part of the student and the instructor as they are important in identifying objects at night.

Instruct the student that to scan effectively, looking from right to left or left to right. Tell the student to begin scanning at the greatest distance an object can be perceived (top) and move inward toward the position of the aircraft (bottom). *[Figure 14-4]* For each stop, an area approximately 30° wide should be scanned. The duration of each stop is based on the

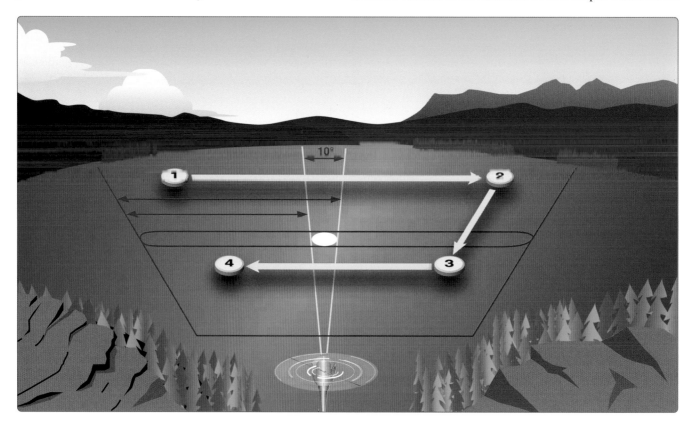

**Figure 14-4.** *Scanning pattern.*

degree of detail that is required, but no stop should last longer than two to three seconds. When moving from one viewing point to the next, the student should overlap the previous field of view by 10°.

Viewing an object using central vision during daylight poses no limitation. If this same technique is used at night, however, the object may not be seen because of the night blind spot that exists during low illumination. Explain to the student that in order to compensate for this limitation, he or she can use off-center vision. This technique requires that an object be viewed by looking 10° above, below, or to either side of the object. In this manner, the peripheral vision can maintain contact with an object. Ensure that the student understands that, with off-center vision, the images of an object viewed longer than 2 to 3 seconds disappear. This occurs because the rods reach a photochemical equilibrium that prevents any further response until the scene changes. This produces a potentially unsafe operating condition. To overcome this night vision limitation, the student must be aware of the phenomenon and avoid viewing an object for longer than 2 or 3 seconds. The peripheral field of vision continues to pick up the object when the eyes are shifted from one off-center point to another. *[Figure 14-5]*

## Visual Illusions

Decreasing visual information increases the probability of spatial disorientation. Ensure the student understands that reduced visual references also create several illusions that can induce spatial disorientation. Many types of visual illusions can occur in flight and it is important that the student becomes familiar with the various types to include:

- Flicker vertigo—much time and research have been devoted to the study of flicker vertigo. A light flickering at a rate between 4 and 20 cycles per second can produce unpleasant and dangerous reactions. Such conditions as nausea, vomiting, and vertigo may occur. On rare occasions, convulsions and unconsciousness may also occur. Flicker vertigo is why the regulations allow pilots to disable the strobe lights and anti-collision beacons while flying in the clouds.

- Fixation—occurs when pilots ignore orientation cues and fix their attention on a goal or an object. This is dangerous because helicopter ground-closure rates are difficult to determine at night.

- False horizons—cloud formations may be confused with the horizon or the ground. While hovering over terrain that is not perfectly level, the pilot might mistake the sloped ground in front of the helicopter for the horizon and cause the helicopter to drift while trying to maintain a stationary position. Another example is a lighted road climbing a mountain side can easily be mistaken for a flat horizon.

- Confusion with ground lights—a common occurrence is to confuse ground lights with stars. When this happens, the pilot unknowingly positions the helicopter in unusual attitudes to keep the ground lights—believed to be stars—above them. *[Figure 14-6]*

- Relative motion—the illusion of relative motion can be illustrated by an example. A pilot hovers a helicopter and waits for hover taxi instructions. Another aircraft hovers alongside. As the other aircraft is picked up in the first pilot's peripheral vision, the pilot senses movement in the opposite direction.

- Altered planes of reference—when approaching a line of mountains or clouds, the pilot may feel the need to climb despite adequate altitude. Also, when flying parallel to a line of clouds, pilots may feel that they need to climb even though their altitude is adequate.

- Structural illusions—these illusions are caused by heat waves, rain, snow, sleet, or other factors that obscure vision. For example, a straight line may appear to be curved when seen through a desert heat wave or a wingtip light may appear to double or move when viewed during a rain shower.

- Height and depth perception illusion—due to lack of visual references when flying over desert, snow, water, or other areas of poor contrast, the pilot may

**Figure 14-5.** *Off-center vision technique.*

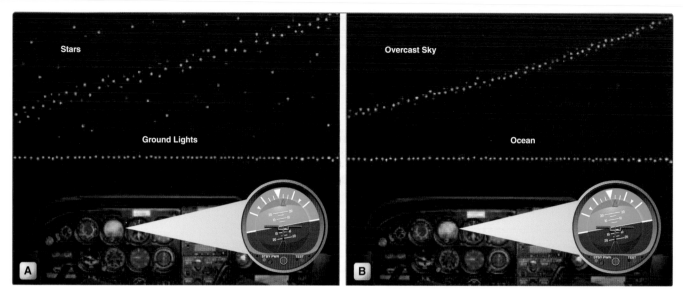

**Figure 14-6.** *This illusion prompts the pilot to place the aircraft in an unusual attitude to keep the misperceived ground lights above them. Isolated ground lights can appear as stars (Part A). When no stars are visible because of overcast conditions, unlighted areas of terrain can blend with the dark overcast to create the illusion that the unlighted terrain is part of the sky (Part B). This illusion can be avoided by referencing the flight instruments and establishing a true horizon and attitude.*

experience the illusion of being higher above the terrain than is actually the case. This illusion may be overcome by dropping an object, such as a chemical light stick or flare, on the ground before landing. Another technique used to overcome this illusion is to monitor shadows cast by nearby objects, such as the skid shadows at a hover. Flight in an area where visibility is restricted by haze, smoke, or fog produces the same illusion.

- Size-distance illusion—results from viewing a source of light that is increasing or decreasing in luminance (brightness). The pilot may interpret the light as approaching or retreating.

- Autokinesis—when the pilot stares at a static light in the dark, the light appears to move. This phenomenon can be readily demonstrated by staring at a lighted cigarette in a dark room. Apparent movement begins in about 8 to 10 seconds. Although the cause of autokinesis is not known, it appears to be related to the loss of surrounding references that normally serve to stabilize visual perceptions. This illusion can be eliminated or reduced by visual scanning, by increasing the number of lights, or by varying the light intensity. The most important of the three solutions is visual scanning. You should not stare at a light or lights for more than 10 seconds. This illusion is not limited to light in darkness. It can occur whenever you stare at a small, bright, still object against a dull, dark or nondescript background. Similarly, it can occur when a small, dark, still object is viewed against a light, structureless environment. Anytime

visual references are not available, pilots are subject to this illusion. Instructors can also relate this illusion to the helicopter's landing light. It becomes very easy to focus on that one beam of light and not see the peripheral vision cues of lateral or rotating movement. Staring into the landing light also tends to ruin much of the accumulated night vision, leaving the pilot nearly blind when the landing light is extinguished

- Reversible perspective illusion—at night, an aircraft may appear to be moving away when it is, in fact, moving toward a second aircraft. This illusion often occurs when an aircraft is flying parallel to another's course. To determine the direction of flight, aircrews should observe aircraft lights and their relative position to the horizon. If the intensity of the lights increases, the aircraft is approaching. If the lights dim, the aircraft is moving away. Also, remembering the "3 Rs" helps identify the direction of travel when other aircraft are encountered. If the red aircraft position lights are on the right, the aircraft is returning (coming toward the observer). *[Figure 14-7]*

## Flight Instruction

Students should receive adequate ground instruction and classroom training prior to advancing to actual flight training. Proper classroom instruction ensures that students are aware of the limitations of night flight and helps ensure a safer operating environment. Once the student has shown proficiency and is comfortable with the demands of night flight, the actual night flying can successfully commence.

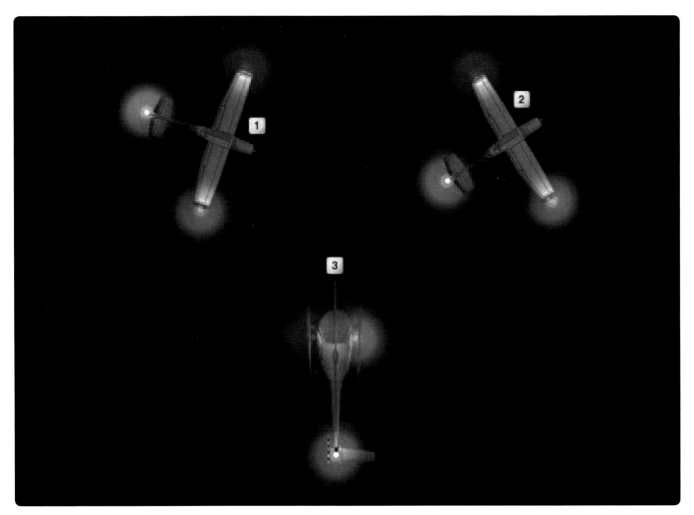

**Figure 14-7.** *Remembering the "3 Rs" helps identify the direction of travel when other aircraft are encountered. If the red aircraft position lights are on the right, the aircraft is returning (moving toward the observer).*

## Preflight Inspection

The aircraft preflight inspection is a critical aspect of safety, and it must comply with the appropriate aircraft operator's manual. Preflight should be scheduled as early as possible, preferably during daylight hours, allowing time for maintenance assistance if necessary. If a night preflight is necessary, a flashlight with an unfiltered lens should be used to supplement lighting. Oil and hydraulic fluid levels and leaks are difficult to detect with blue-green or red lens. Windscreens must be checked to ensure they are clean and relatively free of scratches. Slight scratches are acceptable for day but may not be acceptable for night flight. The searchlight or landing light should be positioned for the best possible illumination during an emergency descent.

Also included in preflight planning, the student should be tasked with deciding how to read the checklist and charts at night. Cockpit organization is a very important chore that must be accomplished before flight. The instructor could have an auxiliary power unit (APU) or power cart connected to the helicopter at night so the student can practice cockpit organization. Small items can become large problems if a flashlight is dropped and blocks the flight controls or a map or checklist is left on the dashboard and blocks the windscreen. The student must learn how to manage a flashlight, map, and checklist while flying at night.

Instructors should review the heliport and airport lighting with the student to include beacons. Since helicopters do not normally use runways, the low blue intensity of airport taxiway lighting should be pointed out. Teach the student that some obstructions around the airport may be lighted but some may not. Pilots should always be looking for wind sock poles, light poles, ASOS/AWOS installations, and other off-runway obstacles. Many of these hazards should be pointed out during the day flights and then again during the night, especially the ones that are not lighted. The student should be prompted to begin developing a hazard map of the area and shown how to properly keep it updated.

Proper preparation of the helicopter for night flight contributes greatly to the success of the flight; however, you should stress to the student that, unless he or she is physically and mentally prepared to participate in the night flight, the flight is considered unsafe. Discuss the following checklist with the student prior to flight to ensure readiness:

- Dark adapt before flight
- Avoid self-imposed stress
- Avoid bright sunlight during the day
- Learn to use the principles of night vision
- Avoid all bright lights after dark adaptation

Participate in frequent night flights to ensure the student can demonstrate the tasks listed below to a satisfactory level:

- Preflight and aircraft lighting
- Proper cockpit lighting
- Engine starting (rotor engagement, if installed)
- Taxi
- Takeoff
- En route procedures
- Specific night emergencies
  - Light failure
  - Alternator failure
  - Loss of orientation
  - Inadvertent IFR
- Collision avoidance
- Approach and landing

Each instructor must determine what type of night training is best for each particular student based on that student's learning style and understanding. It is beneficial to require portions of training with and without aircraft lighting and training with night vision devices, when necessary. Ensure that the training is always based on the appropriate Rotorcraft Flight Manual (RFM) of the particular helicopter being flown and that the training is tailored to the individual's needs. Remember, a successful training session depends on adequate support such as flight hours, equipment, and training areas. See *Figure 14-8* for a sample lesson plan.

Training should begin in high ambient light levels, such as a full moon, and progress to successively lower light levels. Students should first perform maneuvers to prepared surfaces. As proficiency increases, allow the student to progress to lighted sites and then to unlighted sites.

Navigation training should begin with easy routes. As the student becomes more proficient, he or she can fly routes with legs of 50 to 100 nautical miles. More difficult and realistic scenarios should have possible landing areas interspersed along the route.

Communication with the student and actions (sequence or timing) are necessary for students to perform flight tasks efficiently, effectively, and safely. If a particular task is labor intensive and requires additional time from the student, assist in performing that task until the student's confidence and understanding is achieved.

Specific night flight emergencies, such as light failure, alternator failure, and loss of orientation by either too much information over a broad metropolitan area or too few lights over open country, should be discussed, practiced, and evaluated. Depending on the student's status and progress, the first few night flights should be short to allow the student time to adjust and absorb all the new information. Students have a tendency to tire very quickly when starting night training, and it is the job of the instructor to prevent the student from getting to that point as fatigue is detrimental to training. As the student acclimates to the environment, training times can be lengthened to facilitate training.

Note: Discuss with the student that controlled flight into terrain seems to be the major fatal error made by pilots particularly during night flights. Therefore, the student must understand that anything that casts a shadow or appears to be blocking lights, natural or manmade, should be treated as an obstacle.

If weather conditions fall below 1,000 feet and 3 miles while attempting to fly VFR at night, particularly during training, cancelation of the flight would be a wise decision. This also shows the student sound decision making and hopefully that student will carry that forward with him the rest of his flying career. Also, the weather may be greater than a 1,000/3 but no ambient light is visible (no stars are visible); in this case, if VFR night flight is accomplished, use a higher en route altitude (minimum safe altitude from VFR chart). If the weather temporarily limits visibility, such as a strong rain storm, day or night, teach the student to land the helicopter and wait for the weather to pass. Discuss flying at lower airspeeds during limited visibility, as well as the development of IMC conditions, such as fog or condensation. If your visual field suddenly becomes blurry, or difficult to see, it is usually because the weather is changing. This should be an excellent cue to make a decision about aborting the flight.

Teach the student that simply flying direct from one point to the next is fine for day VFR. However, night flights demand more attention to navigate successfully. Using distant towns, towers, or any other lit object is acceptable during night flights. Major roads with traffic also provide a means to navigate during hours of darkness. Remind the student of the blue maximum elevation figures that are published on the sectional charts and their importance in night route planning.

## Lesson One for Night Flight

### Objective

The purpose of this lesson plan is to review with the student all the common terms associated with night flight. The student needs to know what visual deficiencies are. Review with the student the anatomy and physiology of the eye and how they affect night flight. Also, instruct the student in the conditions, hazards, risks, and risk mitigation techniques for night flights. During night flight the student will apply what was learned in the classroom environment.

1. The student will demonstrate a basic knowledge of the terms associated with night flight, to include the anatomy of the eye and the night blind spot.

2. The student will demonstrate a basic knowledge of visual illusions.

3. The student will demonstrate to the instructor that he can comfortably fly the helicopter at night and apply the scanning techniques.

4. The student will demonstrate the tasks listed below to a satisfactory level:

   - Preflight and aircraft lighting
   - Proper cockpit lighting
   - Engine starting (rotor engagement, if installed)
   - Taxi
   - Takeoff
   - En route procedures
   - Specific night emergencies
     1. Light failure
     2. Alternator failure
     3. Loss of orientation
     4. Inadvertent IFR
   - Collision avoidance
   - Approach and landing

### Content

1. Preflight discussion:
   a. Discuss lesson objective and completion standards.

2. Review terms associated with night operations.

3. Instructor actions:
   a. In a classroom environment, review all terms associated with night operations.
   b. Discus with the student the visual deficiencies.
   c. Discus with the student the anatomy of the eye and the night blind spot.
   d. Discus with the student the visual illusions, and how to detect and react to the illusions.
   e. Discuss with the student the proper night scanning techniques.
   f. Discuss with the student the proper lighting inside the helicopter.
   g. Discuss with the student the different techniques for navigating at night.
   h. During the flight portion, the student must demonstrate proficiency and comfort with the demands of night flight.

4. Student actions:
   a. Study the terms associated with night operations.
   b. Be able to discuss with the instructor the visual deficiencies.
   c. Be prepared to discuss with the instructor the anatomy of the eye and how to overcome the night blind spot.
   d. Be prepared to discuss with the instructor your understanding of the night scanning techniques, aircraft lighting requirements, and dark adaptation.
   e. Be prepared to discuss with the instructor the visual illusions.
   f. Be prepared to apply 3(a) through (h) listed above to actual night flight operations.

### Postflight Discussion

Review what was covered during this phase of training. If any problems were encountered during the flight tasks, correct or, if necessary, retrain the student in that particular task(s). Discuss what will be covered on the next lesson plan.

**Figure 14-8.** *Sample lesson plan.*

If the flight is planned to be above the maximum elevations, all obstructions should be cleared by a reasonable margin. Proper flight planning can truly save a life or lives.

If the helicopter is IFR equipped and the student is IFR rated, than teach the student IFR recovery techniques.

## Instructor Tips

- Ensure that the student can recognize and understands the visual deficiencies.

- Ensure the student has a basic understanding of the anatomy and physiology of the eye.

- Point out visual illusions to the student during flight, when possible.

- Discuss with the student the inability to see weather phenomena while flying at night.

- Discuss how unusual attitudes are harder to detect at night without the normal visual references.

- If corrective lenses are prescribed to aircrew members, they must use corrective lenses (glasses) in all modes of flight.

- Be aware that it will take 30 to 45 minutes for the average individual's eyes to reach maximum dark adaptation.

- Use off-center vision when viewing objects under reduced lighting conditions.

- Avoid self-imposed stress.

- Protect night vision by avoiding bright lights once dark adaptation has been achieved.

- Scan using a series of short, regularly spaced eye movements between being still for a second at a time.

## Chapter Summary

This chapter described the basic anatomy and physiology of the eye. Night flying and visual illusions were explained, and how a pilot can overcome them while in flight. Various techniques were described to teach the student safer ways to conduct flight at night. Instructors should ensure that the student has a basic understanding of the requirements and common problems associated with night flying.

# Chapter 15
# Helicopter Operations

## Introduction

This chapter provides information to help the instructor guide the student through various helicopter operations. It includes operational and safety considerations for on the ground and in the air.

Caution

Horizontal Stabilizer

## Collision Avoidance

As discussed in this handbook's Chapter 1, Introduction to Flight Training, as well as in the Aviation Instructor's Handbook, the instructor must ensure the student develops the habit of looking for other air traffic at all times. If a student believes the instructor assumes all responsibility for scanning and collision avoidance procedures, he or she will not develop the habit of maintaining the constant vigilance essential to safety. Remember to establish scan areas and communication practices for keeping the aircraft cleared. Any observed tendency of a student to enter flight maneuvers without first making a careful check for other air traffic must be corrected immediately.

From the first flight, the instructor must make the student aware that it is every pilot's responsibility to see and avoid other aircraft. Explain the blind areas in the helicopter being flown, as well as those in other aircraft. Develop in the student a habit of checking for other aircraft during his or her regular scan pattern. All radio and radar aids should be used to the fullest extent possible, but with the realization that they are only aids, and vigilance should not be relaxed. Radar traffic advisories are very helpful, but there is evidence that indicates some pilots become complacent when in the radar environment and relax their vigil. Also, no turn should ever be made without first looking in the direction of the turn to see that the airspace is clear of other traffic. In the vicinity of an airport, all possible aids should be used and looking for other aircraft should occupy more of the student's time. Landing and anticollision lights should be turned on to make the helicopter more visible, especially in the vicinity of an airport.

Flight instructors should:

- Guard against preoccupation during flight instruction to the exclusion of maintaining a constant vigilance for other traffic.

- Be particularly alert during the conduct of simulated instrument flight in which there is a tendency to "look inside."

- Place special training emphasis on areas of concern in which improvements in pilot education, operating practices, procedures, and techniques are needed to reduce midair conflicts.

- Notify the control tower operator, at airports where a tower is manned, regarding student first solo flights.

- Explain the availability and encourage the use of expanded radar services for arriving and departing aircraft at terminal airports where this service is available, as well as the use of radar traffic advisory services for transiting terminal areas or flying between en route points.

- Understand and explain the limitations of radar that may frequently limit or prevent the issuance of radar advisories by air traffic controllers.

## Runway Incursions

Stress upon the student that, even though helicopters do not regularly use runways for takeoffs and landings, runway incursions need to be understood and discussed. Students need to listen carefully to any clearances and instructions from air traffic control (ATC) and acknowledge them in full. They should also be aware of their position and the position of other aircraft and obstructions at all times. *[Figure 15-1]* During flight training, instructors often use runways to practice maneuvers and procedures. Extra vigilance must be exercised under these circumstances as the instructor and student may become so focused on the particular maneuver or procedure that they become inattentive to the surroundings.

## Safety Considerations

Good manners are an essential part of helicopter operations. If a helicopter is not operated with consideration for nearby persons and property, it can be a nuisance or, even worse, a hazard. A considerate attitude must be cultivated by example and instruction from the beginning of training. Stress to the student that the helicopter's unique ability of landing and taking off near a crowd of people creates downwash that can stir up debris and blow it at high velocity for a considerable distance, causing possible injury to people and damage to property. Remind the student of the potential hazard of someone on the ground walking into turning rotors. *[Figure 15-2]* The tail rotor, in particular, is hard to notice. Therefore, it is mandatory that a student understand the potential hazards to others created by a helicopter and the pilot's responsibility to prevent them. The rotor tip-path plane is not always easy to see, and it may be difficult to judge its distance from fixed objects. A beginning student should be encouraged to maintain more than adequate clearance from all objects and to be constantly aware of both main and tail rotor paths. The instructor should review with the student the pilot handbook and discuss the danger areas of the main and tail rotor clearance distances. Review *Figure 1-5* with student.

## Traffic Patterns

The student must be able to describe the traffic patterns used by both helicopters and fixed-wing aircraft, naming the legs and specifying pattern altitudes. The student must also demonstrate the ability to fly traffic patterns at uncontrolled fields while avoiding the flow of fixed-wing traffic and complying with tower instructions at controlled airports. *[Figure 15-3]*

**Figure 15-1.** *Students should always be aware of possible runway incursions.*

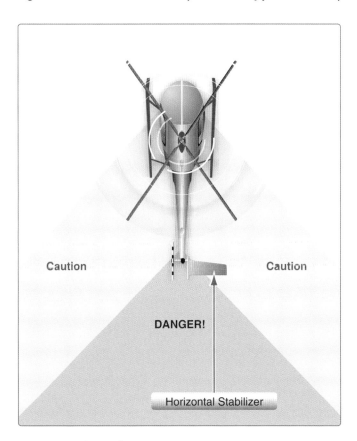

**Figure 15-2.** *Rotor danger zones.*

## Instructional Points

- The student should learn the correct procedures for fixed-wing aircraft at controlled and uncontrolled airports. This knowledge provides the student with an understanding of where to expect fixed-wing aircraft and how to avoid them in the traffic pattern. Teach the student how to search the airport facility guide to check for the fixed-wing traffic pattern in order to be able to avoid the flow of traffic. Instructors should also reference the Helicopter Flying Handbook, which details fixed-wing traffic patterns and how helicopters can avoid the traffic and learn how to operate in it in the event the tower has mixed traffic.

- Advise the student to pay attention to any wind indicators, such as wind socks, flags, and smoke.

- Typically, traffic patterns in a helicopter are flown lower and closer than those flown by fixed-wing aircraft. The typical traffic pattern altitude is 500–800 feet above ground level (AGL) for helicopters, while for most fixed-wing aircraft it is 1,000–1,500 feet AGL.

Note: Always refer to the airport facility guide for traffic pattern altitudes as some airports use different altitudes. By regulation, turbine-powered airplanes should use 1,500 feet for the downwind leg.

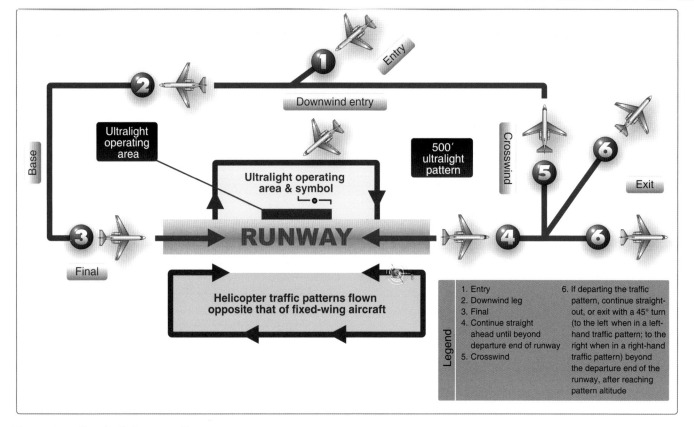

**Figure 15-3.** *Standard airport traffic pattern.*

- For training purposes, a rectangular course should be flown because there is better visibility; the aircraft has a level portion in each leg that facilitates clearing of traffic. A rectangular course also allows the pilot to estimate winds from the amount of crab necessary to offset the wind drift and provides a repeatable profile point to begin the approach. It helps the student to practice good aircraft control profile usable in many other maneuvers.

## Common Student Difficulties
### *Drift Correction*

The student might fail to notice the effect of wind, especially on the downwind and base legs, resulting in a distorted pattern. If this problem persists, it may be necessary to review and practice ground reference and tracking maneuvers.

### *Spacing From Other Aircraft*

It is difficult for beginning students to estimate distances from other aircraft, to estimate the space required to avoid interference, and to decide whether their own aircraft and another are on a collision course. Point out that with converging aircraft, if the other aircraft's relative position is not changing, then both aircraft are on a collision course. In this case, the quickest way to change relative position is to turn toward the other aircraft's tail. When two aircraft are approaching head on, each pilot should alter course to the right. Review with the student Title 14 of the Code of Federal

Regulations (14 CFR) part 91, section 91.113, Right of Way Rules: Except Water Operations.

### *Altitude and Airspeed*

Maintaining a constant altitude and airspeed can be very difficult for the beginning student. Problems can stem from lack of scanning, chasing the instruments or not flying the horizon, and not being able to recognize changes in engine and rotor sounds. Students should be encouraged to use all of their senses to help focus on the entire aircraft and not just one or two instruments at a time. Student trouble with fluctuating airspeeds is often caused by a hand and arm relaxing on the cyclic. Demonstrate how the arm at the cyclic slips slowly back when the pilot is fatigued or overly relaxed, which brings the cyclic back, causing the airspeed to fluctuate and altitude to increase. If the student is not tall enough to rest the elbow on a leg while flying, the arm can tire as the student must hold it up the entire time. Watch the student's arm position throughout the flight. If this does become a problem, place a large sponge or rolled up shirt between the arm and leg to help with the fatigue, which also helps to maintain altitude and airspeed. Failure to scan can also be the cause of altitude and airspeed deviations. Fixating on instruments or the intended landing area can be fixed by reminding the student to keep scanning and to focus on more than just one aspect of the flight.

## Airspace

The helicopter instructor should integrate knowledge of the classifications of airspace throughout the training process from preflight planning to actual flight. Ensure the student understands helicopter regulatory requirements for operations in the various types of airspace based on type of pilot certificate held. Provide a thorough discussion of airspace particularly as it is relevant to helicopters. Instruction should include as a minimum:

- Endorsement requirement for student pilots
- Equipment requirements
- Communication requirements
- Weather minimums

Discuss the pilot's responsibility regarding operations in all airspace and over all types of terrain. *[Figure 15-4]* For example, operations and procedures in Class B airspace differ from operations in Class G airspace, and operations in mountainous terrain differ from operations conducted over open water. Tie in preflight planning, aeronautical decision-making (ADM), risk management, and other topics as they relate to airspace and type of operations. Discuss the ATC System and the services it provides. Refer the student to the Pilot's Handbook of Aeronautical Knowledge, 14 CFR, and the Aeronautical Information Manual (AIM) for additional information.

## Helicopter Turbine and Multiengine Transition

When transitioning a student into either a turbine or multiengine helicopter, the instructor should carefully plan the course of instruction to fully encompass the procedures contained in the Rotorcraft Flight Manual (RFM) for the helicopter being used.

Ensure the student pilot has the opportunity to fly the helicopter at maximum gross weights to learn the characteristics and different aircraft responses when the helicopter is fully loaded. Emergency training should never be conducted while carrying passengers.

The student pilot should fully understand the significance of the helicopter specific airspeeds, such as the takeoff safety speed for Category A Rotorcraft ($V_{TOSS}$) and Category A versus Category B helicopter operations and limitations, powerplant limitations, and possibly transmission limitations for the helicopter being flown. Instrumentation and navigation displays must be understood before flight, as well as the operation of controls such as engine condition levers, governors, and stability augmentation systems.

If a multiengine helicopter requires backing up and climbing some for a Category A takeoff profile, the instructor should ensure that some reasonable and practical procedure is taught and practiced to maintain tail rotor and tail boom obstruction clearance during the maneuver, as discussed in 14 CFR part 25, sections 23.53 through 23.61. In any event, the instructor must ensure that the student gains a thorough understanding of the differences between Category A and Category B operations.

Additionally, if the helicopter is usually flown with a crew of two, then crew resource management should be explained and practiced. Terminology, checklist procedures, crew coordination, flight briefing, and crew position duties and responsibilities should be presented and practiced in detail.

If the student is transitioning to a turbine engine helicopter, the instructor should review the differences in the powerplant response to load changes and power demands, the importance of proper starting procedures, monitoring, limitations, failure modes, and consequences of poor procedures and inattention.

When a student transitions into a turbine or multiengine helicopter, this usually includes their first introduction to power checks (also called health indicator (HIT) checks, engine monitoring, etc.). Power checks allow the pilot to determine if the engine or engines is/are producing rated power before takeoff. Usually, power checks are a function of some type of maintenance program to extend the service time of the powerplant. Turbine engines are very expensive, so any method to safely extend the service time of the power plant is welcome. If the power check values are not within limits or change from one day to the next by a large margin, the pilot should write up the check as a discrepancy and bring it to the attention of maintenance. It is always cheaper to fix a problem before it becomes an airborne emergency. Turbine engines can have a failure mode of disintegrating and sending out parts at high enough speeds to penetrate the other engine, fuel lines, driveline components, and compartments. A poor power check value can be an indication of a worn engine or one that may be ready to fail.

The student should be thoroughly trained to observe temperature and torque limitations. Additionally, the student should be trained how to determine which limitation is for that time and place, as well as why.

## Floats, Wheeled Landing Gear, or Ski Transitions

When transitioning a student to a helicopter equipped with floats, wheeled landing gear, or skis, the instructor should carefully plan the course of instruction to fully encompass

## Airspace

**Objective**

The purpose of this lesson is for the student to learn how to fly from a nontowered airport, through Class G and Class E airspace, to a towered airport.

**Content**

1. Scenario:
   Fly to a specific towered airport.

2. Possible Hazards/Considerations:
   - Ground-based obstructions/hazards
   - Visibility/Ceiling requirements for:
     Class G
     Class E
     Class D
   - Wind conditions
   - Engine-out procedures
   - Communication and equipment requirements for:
     Class G
     Class E
     Class D
   - Airport traffic

3. Fly the Scenario

4. Postflight review:
   This review should include a dialogue between the instructor pilot and the student or transitioning pilot encompassing the flight scenario. Generally, the instructor pilot should lead the discussion with questions that generate reflective thinking on how the overall flight went. The instructor pilot should use this to assist in evaluating the student or transitioning pilot's assessment skills, judgment, and decision-making skills. Typically, the discussion should begin with student self-critique; the instructor pilot enables the student to solve the problems and draw conclusions. Based on this analysis, the student and instructor pilot should discuss methods for improvement, even on those items that were considered successful.

**Postflight Discussion**

- Review what was covered during this phase of training.
- If any problems were encountered during the flight tasks, correct or, if necessary, retrain for that particular task(s).
- Discuss what will be covered on the next lesson plan.

**Figure 15-4.** *Sample lesson plan.*

the procedures contained in the Rotorcraft Flight Manual (RFM) for the helicopter being used.

## Floats

Fixed inflated type floats must be checked daily for inflation and may limit the airspeeds and maneuvering capability of the helicopter. Water landings can be uneventful or very demanding, depending on the winds and waves. Autorotations to the water can be very challenging if the water is smooth and calm. Rotor wash disturbs the water surface, which can make hovering over a position most difficult. Should the pilot need to shut down on floats and then restart, uncontrolled

turning should be expected until sufficient rpm is gained to allow heading control with the pedals.

A pilot flying over the water should be taught to carefully observe the water's surface for wind direction and swell parameters. Due to the lack of reference points, navigation over large bodies of water is somewhat different than land navigation. For example, the wind direction does not usually vary as much over land due the lack of surface friction, but thunderstorms tend to form more at night over larger bodies of water, which makes the occurrence of fog more likely. Haze can also be dense enough to restrict visibility to 3 miles.

Many helicopters sit somewhat low in the water on fixed floats, so tail rotor clearance can be a hazard in very small waves. Landing close to boats or ships exposes the helicopter to ropes, lines, and cranes on the larger vessels. All of these can constitute deadly hazards to a helicopter and should be discussed in detail.

Work on floats involves flight over water so the crew and passengers must have life jackets. The pilot must ensure the passengers are equipped with life jackets and briefed on how the equipment works and what the best course of action is in the event they are required to land in water.

Discuss with the student that water operations are much more demanding on the maintenance crews because of corrosion control, to include engine washes are major items. In most operations near salt water, the pilot performs a daily engine wash, while the maintenance crew performs more extensive washes periodically.

### Wheeled Landing Gear

Wheeled landing gear must be inflated properly to prevent ground resonance. If the wheels are retractable, the pilot must follow a checklist to ensure gear extension before landing. Usually, if the gear retracts, there are emergency landing gear extension procedures for the student to learn. There may be maximum airspeeds for landing gear operation, retraction, and extension.

### Skis

Skis, or snow pads, settle into the snow over time and become almost glued to the surface. Skis also slip under tree limbs and other obstructions very easily. Always ensure the skis or snow pads are free of the surface before lifting all of the way to a hover. Snow operations include other hazards, such as whiteouts and loss of the horizon, as the snow blends into a white sky. Snow can hide depressions and in itself be unstable if near a high area. Always maintain flight rpm until you are certain that the snowy surface fully supports the helicopter and is stable under the added weight of the helicopter.

## External Loads

14 CFR part 133 and Federal Aviation Administration (FAA) Advisory Circular (AC ) 133-1 provide information for rotorcraft external-load operations. No person subject to this part may conduct rotorcraft external-load operations within the United States without a Rotorcraft External-Load Operator Certificate issued by the FAA in accordance with 14 CFR part 133, section 133.17. Additionally, the pilot must have the exclusive use of at least one rotorcraft that—

1.  Was type certificated under and meets the requirements of 14 CFR part 27 or 29 (but not necessarily with external-load-carrying attaching means installed) or of

14 CFR part 21, section 21.25, for the special purpose of rotorcraft external-load operations;

2.  Complies with the certification provisions in 14 CFR part 133, subpart D that apply to the rotorcraft-load combinations for which authorization is requested;

3.  Has a valid standard or restricted category airworthiness certificate.

For the purposes of this section, a person has exclusive use of a rotorcraft if he or she has the sole possession, control, and use of it for flight, as owner, or has a written agreement (including arrangements for the performance of required maintenance) giving him or her that possession, control, and use for at least six consecutive months.

### Personnel

The pilot must hold a current commercial or airline transport pilot certificate issued by the FAA, with a rating appropriate for the rotorcraft as prescribed in 14 CFR part 133, section 133.19. One pilot, who may be the applicant, must be designated as chief pilot for rotorcraft external-load operations. The applicant also may designate qualified pilots as assistant chief pilots to perform the functions of the chief pilot when the chief pilot is not readily available. The chief pilot and assistant chief pilots must be acceptable to the FAA and each must hold a current Commercial or Airline Transport Pilot Certificate, with a rating appropriate for the rotorcraft as prescribed in 14 CFR section 133.19.

The holder of a Rotorcraft External-Load Operator Certificate must report any change in designation of chief pilot or assistant chief pilot immediately to the FAA certificate-holding office. The new chief pilot must be designated and must comply with 14 CFR section 133.23, within 30 days or the operator may not conduct further operations under the Rotorcraft External-Load Operator Certificate unless otherwise authorized by the FAA certificate-holding office.

### Knowledge and Skill

The applicant, or the chief pilot designated in accordance with 14 CFR part 133, section 133.21(b), must demonstrate to the FAA satisfactory knowledge and skill regarding rotorcraft external-load operations. The test of knowledge (which may be oral or written, at the option of the applicant) covers the following subjects:

1.  Steps to be taken before starting operations, including a survey of the flight area.

2.  Proper method of loading, rigging, or attaching the external load.

3.  Performance capabilities, under approved operating procedures and limitations, of the rotorcraft to be used.

4. Proper instructions of flight crew and ground workers.

5. Appropriate rotorcraft-load combination flight manual.

The test of skill requires appropriate maneuvers for each class requested. The appropriate maneuvers for each load class must be demonstrated in the rotorcraft as prescribed in 14 CFR part 133, section 133.19. These include:

1. Takeoffs and landings.

2. Demonstration of directional control while hovering.

3. Acceleration from a hover.

4. Flight at operational airspeeds.

5. Approaches to landing or working area.

6. Maneuvering the external load into the release position.

7. Demonstration of winch operation, if a winch is installed to hoist the external load.

Before attempting external loads, the student must be familiar with helicopter performance and the procedures outlined in the RFM. Ensure the student is aware that preflight planning is not complete until the ground crew is briefed on essential safety criteria, such as signals and emergency procedures. Discuss load pickup, en route, and load release procedures and the fact that operations need to be at an altitude that ensures the load clears all obstacles. When possible, plan the route of flight through an area that is not densely populated. Emphasize the differences in helicopter handling characteristics for external loads and internal loads.

### Emergency Procedures

Instruct the student in emergency load release procedures and point out that unnecessary overflight of populated areas should be avoided. Ensure the student is aware that, during pickup and normal release of a load, the helicopter is usually operating in the danger area of the height-velocity diagram.

Depending on the operation and configuration, great care should be exercised to isolate the external load release from the radio or intercom transmit button to prevent inadvertent load release. If at all possible, the student should exercise the emergency or manual load release system during flight to build the habit pattern. If the helicopter does not have dual or mutually accessible emergency cargo release controls, the instructor should develop a procedure and brief the student on the emergency actions to be accomplished by each crewmember in the case of an actual emergency. Some version of cargo hook arming and safing procedures should be practiced.

## Instructor Tips

- Remind the student that all aircraft have blind spots whether they are in the air or on the ground, and pilots must maintain a continuous scan to keep the helicopter clear.

- During external load training, remind the student that not only the helicopter needs to clear an obstacle on takeoff, sufficient altitude must also be attained in order for the load to clear the obstacle.

- Stay close to the controls at all times and always be ready to take control of the aircraft. Be prepared for both the expected and the unexpected.

## Chapter Summary

This chapter presented some training techniques and instructional points that can be used to familiarize the student with helicopter operations in and around airports. It also briefly discussed transition of students into turbine and multiengine helicopters, as well as points to emphasize during external load training.

## Chapter 16

# Practical Examination and Preparation for Flight Review

## Introduction

The instructor must be able to prepare the student for a practical examination as well as the flight review.

## Documentation

Every instructor must have a thorough understanding of Title 14 of the Code of Federal Regulations (14 CFR) part 61. For instance, 14 CFR part 61, section 61.3, provides information on certificates, ratings, and authorizations. Prior to recommending the student pilot for the practical examination or prior to conducting a flight review, the student must have documentation of experience and training requirements. Part 61 also provides regulatory guidance for the requirements and documentation needed for the knowledge test (14 CFR section 61.35), practical test (14 CFR section 61.39), and flight review (14 CFR section 61.56), as well as other pertinent information.

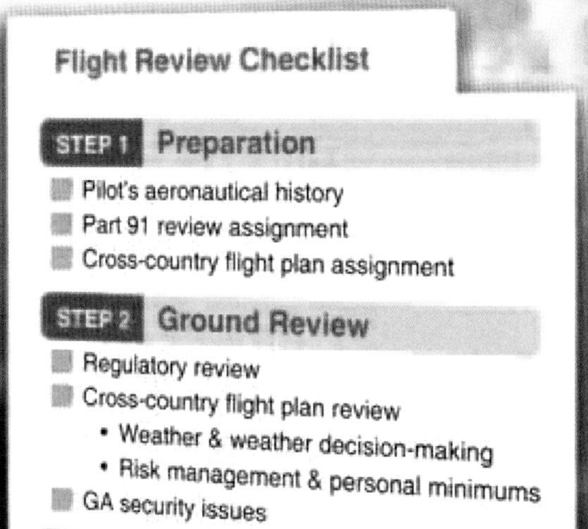

**Flight Review Checklist**

**STEP 1** **Preparation**
- Pilot's aeronautical history
- Part 91 review assignment
- Cross-country flight plan assignment

**STEP 2** **Ground Review**
- Regulatory review
- Cross-country flight plan review
  - Weather & weather decision-making
  - Risk management & personal minimums
- GA security issues

**STEP 3** **Flight Activities**
- Physical airplane (basic skills)
- Mental airplane (system knowledge)
- Aeronautical decision-making

**STEP 4** **Postflight Discussion**
- Replay, reflect, reconstruct, redirect
- Questions

**STEP 5** **Aeronautical Health Maintenance & Improvement Plan**
- Personal minimums checklist
- Personal proficiency practice plan
- Training plan (if desired)

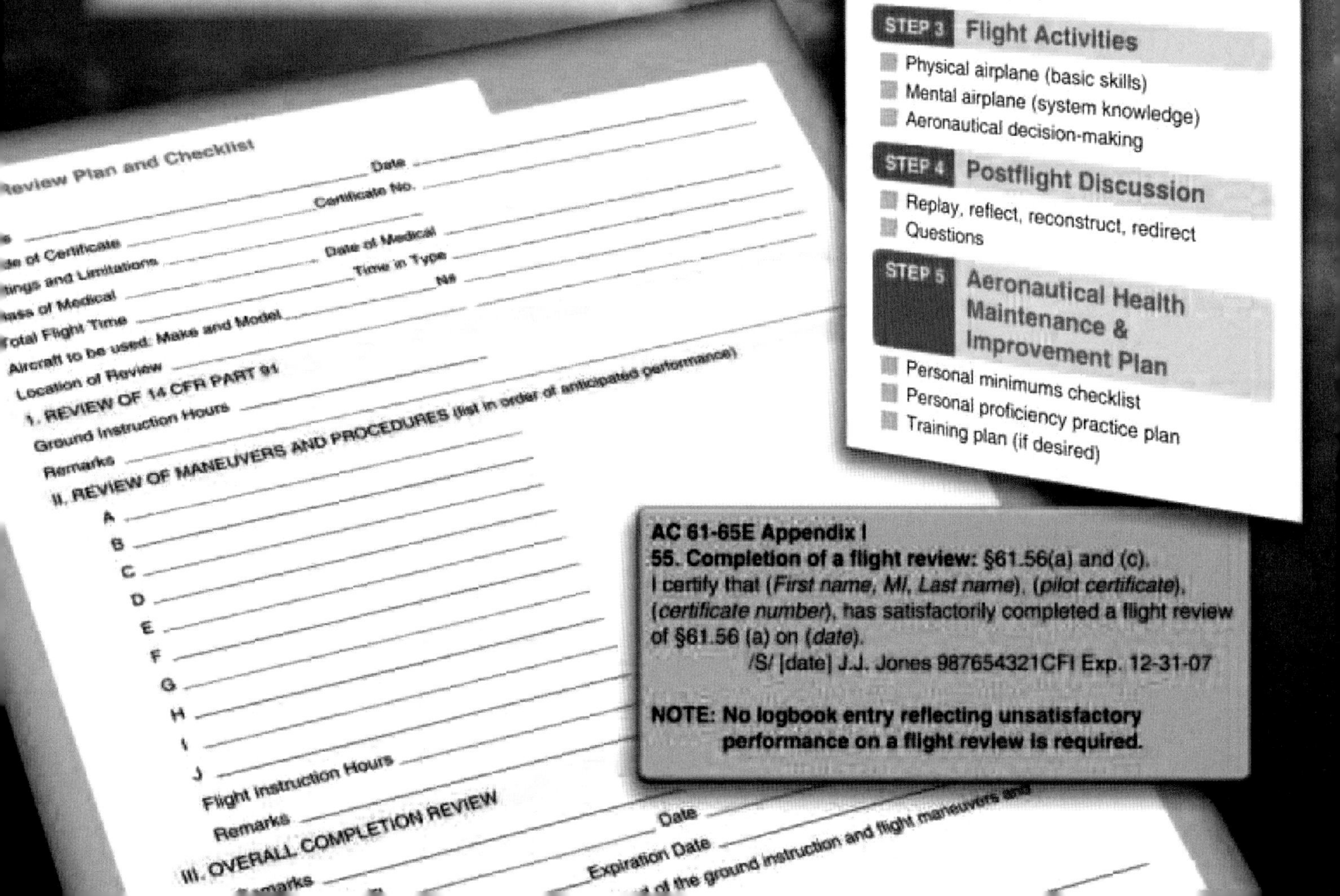

Review Plan and Checklist

Date _____
Certificate No. _____

de of Certificate _____ Date of Medical _____
ings and Limitations _____ Time in Type _____
ass of Medical _____ N# _____
otal Flight Time _____
Aircraft to be used: Make and Model _____
Location of Review _____
1. REVIEW OF 14 CFR PART 91
Ground Instruction Hours _____
Remarks _____
II. REVIEW OF MANEUVERS AND PROCEDURES (list in order of anticipated performance)
A _____
B _____
C _____
D _____
E _____
F _____
G _____
H _____
I _____
J _____
Flight Instruction Hours _____
Remarks _____
III. OVERALL COMPLETION REVIEW
Date _____
Expiration Date _____
of the ground instruction and flight maneuvers and

**AC 61-65E Appendix I**
**55. Completion of a flight review: §61.56(a) and (c).**
I certify that (*First name, MI, Last name*), (*pilot certificate*), (*certificate number*), has satisfactorily completed a flight review of §61.56 (a) on (*date*).
/S/ [date] J.J. Jones 987654321CFI Exp. 12-31-07

**NOTE: No logbook entry reflecting unsatisfactory performance on a flight review is required.**

Requirements differ; therefore, it is important for the instructor to become familiar with the prerequisites for the specific training and test to be conducted. Documentation for each type of certificate generally includes:

- Current student or pilot certificate,
- Medical certificate appropriate for the rating sought,
- Instructor's endorsement to take the practical exam,
- Experience requirements,
- Airman's application form, and
- Photo identification.

## Preparing the Student

A student pilot is likely to be anxious and apprehensive when preparing for the knowledge test, practical exam, and flight review. Help the student understand the process and intent of each.

### Last Training Flight

The last few training sessions should attempt to mirror what is to be expected during the actual academic or flight evaluation. Replicate the evaluation and allow the student to perform all maneuvers or discuss all subjects without assistance. This should provide the student with confidence going into the evaluation.

Describe the difference between a FAA inspector and designated pilot examiner (DPE). Remind the student that each evaluator has different techniques, but that the same standards apply regardless of the evaluator.

### Application and Testing Preparation

Assist the student by walking them through the application process and explaining what is to be expected at the testing locations. Questions, such as, how long the test will be, what to bring, what items they may need to bring are most common. Several FAA websites are available to help the student prepare for the exam or flight review process:

- http://www.faa.gov/training_testing/ (provides a list of resources and guides to help the student)
- http://www.faa.gov/pilots/testing/ (provides testing information and locations)
- http://av-info.faa.gov/DesigneeSearch.asp (provides a directory for Pilot Examiners)

## Preparation for a Practical Exam

### Training

As previously discussed in earlier chapters, the academic and flight training leading up to this point should have scenarios included. These scenarios start with basic fundamentals and develop into more complex and comprehensive task-integrated scenarios. In addition to teaching specific maneuvers and knowledge required for the exams, a large part of ensuring a pilot is ready for pilot-in-command (PIC) duties is developing the student's ability to exercise good judgment and mature decision-making skills.

A scenario based flight training program should provide the student with the skills and knowledge necessary to meet or exceed the standards outlined in the Practical Test Standards (PTS) found at http://www.faa.gov/pilots/testing for the rating being sought. The PTS is published by the Federal Aviation Administration (FAA) to establish the standards for the practical test and pilot certification. FAA inspectors and designated pilot examiners (DPE) conduct practical tests in compliance with these standards. Flight instructors and applicants should find these standards helpful during training and when preparing for the practical test. By incorporating the PTS into scenario-based instructional programs, instructors can ensure the student not only meets or exceeds the standards but understands the practical application of those standards.

An instructor should not recommend a student for the practical exam unless the student has demonstrated both the knowledge and the skills required to meet the PTS. The instructor should not use the PTS as a training document because the PTS contains only the minimum standards to meet to obtain a certificate. Training that only meets the minimum standard of the PTS is poor training. Instructors should determine the student's goals and focus the training towards those goals, developing training that challenges the student to exceed the minimum standards.

### Flight Review

The flight review is not a test, but rather a mandated opportunity to receive updated information and instruction concerning the national airspace system, the aircraft to be used during the instruction, safety policies, and procedures. Additionally, the flight review is an instructional service designed to assess the pilot's knowledge, skills, and proficiency. For more in-depth discussion of the conduct of flight reviews, refer to Conducting an Effective Flight Review (a downloadable PDF handbook available at www.faa.gov); FAA Advisory Circular (AC) 61-98, Currency and Additional Qualification Requirements for Certificated Pilots; and 14 CFR part 61, section 61.56, Certification: Pilots, Flight Instructors, and Ground Instructors. Appendices in these documents include such items as a sample flight review plan and checklist, a sample list of flight review knowledge, maneuvers and procedures, and personal minimums worksheets. [Figure 16-1]

## Flight Review Checklist

**STEP 1  Preparation**

- Pilot's aeronautical history
- Part 91 review assignment
- Cross-country flight plan assignment

**STEP 2  Ground Review**

- Regulatory review
- Cross-country flight plan review
  - Weather & weather decision-making
  - Risk management & personal minimums
- General aviation security issues

**STEP 3  Flight Activities**

- Physical airplane (basic skills)
- Mental airplane (system knowledge)
- Aeronautical decision-making

**STEP 4  Postflight Discussion**

- Replay, reflect, reconstruct, redirect
- Questions from applicant

**STEP 5  Aeronautical Health Maintenance & Improvement Plan**

- Personal minimums checklist
- Personal proficiency practice plan
- Training plan (if desired)

**Figure 16-1.** *Sample flight review checklist.*

It should be noted that in a flight review, the two possible outcomes are a sign-off in the logbook for successful completion or an opportunity to return and practice more to regain or sharpen certain aircraft skills. There is no possibility of failure in a flight review. At the conclusion of a successful flight review, the logbook of the pilot should be endorsed as recommended by AC 61-65, Certification: Pilots and Flight and Ground Instructors. *[Figure 16-2]*

The purpose of the flight review is to provide for an independent evaluation of pilot skills and aeronautical knowledge. According to the regulation, it is also intended to offer pilots the opportunity to design a personal currency and proficiency program in consultation with a certificated flight instructor (CFI). In effect, the flight review is the aeronautical equivalent of a regular medical checkup and ongoing health improvement program.

**AC 61-65E Appendix I**
**55. Completion of a flight review:** §61.56(a) and (c).
I certify that (*First name, MI, Last name*), (*pilot certificate*), (*certificate number*), has satisfactorily completed a flight review of §61.56 (a) on (*date*).
    /S/ [date] J.J. Jones 987654321CFI Exp. 12-31-07

**NOTE: No logbook entry reflecting unsatisfactory performance on a flight review is required.**

**Figure 16-2.** *Endorsement after successful flight review.*

The conduct of flight reviews for certificated pilots is a responsibility of the flight instructor, and is also an excellent opportunity for the instructor to expand his or her professional services. The flight review is intended to be an industry-managed, FAA-monitored currency program. As stated in 14 CFR part 61, no person may act as pilot in command (PIC) of an aircraft unless a flight review has been accomplished within the preceding 24 calendar months.

Effective pilot refresher training must be based on specific objectives and standards. The objectives should include a thorough checkout appropriate to the pilot certificate and aircraft ratings held, and the standards should be at least those required for the issuance of that pilot certificate. Before beginning any training, the pilot and the instructor should agree fully on these objectives and standards, and, as training progresses, the pilot should be kept appraised of progress toward achieving those goals.

A flight review is an excellent opportunity for an instructor to review pilot decision-making skills. To get the information needed to evaluate aeronautical decision-making (ADM) skills, including risk management, give the pilot multiple opportunities to make decisions and ask questions about those decisions. For example, ask the pilot to explain why the alternate airport selected for the diversion exercise is a safe and appropriate choice. What are the possible hazards and what can the pilot do to mitigate them? Be alert to the pilot's information and automation management skills as well. For example, does the pilot perform regular "common sense cross-checks?" For more ideas on generating scenarios that teach risk management, visit www.faa.gov.

In addition to the required maneuvers conducted during the flight review, flight instructors should also review and discuss those special emphasis items listed in the flight instructor PTS.

## Who Needs a Flight Review?

Pilots require a flight review every 24 months, with the following exceptions:

1. A person who has a pilot proficiency check conducted by an examiner, an approved pilot check airman, or is in the U.S. Armed Forces, for pilot certificate, rating, or operating privilege.

2. A person who has satisfactorily completed one or more phases of an FAA-sponsored pilot proficiency award program.

3. A student pilot need not accomplish the flight review provided that student pilot is undergoing training for a certificate and has a current solo flight endorsement as required under 14 CFR part 61, section 61.87.

4. A person who has passed the required PIC proficiency check under 14 CFR part 61, section 61.58, or part 121, 135, or 141.

5. A person who holds a current flight instructor certificate and has satisfactorily completed renewal of a flight instructor certificate under the provisions of 14 CFR part 61, section 61.197, need not complete the one hour of ground training listed under the requirements of the flight review.

## Flight Review Requirements

14 CFR part 61, section 61.56 stipulates that a flight review must contain at least one hour of ground instruction and one hour of flight instruction. The instruction must include a review of the general operating and flight rules of 14 CFR part 91 and a review of those maneuvers and procedures that, at the discretion of the instructor giving the review, are necessary for the pilot to demonstrate the safe exercise of the privileges of the pilot certificate held. *[Figure 16-3]* Instructors should tailor the review of general operating and flight rules to the needs of the pilot being reviewed to ensure the pilot can comply with all regulatory requirements and operate safely. Flight instructors need to keep in mind that a flight review can only end with a biennial flight review endorsement or dual given. There are no failures on a flight review.

## Preparation for the Flight Review

The flight review gives pilots the opportunity to ride with a flight instructor of their own choosing for an appraisal of their flying skills and proficiency, and to get further assistance and guidance in any area(s) in which they are deficient. Accident rates suggest that, among other things, some instructors administering flight reviews may not sufficiently recognize and correct poor pilot technique or decision-making capabilities. Since the maneuvers and procedures performed are at the discretion of the instructor giving the review, it is important that instructors adequately prepare for the review. The instructor can make the most of a flight review by beginning with an interview of the pilot to determine the nature of his or her flying and operating requirements. Ask the pilot what he or she wishes to refresh or relearn. This helps the pilot become motivated and to accept the flight review evaluation. AC 61-98 suggests some of the elements to consider during this interview. Flight instructors should take into consideration the typical flight areas of the flight review pilot and attempt to tailor the review towards their expected hazardous conditions and maneuvers. For example, if flights are in high elevations, then particular attention should be focused on high density altitude flight planning and mountain flying to include box canyons. Flight instructors should also be aware of the local accident rate and causative factors.

### Model or Type of Helicopter Flown

For the purposes of giving a flight review, the regulations do not require an instructor to have any minimum amount of time in a particular make and model of helicopter. To ensure the review is safely conducted within the operating limitations of the helicopter to be used, it is a good idea to be familiar with the helicopter. An instructor conducting a flight review must hold a category, class, and, if appropriate, the type rating on the pilot certificate, as well as a category and class rating on the flight instructor certificate appropriate to the aircraft in which the review is to be conducted.

### Nature of Flight Operations

An instructor giving a flight review should consider the type of flying usually done by the pilot before deciding how to conduct a review. Most pilots may want to review only emergency procedures, but other pilots may want to concentrate on areas of operation in which they lack experience or feel deficient.

Special Federal Aviation Regulation No. 73 (SFAR 73) has additional requirements for pilots to act as PIC in Robinson R-22 or R-44 helicopters. To act as PIC in either of these helicopters, the pilot must complete the flight review in the specific model helicopter.

### Recency of Flight Experience

The instructor should review the pilot's logbook to determine total flight time and recency of experience. This allows the instructor to evaluate the need for particular maneuvers and procedures in the flight review. Pilots who have not flown in several years may require an extensive review of the basic maneuvers and a more extensive review of 14 CFR part 91, as well as airspace and other operating requirements. More experienced and current pilots may want to review advanced

## Flight Review Plan and Checklist

Name _____ Date _____

Grade of Certificate _____Certificate No. _____

Ratings and Limitations _____

Class of Medical _____ Date of Medical _____

Total Flight Time _____Time in Type _____

Aircraft to be used: Make and Model_____N# _____

Location of Review _____ _____

### 1. REVIEW OF 14 CFR PART 91

Ground Instruction Hours _____ _____

Remarks _____

### II. REVIEW OF MANEUVERS AND PROCEDURES (list in order of anticipated performance)

A _____

B _____

C _____

D _____

E _____

F _____

G _____

H _____

I _____

J _____

Flight Instruction Hours _____

Remarks _____

### III. OVERALL COMPLETION REVIEW

Remarks _____

Signature of CFI _____ Date _____

Certificate No. _____Expiration Date _____

I have received a flight review which consisted of the ground instruction and flight maneuvers and procedures noted above.


Signature of the Pilot_____ Date _____

**Figure 16-3.** *Sample plan for a flight review.*

maneuvers. Regardless of flight experience, the flight review should include all areas the instructor deems necessary for future safe operations.

## Chapter Summary

This chapter discussed the instructor's role in a student pilot's preparation for a practical examination. It also outlined recommended procedures for conducting a flight review.

## Chapter 17

# Single-Pilot Resource Management, Aeronautical Decision-Making, and Risk Management

## Introduction

The accident rate for helicopters has traditionally been higher than the accident rate for airplanes, probably due to the helicopter's unique abilities to fly and land in more diverse situations than airplanes. With no significant improvement in helicopter accident rates for the last 20 years, the Helicopter Association International (HAI) set the goal of an 80 percent reduction in helicopter accidents by the year 2016. The Federal Aviation Administration (FAA) has joined HAI and other members of the helicopter community in accepting the challenge of improving the safety of helicopter operations.

According to the National Transportation Safety Board (NTSB) statistics, approximately 80 percent of all aviation accidents are caused by "pilot error" or the human factor. *[Figure 17-1]* Many of these accidents are the result of the tendency of instructors to focus flight training on teaching the student pilot the physical aspects of flying the aircraft and only enough aeronautical knowledge to pass the written and practical tests. Today's instructor must incorporate aeronautical decision-making (ADM) into flight training.

ADM includes single-pilot resource management (SRM), risk management, situational awareness, task management, and controlled flight into terrain (CFIT) awareness. Ignoring these safety issues can have fatal results. The flight instructor who integrates SRM into flight training teaches aspiring pilots how to be more aware of potential risks in flying, how to clearly identify those risks, and how to manage them successfully.

An instructor's first priority should be the student. Too often instructors rush through academics, flight planning, and preflight to get into the helicopter and begin logging time. This is a very dangerous first impression for the student. Instructors learn the Law of Primacy, yet so often this is the first principle violated. The first flight should be preceded by very thorough preflight planning, encompassing every facet of aeronautical decision-making. While we do not want to overwhelm the student, it is very important to instill a sense of methodical decision-making, ensuring the student understands that safety should never take a back seat to "getting off the ground." What will be the first impression of a student who is rushed through preflight planning for the purpose of getting off the ground? If not shown the proper procedure from the very first flight, students most likely accomplish only what they remember and what was first learned. To go backward to earlier steps that were not discussed or conducted improperly leads to confusion or rules misinterpreted by the student. We often hear a student state, "We didn't check weather last flight or do performance planning, but today the instructor says it looks marginal." In other words, the student is thinking, "I guess we just check the weather if it looks bad." This could be the first link in the accident chain, and a prime example of poor decision-making as a result of poor instructional practices.

The helicopter instructor should include the student in all flight decisions made during the course of instruction. By discussing the mental processes used to determine whether or not to fly, the student learns SRM, crew resource management, and human factors from the beginning of training. It is important for the instructor to model a high standard of professionalism which provides the student with a good role model and helps develop safe flying habits from the start of training.

During dual instruction, the instructor should practice good crew resource management (CRM). As the student approaches solo proficiency, the instructor should begin discussing what actions and resources the student will have

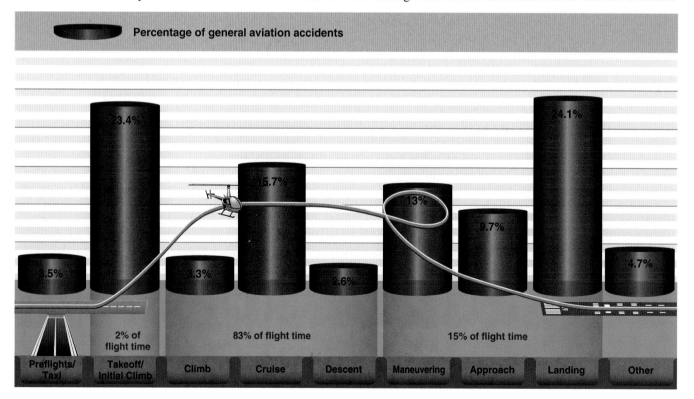

**Figure 17-1.** *The percentage of aviation accidents by different phases of flight. Note that the greatest percentage of accidents take place during a minor percentage of the total flight.*

available during solo flights. Good preflight briefings review past performances, provide suggestions for improvement, and define the flight lessons to be demonstrated and performed during this training period. Performance of any new maneuvers should be understood by the student prior to leaving the briefing room.

An effective instructor of ADM guides the student through the decision process by:

- Posing a question or situation that engages the student pilot in some form of decision-making activity.

- Examining the decisions made.

- Exploring other ways to solve the problem.

- Evaluating which way is best.

For example, if the student is going to practice simulated engine failures, prior to flying and then during the debrief, the instructor might ask questions such as: "Where are you going to land?" "Why did you pick that place to land?" "Is there a better choice?" "Which place is the safest?" or "Why?" These questions force the student to focus on the decision process. This accelerates the acquisition of improved judgment, which is simply the decision-making process resulting from experience. By introducing decision-making opportunities into routine training lessons, instructors speed up acquisition of experience, thus enhancing judgment.

## Origins of ADM and SRM

For over 25 years, the importance of good pilot judgment, or ADM, has been recognized as critical to the safe operation of aircraft, as well as accident avoidance. Motivated by the need to reduce accidents caused by human factors, the airline industry developed the first training programs based on improving ADM. Called crew resource management (CRM), these programs focused on training flight crews on the effective use of all available resources: human resources, hardware, and information supporting ADM to facilitate crew operation and improve decision-making. The effectiveness of this training prompted the FAA to incorporate these concepts into training directed at improving the decision-making of pilots. It also led to current FAA regulations that require decision-making be taught as part of pilot training curriculum.

The effectiveness of ADM training has been validated in independent studies in which student pilots received ADM training in conjunction with the standard flying curriculum. When tested, the pilots who received ADM training made fewer in-flight errors than those who had not. Contrary to popular opinion, good judgment can be taught. Building upon the foundation of conventional decision-making, ADM enhances the process to decrease the probability of human error and increase the probability of a safe flight.

Many of the concepts utilized in CRM have been successfully applied to single-pilot operations which led to the development of SRM. Defined as the art and science of managing all the resources (both onboard the aircraft and from outside resources) available to a single pilot (prior to and during flight), SRM ensures the successful outcome of the flight. As mentioned earlier, this includes ADM, risk management, situational awareness (SA), task management, and CFIT awareness.

SRM is about helping pilots learn how to gather information, analyze it, and make decisions. It helps the pilot assess and manage risk accurately and make accurate and timely decisions. For instance, give the new student tasks such as checking weather, determining weather sources and there content, and conduct performance planning for current and forecast conditions. Having the student perform these vital tasks is likely to instill confidence and, when reviewed by the instructor, provide a solid ADM foundation.

Help the student identify hazards, report them, and explain why and under which circumstances they are applicable. After training, the student may not report them to anyone, but rather acknowledge their existence and make seasoned, rational risk management decisions concerning those hazards. This is the essence of risk management. A pilot should acknowledge hazards, determine risk factors, and develop risk mitigation plans. For example, a pilot notices a large rock has surfaced in the hover training area. The pilot should recognize the rock could cause a dynamic rollover accident. In this case, the pilot should report the rock to the airport authority and avoid the area until the rock is removed and the hole filled. The pilot should also remember the rock can be a hazard during mowing seasons. During mowing, pieces of the rock can be ejected from the mower like small missiles, damaging thin helicopter parts.

In effect, you lead the student through the risk management process, which is much easier to comprehend at the planning table than in the air. At some point in the training, the instructor should begin to ask the student to evaluate situations, and determine if hazards are present and what results risk management analysis would yield. For aeronautical decision-making skills to be gained and developed, conclusions must be made and results determined. That is called experience.

### The Decision-Making Process
The instructor needs to understand the basic concepts of the decision-making process in order to provide the student with a foundation for developing ADM skills. It is important to teach students how to respond to emergency situations, such as an engine failure, which requires an immediate response using established procedures with little time for detailed

analysis. This type of decision-making is called automatic decision-making and is based upon training, experience, and recognition. Traditional instruction trains students to react to emergencies, but does not prepare the student to make decisions requiring a more reflective response through greater analysis. There is usually time during a flight to examine any changes that occur, gather information, and assess risk before reaching a decision. The steps leading to this conclusion constitute the decision-making process. In many cases, decision-making is the filtering of options and accurate perception of the true conditions. For example, if the weather is at all questionable, cancel the flight and explain to the student that their lives are worth much more than getting training done in questionable weather and possibly risking an accident due to the conditions.

Instructors can demonstrate decision-making skills through emergency training, required maneuvers or table discussion. Too often instructors "check the block;" that is, they have the student perform a series of tasks–without ever seeing the student exercise problem-solving skills. Have you ever participated in an academic evaluation with an evaluator asking, "What is the emergency procedure for an engine failure, for a hydraulic failure, for a high engine oil temp indication…" and so on? What is being evaluated? The answer is simple—the student's ability to memorize emergency procedures and limitations.

Provide system or mechanical indicators as part of a flight profile scenario that can be associated with a specific emergency or maneuver that is likely to be encountered. One example is teaching students to abort takeoffs. Although quick stops are taught as to how to stop, aborted takeoffs are rarely practiced with the judgment factor exercised. Instructors can evaluate or train this maneuver while evaluating the student's decision-making process.

Instructors should challenge and evaluate the student's decision making abilities. The intent is not to trick the student or demonstrate your superior knowledge, but rather to increase student ability to gather information, assess a situation, determine options, and choose a course of action.

### Defining the Problem
The first step in the decision-making process is to define the problem, which begins with recognizing that a change has occurred or that an expected change did not occur. A problem is perceived first by the senses, then is distinguished through insight and experience. One critical error that can be made during the decision-making process is incorrectly defining the problem.

Teach students to establish scanning or cross-checks. Do not allow an unusual finding to become a flight distraction, but do not ignore what may be an abnormal condition.

For example, a rapidly falling engine tachometer reading could indicate the engine has failed, and an autorotation needs to be entered immediately. It could mean the engine tachometer has failed. The actions to be taken in each of these circumstances would be significantly different. One requires an immediate decision based upon training, experience, and evaluation of the situation; the latter decision is based upon an analysis. It should be noted that the same indication could result in two different actions depending upon other influences.

Given this set of circumstances, we should immediately cross-check other instruments to verify our tachometer reading. By teaching our students to gather, identify and assess the information quickly we expect them to make the appropriate decision. In this case, if the engine rpm indication is decaying, but the rotor rpm is within the normal range, then the problem is the gauge. If the rotor rpm is decaying, then it is a power problem. Gathering all available information in a timely manner is a very important factor in defining the problem.

In the first instance, if the engine failed, then the proper action would be to autorotate. On the other hand, if the rapidly falling engine tachometer reading was due to a failure of the engine tachometer, then the engine is still running. The helicopter is under power, and a landing under power could be accomplished.

### Choosing a Course of Action
After the problem has been identified, the student must evaluate the need to react to it and determine the actions that may be taken to resolve the situation in the time available. The expected outcome of each possible action should be considered and the risks assessed before selecting a response to the situation.

### Implementing the Decision and Evaluating the Outcome
Although a decision may be reached and a course of action implemented, the decision-making process is not complete. The decision-making process continues, eliminating some options and recognizing new options as conditions change. It is important that the instructor teach the student how to think ahead and determine how the decision could affect other phases of flight. As the flight progresses, encourage the student to continue to evaluate the outcome of the decision to ensure it is producing the desired result.

## Improper Decision-Making Outcomes

Pilots sometimes get in trouble not because of deficient basic skills or system knowledge, but because of faulty decision-making skills. One realistic scenario involves a fuel gauge suddenly indicating near zero or empty. Should the pilot land and visually check the fuel tanks in the next safe area to land, or continue as planned and just consider it a gauge malfunction? The instructor should be teaching the pilot to evaluate the risk in landing to check the fuel state visually, instead of continuing the flight uninterrupted and just hoping that it is only the gauge system failing. What could be the consequences if the gauge were correct versus being failed?

Although aeronautical decisions may appear to be simple or routine, each individual decision in aviation often defines the options available for the next decision the pilot must make and the options, good or bad, they provide. Therefore, it is important for the instructor to stress to the student that a poor decision early in a flight can compromise the safety of the flight at a later time. Emphasize the importance of making accurate decisions. Good decision-making early in an emergency provides greater latitude for options later on.

## FAA Resources

FAA resources offer the instructor a variety of structured frameworks for decision making that provide assistance in organizing the decision process. These models include but are not limited to the 3P, the 5P, the OODA, and the DECIDE models. All these models and their variations are discussed in detail in the Pilot's Handbook of Aeronautical Knowledge chapter on aeronautical decision-making. *[Figure 17-2]*

Whatever model is used, the instructor wants to ensure the student learns how to define a problem, recognize all feasible options available, choose a course of action, implement the decision, and evaluate the outcome (continuing the process if necessary). Remember, there is no one right answer in this process. Each student is expected to analyze the situation in light of experience level, personal minimums, and current physical and mental readiness level, and make his or her own decision.

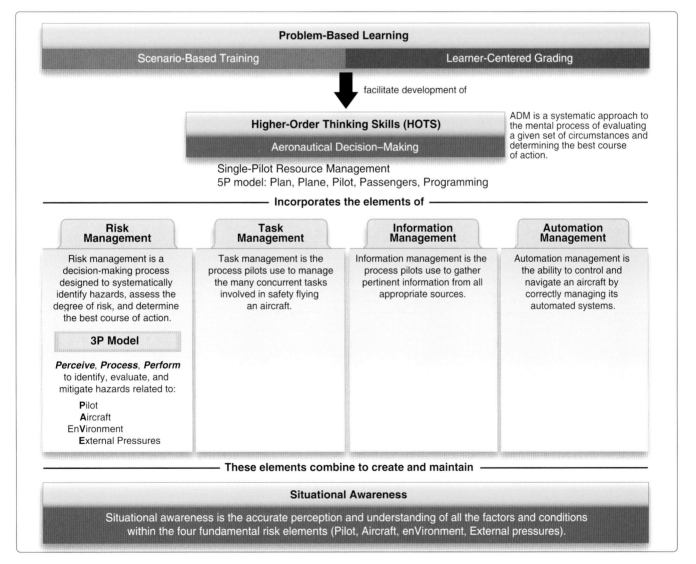

**Figure 17-2.** *Teach students that various models of decision-making are used in problem solving.*

## Human Factors

### Curiosity: Healthy or Harmful?

Curiosity is another human trait that kills. Airmen should be taught to control their curiosity until they land unless they wish to be test pilots and perform the standard risk management operations prior to the test flight!

Children tend to reply, "I don't know" when asked why they did something foolish because they really do not know that they succumbed to a burning curiosity about what would happen if they took the action. Humans are born curious and are always looking to see what is over the next hill or wave. That innate drive can make us bored or unhappy with what we have because we do not know what is "over there." Whether it is a new country, planet, or just a different recipe for a new taste, we are curious explorers.

Restrained curiosity can be healthy. Spontaneous excursions in aviation can be deadly. An airman musing "Let's see how this works" should not tinker with the object of curiosity while in the air. Airmen should abide by established procedures until proper hazard assessment and risk management are complete.

Pilot curiosity crashes modern aircraft just as it did early aircraft. A pilot may wonder how much he or she can fly or haul in one day. That curiosity leads to a personal challenge that may in turn lead to overloading the machine. The pilot may think of a method to improve production, but may neglect to factor in the design criteria for the machine, leading to unsafe overloads or fatigue. A pilot may decide not to act upon curiosity about something only after learning that another pilot's action stemming from the same curiosity ended in tragedy.

Instructors must be aware of not only their own curiosity but also the attitude and tendencies of their students. The tone of a student's question can often be interpreted. For instance, one student may ask, "Is the maximum weight limit of this aircraft due to design or power limits?" This is a question that promotes further discussion or correlation of the limiting factors. Another student may ask, "You can exceed the maximum weight limit if you really need to, can't you?" This student wants to push the envelope, and his or her curiosity may not be of a healthy nature. if left unchecked, curiosity can kill the pilot.

## Risk Management

Risk management is a formalized way of dealing with hazards. It is the logical process of weighing the potential costs of risks against the possible benefits of allowing those risks to stand uncontrolled. It is a decision-making process designed to help the pilot identify hazards systematically,

assess the degree of risk, and determine the best course of action. Once risks are identified, they must be assessed. The risk assessment determines the degree of risk (negligible, low, medium, or high) and whether the degree of risk is worth the outcome of the planned activity. If the degree of risk is "acceptable," the planned activity may then be undertaken. Once the planned activity is started, consideration must then be given whether to continue. Pilots must have preplanned, viable alternatives available in the event the original flight cannot be accomplished as planned.

Instructors play a critical role in developing the decision-making skills of new pilots. Observe various levels of and rates at which students acquire these skills. Some students seem very aware of their surroundings, and others focus solely on the task at hand. Additionally, all new pilots lack the experience base to identify potential hazards (such as a buzz in the pedals) or the options available to them. Instructors must share their knowledge and discuss options available to the student pilot.

Hazard and risk mitigation are key terms in risk management. Define those terms for the student:

- Hazard—present condition, event, object, or circumstance that could lead to or contribute to an unplanned or undesired event—like an accident. It is a source of danger. For example, binding in the antitorque pedals represents a hazard.

- Risk—the future impact of a hazard that is not controlled or eliminated. It is the possibility of loss or injury. The level of risk is measured by the number of people or resources affected (exposure), the extent of possible loss (severity), and likelihood of loss (probability).

- Mitigation—the effort to reduce loss of life and property by lessening the impact of disasters through proper planning and developing and implementing procedures.

A hazard can be a real or perceived condition, event, or circumstance that a pilot encounters. Teaching the student how to identify hazards, assess the degree of risk they pose, and determine the best course of action form an important element of today's flight training programs. For more information on risk management, refer the student to the Risk Management Handbook, FAA-H-8083-2.

### Assessing Risk

It is important for the flight instructor to teach the student how to assess risk. Before the student can begin to assess risk, he or she must first perceive the hazard and attendant risk(s). Experience, training, and education help a pilot learn how to spot hazards quickly and accurately.

Valuable information for instructors can be found on the NTSB web site, http://www.ntsb.gov/aviation/aviation.htm. Researching an accident and discussing the events that preceded it provide a real opportunity for an instructor to impart knowledge and insight to the new student pilot. Many accidents are due to the failure of the pilot to properly and/or quickly assess the risk of a hazard. Additionally, applying the applicable circumstances that led up to the accident will aid in building the new pilot's knowledge base. Associating the events and circumstances that led to an accident helps to build the new pilot's knowledge base.

Once a hazard is identified, determining the probability and severity of an accident (level of risk associated with it) becomes the next step. For example, the hazard of binding in the antitorque pedals poses a risk only if the helicopter is flown. If the binding leads to a loss of directional control, there is a high risk it could cause catastrophic damage to the helicopter and the passengers. The instructor helps the student identify hazards and how to deal with them by incorporating risk assessment into the training program.

Every flight has hazards and some level of risk associated with those hazards. It is critical that students are able to:

- Differentiate in advance between a low-risk flight and a high risk flight.

- Establish a review process and develop risk mitigation strategies to address flights throughout the low to high risk range.

- Determine low risk versus high risk by being educated on the primary causes, reactions, and final outcomes of failures caused by weather and aerodynamics.

## Using the 3P To Form Good Safety Habits

As is true for other flying skills, risk management thinking habits are best developed through repetition and consistent adherence to specific procedures. The 3P model, while similar to other methods, offers three good reasons for its use:

1. It is fairly simple to teach and remember.

2. It gives students a structured, efficient, and systematic way to identify hazards, assess risk, and implement effective risk controls.

3. Practicing risk management needs to be as automatic as basic aircraft control. *[Figure 17-3]*

To assist the student pilot in using the 3P process, develop scenarios that use the building block theory. Introduce a simple circumstance that requires the student to progress through the

**Figure 17-3.** *3P Model (perceive, process, and perform).*

perceive, process, and perform functions. For instance, during a hover power check the predicted value is exceeded. Have the student first go through the steps determining a course of action, then follow that course of action with continuous reassessment. Further examples can be made more complicated to augment the student's decision-making ability.

## Stressors Affecting Decision-Making

Many factors, or stressors, can increase a pilot's risk of making a poor decision that affects the safety of the flight. Stressors are generally divided into three categories: environmental, physiological, and psychological. *[Figure 17-4]* Reduction of identifiable stressors can be seen in the simplification of instrumentation, clear procedures, and redundant systems.

| **Stressors** |
| --- |
| **Environmental**<br>Conditions associated with the environment, such as temperature and humidity extremes, noise, vibration, and lack of oxygen. |
| **Physiological Stress**<br>Physical conditions, such as fatigue, lack of physical fitness, sleep loss, missed meals (leading to low blood sugar levels), and illness. |
| **Psychological Stress**<br>Social or emotional factors, such as a death in the family, a divorce, a sick child, or a demotion at work. This type of stress may also be related to mental workload, such as analyzing a problem, navigating an aircraft, or making decisions. |

**Figure 17-4.** *Environmental, physiological, and psychological stress affect decision-making skills. These stressors have a profound impact on the pilot, especially during periods of high workload.*

By making aviation as simple and predictable as possible, its stressfulness is reduced. ADM attempts to prevent the effects of stress and increase flight safety. Discuss stressors with the student and how stressors affect flight decision-making.

### Pilot Self-Assessment

Review the IMSAFE checklist with the student. Stress its importance as one of the best ways single pilots can mitigate risk by determining physical and mental readiness for flying. *[Figure 17-5]*

**Figure 17-5.** *I'M SAFE Checklist.*

### The PAVE Checklist

Explain to the student that mitigation of risk begins with perceiving hazards. By incorporating the PAVE checklist into preflight planning, the instructor teaches the student how to divide the risks of flight into four categories: Pilot, Aircraft, enVironment, and External pressures (PAVE). Discuss with the student how these categories form part of a pilot's decision-making process. *[Figure 17-2]*

The PAVE checklist provides the student with a simple way to remember each category to examine for risk prior to each flight. Once the student identifies the risks of a flight, he or she needs to decide whether the risk or combination of risks can be managed safely and successfully. Stress to the student that the PIC is responsible for making the decision of whether or not to cancel the flight. Explain that if the pilot decides to continue with the flight, he or she should develop strategies to mitigate the risks.

Encourage the student to learn how to control the risks by setting personal minimums for items in each risk category. Emphasize that these are limits unique to that individual pilot's current level of experience and proficiency, and should be reevaluated periodically based upon experience and proficiency.

Incorporate ongoing discussions of hazards, risk assessment, and risk mitigation into training to reinforce the student's decision-making skills.

### Recognizing Hazardous Attitudes

As discussed in the Aviation Instructor's Handbook, it is not necessary for a flight instructor to be a certified psychologist, but it is helpful to be aware of student behavior before and during each flight. If the instructor notices a hazardous attitude (which contributes to poor pilot judgment), he or she can counteract it effectively by converting that hazardous attitude into a positive attitude.

Since recognizing a hazardous attitude is the first step toward neutralizing it, it is important for the student to learn the hazardous attitudes and the corresponding antidotes. The antidote for each of the hazardous attitudes should be memorized so it automatically comes to mind when needed. *[Figure 17-6]*

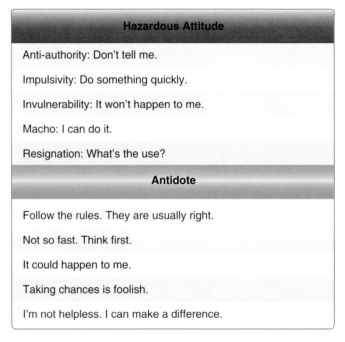

| Hazardous Attitude |
| --- |
| Anti-authority: Don't tell me. |
| Impulsivity: Do something quickly. |
| Invulnerability: It won't happen to me. |
| Macho: I can do it. |
| Resignation: What's the use? |
| **Antidote** |
| Follow the rules. They are usually right. |
| Not so fast. Think first. |
| It could happen to me. |
| Taking chances is foolish. |
| I'm not helpless. I can make a difference. |

**Figure 17-6.** *The antidotes to five hazardous attitudes.*

When reading *Figure 17-6*, keep in mind that each hazardous attitude relates directly to a potential incident or accident. So many of the regulations we abide by are in response to an increase of the aviation accident rate. Therefore, discuss specifics with the student that correlate to each.

The anti-authority hazardous attitude explains numerous accidents involving weather-related decisions. We have ceiling and visibility minimums not just for you, but to allow other pilots to see and avoid you. Or, that old adage, "I don't need my landing light on at night, I can see the lit runway just fine." Turn the landing lights on at night so other pilots can see *you*. Each hazardous attitude has an explanation and an antidote as seen in the figure.

## Stress Management

Stress is the body's response to physical and psychological demands placed upon it. While a certain level of stress is necessary to perform optimally, too little stress can have as much of an adverse affect as too much stress. If the student is under too little stress, the thinking processes tend to wander to non-related thoughts and activities. For instance, the "sterile cockpit" rule (14 CFR part 121, section 121.542) resulted from numerous accidents where the crew seemed to exhibit no stress and their attention wondered from their flight duties. Too much stress and the thinking processes seem to stagnate, resulting in a sensory overload and subsequent mental shutdown.

The causes of student stress can range from poor performance of flight maneuvers to personal issues unrelated to flying. Stress is an inevitable and necessary part of life that can add motivation and heighten an individual's response to meet a challenge.

The effects of stress are cumulative and, if the student does not cope with them in an appropriate way, they can eventually add up to an intolerable burden. Performance generally increases with the onset of stress, peaks, and then falls off rapidly as stress levels exceed a person's ability to cope. At this point, a student's performance begins to decline and judgment deteriorates. Complex or unfamiliar tasks are more subject to the adverse effects of increasing stress than simple or familiar tasks.

Stress falls into two broad categories, acute (short term) and chronic (long term). Acute stress involves an immediate threat that is perceived as danger. This is the type of stress that triggers a "fight or flight" response in an individual, whether the threat is real or imagined. Normally, a healthy person can cope with acute stress and prevent stress overload. However, ongoing acute stress can develop into chronic stress.

Chronic stress can be defined as a level of stress that presents an intolerable burden, exceeds the ability of an individual to cope, and causes individual performance to fall sharply. Unrelenting psychological pressures, such as loneliness, financial worries, and relationship or work problems can produce a cumulative level of stress that exceeds a person's ability to cope with a situation. When stress reaches this level, performance falls off rapidly. The instructor should make the student aware that pilots experiencing this level of stress are not safe and should not exercise their airman privileges.

The indicators of excessive stress often show as three types of symptoms: emotional, physical, and behavioral. Emotional symptoms may surface as overcompensation, denial, suspicion, paranoia, agitation, restlessness, or defensiveness. Physical stress can result in acute fatigue. Behavioral degradation is manifested as sensitivity to criticism, tendency to be argumentative, arrogance, and hostility. Instructors need to learn to recognize the symptoms of stress in students.

There are several techniques an instructor can suggest to a student to help manage the accumulation of life stresses and prevent stress overload. For example, to help reduce stress levels, suggest the student set aside time for relaxation each day or maintain a program of physical fitness. To prevent stress overload, encourage the student to learn to manage time more effectively to avoid pressures imposed by getting behind schedule and not meeting deadlines.

For a more in-depth discussion of stress and ways to deal with it, see the Pilot's Handbook of Aeronautical Knowledge, Chapter 16, Aeromedical Factors, and Chapter 17, Aeronautical Decision-Making.

## Use of Resources

To make informed decisions during flight operations, a student must be introduced to and learn how to use all the resources found inside and outside the cockpit. Since useful tools and sources of information may not always be readily apparent, an essential part of ADM training is learning to recognize these resources. The instructor must not only identify the resources, but also help the student develop the skills to evaluate whether there is time to use a particular resource and the impact its use has upon the safety of flight.

Remember to point out to your students the most valuable resource or option as a helicopter pilot is the ability to land the helicopter almost anywhere. Whether at an airport or any suitable landing area along the flight path, the option to land the aircraft is almost always available. Too many fatalities have occurred when this most basic helicopter option is overlooked. A controlled landing under power is always preferable to an emergency power off landing. It is easy to land in an open field or lot and call for fuel or maintenance rather than crash and never get the chance again to make any calls. Getting the helicopter safely on the ground allows time to process other options without endangering the crew and/or passengers.

The assistance of air traffic control (ATC) may be very useful if a student becomes lost; but in an emergency situation, there may be no time available to contact ATC.

Cockpit management is also a key resource for preventing a potential accident from happening. Students must learn to manage avionics, computer messages, radios, transponders, and checklists while flying safely and under all conditions, VFR, IFR, and at night. Proper management in the cockpit helps the student to organize and learn to safely multitask.

### Internal Resources

Point out to the student that the person in the other seat can be an important resource even if that person has no flying experience. When appropriate, passengers can assist with tasks, such as watching for traffic or reading checklist items aloud.

Emphasize to the student the importance of verbal communication. It has been established that verbal communication reinforces an activity. Touching an object while communicating further enhances the probability an activity has been accomplished. For this reason, many solo pilots read checklists out loud. When they reach a critical item, they touch the switch or control. For example, to ascertain the force trim is on, the student can read the checklist. But, if he or she touches the force trim switch during the process, the checklist action is confirmed.

Explain to the student that it is necessary for a pilot to have a thorough understanding of all the equipment and systems in the aircraft being flown. Discuss with the student that a lack of aircraft systems knowledge, for example, can lead to a tragic error. For instance, if a new pilot is unaware of the mechanical differences between a direct reading gauge (wet line) versus a gauge operating from sensors, the student pilot may not associate the oil residue under the center console to a loose fitting on the affected gauge. Without full awareness and understanding of the wet line system, the student may choose to sidestep the issue, thinking it to be out of place and of no concern. The instructor's role in expanding the student's knowledge may directly lead to a correct assessment, evaluation of available options and, ultimately, a good decision.

Explain to the student that it is necessary for a pilot to have a thorough understanding of all equipment and systems in the aircraft being flown. Lack of knowledge such as whether the oil pressure gauge is a direct reading or uses a sensor is the difference between making a good decision or a poor one that leads to a tragic error.

Checklists are essential internal resources used to verify the aircraft instruments and systems are checked, set, and operating properly. They also ensure proper procedures are performed when there is a system malfunction or in-flight emergency. One bad habit is reading a step on the checklist and moving on without having thoroughly performed the step. Complacency with the routine can allow us to lose focus on the task at hand. The result is a failure to complete a step or to skip it entirely. Either failure may lead to aircraft damage.

Additionally, students should be taught not just the step or procedure, but the reason for the step or procedure. In essence, each step involves an aircraft system or procedure. The student should be taught why the step is performed, what indications or system settings are affected and what potential hazards may occur if the checklist is not properly used. While instructing, notice if the student is providing verbal response out of habit, or if that student is actually comprehending and performing the checks. Verbal response is commonly used under two-pilot situations; however, many pilots of single-pilot aircraft also verbalize the checks for confirmation.

Another internal resource is the Pilot's Operating Handbook (POH). Instructors should be teaching the student pilot the contents of the POH, as well as interpreting and validating the information found within the POH. Stress to the student that certain emergencies require immediate action without referencing the POH or checklist. Emphasize to the student that the POH:

- Is required to be carried on board the aircraft.
- Is indispensable for accurate flight planning.
- Plays a vital role in the resolution of in-flight equipment malfunctions when time allows.

Workload management (page 17-11) is also a valuable internal resource.

### External Resources

Discuss with the student the role of air traffic controllers and flight service specialists, the best external resources during flight. To promote the safe, orderly flow of air traffic around airports and along flight routes, ATC provides pilots with traffic advisories, radar vectors, and assistance in emergency situations.

Explain that it is the pilot's responsibility to make the flight as safe as possible, but a pilot with a problem can request assistance from ATC. For example, if a pilot needs to be given a vector, ATC assists and becomes integrated as part of the crew. Stress to the student that the services provided by ATC can not only decrease pilot workload, but also help pilots make informed in-flight decisions.

Discuss the role of the Flight Service Station (FSS)/ Automated Flight Service Station (AFSS) with the student:

- Air traffic facilities that provide pilot briefing, en route communications, and visual flight rules (VFR) search and rescue services; assist lost aircraft and aircraft in emergency situations; relay ATC clearances, originate Notices to Airmen (NOTAM); broadcast aviation weather and National Airspace System (NAS) information; receive and process instrument flight rules (IFR) flight plans; and monitor navigational aids (NAVAIDs).

- At selected locations, an FSS/AFSS provides En Route Flight Advisory Service (Flight Watch), issues airport advisories, and advises Customs and Immigration of transborder flights.

- Selected FSS/AFSS in Alaska also provide transcribed weather broadcast (TWEB) recordings and take weather observations.

Helicopters often operate in locations where radio reception is poor or contact with ATC is not possible. Prepare the student for this likelihood by providing scenarios in which it may be possible to use other devices or NAVAIDS to communicate. Cellular phones or satellite phones are of great benefit.

## Workload Management

Humans have a limited capacity for information. Once information flow exceeds the person's ability to mentally process the information, any additional information becomes unattended or displaces other tasks and information already being processed. Once this situation occurs, only two alternatives exist: shed the unimportant tasks or perform all tasks at a less than optimal level. Information overload for the pilot is like an overloaded electrical circuit; either the consumption is reduced or a circuit failure is experienced. *[Figure 17-7]*

Teaching a student effective workload management ensures essential operations are accomplished by planning, establishing a priority for the tasks, and then placing them in a sequence that avoids work overload. As the student gains experience, he or she learns to recognize future workload requirements and can prepare for high workload periods during times of low workload.

It is important for the instructor to model good workload management techniques. For example, review the appropriate chart and set radio frequencies well in advance of when they are needed to reduce workload as the flight nears the airport. In addition, listen to the Automatic Terminal Information Service (ATIS), Automated Surface Observing System (ASOS), or Automated Weather Observing System (AWOS), if available, and then monitor the tower frequency or Common Traffic Advisory Frequency (CTAF), explaining to the student that these external resources give a pilot a good idea of what traffic conditions to expect.

Another tool of workload management (that complements the use of these systems) is simple navigation and landmark referencing. By planning ahead and using visual landmarks, the student learns to maintain situational awareness and use these landmarks to prompt a necessary call or to alert the pilot of upcoming controlled or special use airspace.

Remind students that checklists should be performed well in advance so there is time to focus on traffic and ATC instructions. Emphasize to the student that these procedures are especially important prior to entering a high density

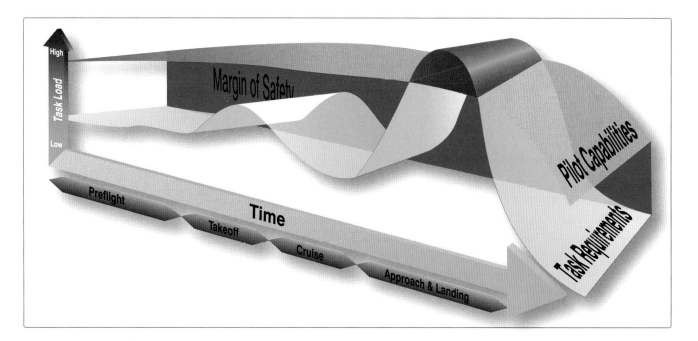

**Figure 17-7.** *The pilot has a certain capacity of doing work and handling tasks. However, there is a point at which the tasking exceeds the pilot's capability. When this happens, tasks are either not done properly or some are not done at all.*

traffic area, such as Class B airspace. Discuss workload management with the student:

1. Recognition of a work overload situation is an important component of managing workload.

   • The first effect of high workload: the pilot may be working harder but accomplishing less.

   • As workload increases, attention cannot be devoted to several tasks at one time and the pilot may begin to focus on one item.

   • When a pilot becomes task saturated, there is no awareness of input from various sources, so decisions may be made on incomplete information, and the possibility of error increases.

2. When a work overload situation exists, a pilot needs to:

   • Stop,

   • Think,

   • Slow down, and

   • Prioritize.

It is important for the student to understand how to decrease workload. Encourage him or her to learn how to place a situation in the proper perspective, remain calm, and think rationally. Explain that these are the key elements in reducing stress and increasing the capacity to fly safely. Remind the student this ability depends upon experience, discipline, and the training that they are in the process of receiving.

## Situational Awareness

Many definitions of situational awareness exist. For the beginning pilot it is most likely placing a term on a preexisting, subconscious practice. No two individuals share exactly the same degree of situational awareness. Some people seem to be aware of almost everything that is going on around them, while others seem oblivious to anything except the single task at hand. As instructors, we try to instill safety by teaching the student situational awareness through heightening of their senses and broadening student awareness before, during, and after flight.

While there are many techniques and learning tools available, one of the most commonly used is found in the Pilot's Handbook of Aeronautical Knowledge (PHAK). The PHAK provides an aviation directed tool that instructors may use for new pilots: the acronym PAVE. PAVE is composed of four fundamental risk elements for flight: Pilot, Aircraft, EnVironment, and External pressures.

Explain that situational awareness is knowing what is going on during the flight. Use the monitoring of radio communications for traffic, weather discussion, and ATC communication to demonstrate to the student how these resources enhance situational awareness by helping the pilot develop a mental picture of what is happening.

Discuss with the student how maintaining situational awareness requires an understanding of the relative significance of all flight-related factors and their future impact on the flight. When a student understands what is going on and has an overview of the total operation, he or she is not fixated on one perceived significant factor. Stress that it is important not only to know the aircraft's geographical location but also to understand what is happening around it. Provide the student with scenario-based training, which enhances the student's ability to maintain situational awareness and uses all of the skills involved in ADM.

### Obstacles to Maintaining Situational Awareness

Explain to the student that fatigue, stress, complacency, and work overload can cause a pilot to fixate on a single item perceived as important and reduce overall situational awareness of the flight. Discuss how a factor contributing to many accidents is distraction that diverts the pilot's attention from monitoring the instruments or scanning outside the aircraft. NTSB accident records offer flight instructors many examples of loss of situational awareness that can be used for training purposes.

Impress upon the student how easily a minor problem, such as a gauge that is not reading correctly, can result in an accident. The pilot diverts attention to the perceived problem and neglects to properly control the aircraft.

### Operational Pitfalls

Operational pitfalls [Figure 17-8] are routinely underemphasized. Instructors have the ability to influence and impact future generations of aviators through their instructional techniques and practices. While we frequently refer to the operational pitfalls as present for most aviation occupations, we fail to highlight the results of making these mistakes.

The NTSB database (http://ntsb.gov/ntsb/query.asp) is one of the greatest tools available to instructors when discussing operational pitfalls with new pilots. Conducting queries of helicopter accidents and incidents provides factual, aviation-related events that can be directly associated with the specific pitfalls outlined in Figure 17-8.

For instance, when discussing loss of positional or situational awareness, a query of the NTSB database presents an incident that occurred in January 2004 (NTSB Identification: MIA04CA048). While taxiing a helicopter for an upcoming early morning flight, the helicopter's left skid contacted a

| Operational Pitfalls |
|---|

**Peer Pressure**
Poor decision-making may be based upon an emotional response to peers, rather than evaluating a situation objectively.

**Mind Set**
A pilot displays mind set through an inability to recognize and cope with changes in a given situation.

**Get-there-itis**
This disposition impairs pilot judgment through a fixation on the original goal or destination, combined with a disregard for any alternative course of action.

**Duck-Under Syndrome**
A pilot may be tempted to arrive at an airport by descending below minimums during an approach. There may be a belief that there is a built-in margin of error in every approach procedure, or a pilot may not want to admit that the landing cannot be completed and a missed approach must be initiated.

**Scud Running**
This occurs when a pilot tries to maintain visual contact with the terrain at low altitudes while instrument conditions exist.

**Continuing Visual Flight Rules (VFR) in Instrument Conditions**
Spatial disorientation or collision with ground/obstacles may occur when a pilot continues VFR into instrument conditions. This can be even more dangerous if the pilot is not instrument rated or current.

**Getting Behind the Aircraft**
This pitfall can be caused by allowing events or the situation to control pilot actions. A constant state of surprise at what happens next may be exhibited when the pilot is getting behind the aircraft.

**Loss of Positional or Situational Awareness**
In extreme cases, when a pilot gets behind the aircraft, a loss of positional or situational awareness may result. The pilot may not know the aircraft's geographical location, or may be unable to recognize deteriorating circumstances.

**Operating Without Adequate Fuel Reserves**
Ignoring minimum fuel reserve requirements is generally the result of overconfidence, lack of flight planning, or disregarding applicable regulations.

**Descent Below the Minimum En Route Altitude**
The duck-under syndrome, as mentioned above, can also occur during the en route portion of an IFR flight.

**Flying Outside the Envelope**
The assumed high performance capability of a particular aircraft may cause a mistaken belief that it can meet the demands imposed by a pilot's overestimated flying skills.

**Neglect of Flight Planning, Preflight Inspections, and Checklists**
A pilot may rely on short- and long-term memory, regular flying skills, and familiar routes instead of established procedures and published checklists. This can be particularly true of experienced pilots.

**Figure 17-8.** *Typical operation pitfalls requiring pilot awareness.*

hedge bush row, causing the helicopter to roll onto its left side (dynamic rollover). Additional investigation discovered an expired biennial review.

This simple, yet costly, error provides the opportunity to discuss many aviation topics. Was the accident due to lack of situational awareness or the mindset of the pilot? Perhaps there was an element of fatigue involved or a perceived sense of urgency or get-there-itis. Follow-up discussion can include dynamic rollover and biennial review requirements. This one example is likely to leave a lasting impression on a new pilot.

Instructors must take advantage of resources available to them not only to discuss these topics with new students, but to instill safety awareness in the minds of new pilots. If a climate of safety awareness is achieved, perhaps we can reduce the number of or even eliminate incidents/accidents and needless fatalities in helicopters with pilots of all experience levels.

While discussing each pitfall, begin with the classic behavioral traps, into which pilots have been known to fall, that lead to accidents. Talk about the tendency of pilots, particularly those with considerable experience, who always try to complete a flight as planned, please passengers, and meet schedules. Warn the student that the desire to meet these goals can have an adverse effect on safety and contribute to an unrealistic assessment of piloting skills under stressful conditions. Encourage the student to learn how to identify and eliminate these operational pitfalls. *[Figure 17-8]*

An instructor must develop the student's awareness of and teach the students how to avoid operational pitfalls by ensuring effective ADM training is given. *[Figure 17-9]*

## Instructor Tips

- NTSB accident records offer many accident/incident reports that can be tailored to test the student's ADM knowledge.

- A student will attempt to imitate instructor actions. Do not take shortcuts. Instill safety from the first day.

- Questions or situations posed by the instructor must be open ended (rather than requiring only rote or one-line responses).

## Chapter Summary

This chapter provided the flight instructor with a review of ADM and SRM. It also offered recommendations on how ADM and SRM can be integrated into training

---

### Single-Pilot Resource Management, Aeronautical Decision-Making, and Risk Management

**Objective**

The purpose of this lesson is for the student to learn how to assess risk effectively when given a flight scenario. The student will demonstrate the ability to assess risk effectively when given a flight scenario.

**Content**

1. Preflight Discussion
   a. Discuss lesson objective and completion standards.
   b. Review the elements of SRM.

2. Instructor Actions
   a. Instructor briefs the student on an incident/accident scenario taken from the NTSB.

   For example:
   A pilot receiving instruction is practicing autorotations. The helicopter touches down on the aft part of the landing skids and starts roll motion forward. What action(s) taken by the pilot could cause one of the main rotor blades to strike the tail boom?

3. Student Actions
   a. Student assesses risk factors that could lead to this type of accident.

**Postflight Discussion**

Review the flight scenario. Preview the next lesson, and assign *Helicopter Flying Handbook*, Chapter 16, Airport Operations.

**Figure 17-9.** *Sample lesson plan.*

# Glossary

**Absolute altitude.** The actual distance an object is above the ground.

**AD.** See Airworthiness Directive.

**Advancing blade.** The blade moving in the same direction as the helicopter. In rotorcraft that have counterclockwise main rotor blade rotation as viewed from above, the advancing blade is in the right half of the rotor disk area during forward movement.

**Agonic line.** A line along which there is no magnetic variation.

**Air density.** The density of the air in terms of mass per unit volume. Dense air contains more molecules per unit volume than less dense air. The density of air decreases with altitude above the surface of the earth and with increasing temperature.

**Aircraft pitch.** The movement of an aircraft about its lateral, or pitch, axis. Movement of the cyclic forward or aft causes the nose of the helicopter to pitch up or down.

**Aircraft roll.** The movement of the aircraft about its longitudinal, or roll, axis. Movement of the cyclic right or left causes the helicopter to tilt in that direction.

**Airfoil.** Any surface designed to obtain a useful reaction of lift, or negative lift, as it moves through the air.

**Airworthiness Directive (AD).** A document issued by the FAA to notify concerned parties of an unsafe condition in an aircraft and to describe the appropriate corrective action.

**Altimeter.** An instrument that indicates flight altitude by sensing pressure changes and displaying altitude in feet or meters.

**Angle of attack.** The angle between an airfoil's chord line and the relative wind.

**Antitorque pedal.** The pedal used to control the pitch of the tail rotor or air diffuser in a NOTAR® system.

**Antitorque rotor.** See tail rotor.

**Articulated rotor.** A rotor system in which each of the blades is connected to the rotor hub in such a way that it is free to change its pitch angle, and move up and down and fore and aft in its plane of rotation.

**Autopilot.** Those units and components that furnish a means of automatically controlling the aircraft.

**Autorotation.** The condition of flight during which the main rotor is driven only by aerodynamic forces with no power from the engine.

**Axis of rotation.** The imaginary line about which the rotor rotates. It is represented by a line drawn through the center of, and perpendicular to, the tip-path plane.

**Basic empty weight.** The weight of the standard rotorcraft, operational equipment, unusable fuel, and full operating fluids, including full engine oil.

**Blade coning.** An upward sweep of rotor blades, resulting from a combination of lift and centrifugal force.

**Blade damper.** A device attached to the drag hinge to restrain the fore and aft movement of the rotor blade.

**Blade feather or feathering.** The rotation of the blade around the spanwise (pitch change) axis.

**Blade flap.** Rotor blade movement in a vertical direction. Blades may flap independently or in unison.

**Blade grip.** The part of the hub assembly to which the rotor blades are attached, sometimes referred to as blade forks.

**Blade lead or lag.** The fore and aft movement of the blade in the plane of rotation. It is sometimes called hunting or dragging.

**Blade loading.** The load imposed on rotor blades, determined by dividing the total weight of the helicopter by the combined area of all the rotor blades.

**Blade root.** The part of the blade that attaches to the blade grip.

**Blade span.** The length of a blade from its tip to its root.

**Blade stall.** The condition of the rotor blade when it is operating at an angle of attack greater than the maximum angle of lift.

**Blade tip.** The farthermost part of the blade from the hub of the rotor.

**Blade track.** The relationship of the blade tips in the plane of rotation. Blades that are in track will move through the same plane of rotation.

**Blade tracking.** The mechanical procedure used to bring the blades of the rotor into a satisfactory relationship with each other under dynamic conditions so that all blades rotate on a common plane.

**Blade twist.** The variation in the angle of incidence of a blade between the root and the tip.

**Blowback.** The tendency of the rotor disk to tilt aft in forward flight as a result of flapping.

**Calibrated airspeed (CAS).** The indicated airspeed of an aircraft, corrected for installation and instrumentation errors.

**Center of gravity.** The theoretical point where the entire weight of the helicopter is considered to be concentrated.

**Center of pressure.** The point where the resultant of all the aerodynamic forces acting on an airfoil intersects the chord.

**Centrifugal force.** The apparent force that an object moving along a circular path exerts on the body constraining the object and that acts outwardly away from the center of rotation.

**Centripetal force.** The force that attracts a body toward its axis of rotation. It is opposite centrifugal force.

**Chip detector.** A warning device that alerts you to any abnormal wear in a transmission or engine. It consists of a magnetic plug located within the transmission. The magnet attracts any ferrous metal particles that have come loose from the bearings or other transmission parts. Most chip detectors have warning lights located on the instrument panel that illuminate when metal particles are picked up.

**Chord.** An imaginary straight line between the leading and trailing edges of an airfoil section.

**Chordwise axis.** A term used in reference to semirigid rotors describing the flapping or teetering axis of the rotor.

**Coaxial rotor.** A rotor system utilizing two rotors turning in opposite directions on the same centerline. This system is used to eliminated the need for a tail rotor.

**Collective pitch control.** The control for changing the pitch of all the rotor blades in the main rotor system equally and simultaneously and, consequently, the amount of lift or thrust being generated.

**Coning.** See blade coning.

**Coriolis effect.** The tendency of a rotor blade to increase or decrease its velocity in its plane of rotation when the center of mass moves closer or further from the axis of rotation.

**Cyclic feathering.** The mechanical change of the angle of incidence, or pitch, of individual rotor blades independent of other blades in the system.

**Cyclic pitch control.** The control for changing the pitch of each rotor blade individually as it rotates through one cycle to govern the tilt of the rotor disk and, consequently, the direction and velocity of horizontal movement.

**Delta hinge.** A flapping hinge with an axis that is skewed so that the flapping motion introduces a component of feathering that results in a restoring force in the flapwise direction.

**Density altitude.** Pressure altitude corrected for nonstandard temperature variations.

**Deviation.** A compass error caused by magnetic disturbances from the electrical and metal components in the aircraft. The correction for this error is displayed on a compass correction card place near the magnetic compass of the aircraft.

**Direct control.** The ability to maneuver a rotorcraft by tilting the rotor disk and changing the pitch of the rotor blades.

**Direct shaft turbine.** A shaft turbine engine in which the compressor and power section are mounted on a common driveshaft.

**Disk area.** The area swept by the blades of the rotor. It is a circle with its center at the hub and has a radius of one blade length.

**Disk loading.** The total helicopter weight divided by the rotor disk area.

**Dissymmetry of lift.** The unequal lift across the rotor disk resulting from the difference in the velocity of air over the advancing blade half and retreating blade half of the rotor disk area.

**Drag.** An aerodynamic force on a body acting parallel and opposite to the relative wind.

**Dual rotor.** A rotor system utilizing two main rotors.

**Dynamic rollover.** The tendency of a helicopter to continue rolling when the critical angle is exceeded, if one gear is on the ground, and the helicopter is pivoting around that point.

**Feathering.** The action that changes the pitch angle of the rotor blades by rotating them around their feathering (spanwise) axis.

**Feathering axis.** The axis about which the pitch angle of a rotor blade is varied, sometimes referred to as the spanwise axis.

**Feedback.** The transmittal of forces, which are initiated by aerodynamic action on rotor blades, to the cockpit controls.

**Flapping hinge.** The hinge that permits the rotor blade to flap and thus balance the lift generated by the advancing and retreating blades.

**Flapping.** The vertical movement of a blade about a flapping hinge.

**Flare.** A maneuver accomplished prior to landing to slow a rotorcraft.

**Free turbine.** A turboshaft engine with no physical connection between the compressor and power output shaft.

**Freewheeling unit.** A component of the transmission or power train that automatically disconnects the main rotor from the engine when the engine stops or slows below the equivalent rotor rpm.

**Fully articulated rotor system.** See articulated rotor system.

**Gravity.** See weight.

**Gross weight.** The sum of the basic empty weight and useful load.

**Ground effect.** A usually beneficial influence on rotorcraft performance that occurs while flying close to the ground (within one rotor diameter). It results from a reduction in upwash, downwash, and blade-tip vortices, which provide a corresponding decrease in induced drag.

**Ground resonance.** Self-excited vibration occurring whenever the frequency of oscillation of the blades about the lead-lag axis of an articulated rotor becomes the same as the natural frequency of the fuselage.

**Gyroscopic procession.** An inherent quality of rotating bodies, which causes an applied force to be manifested 90° in the direction of rotation from the point where the force is applied.

**Human factors.** The study of how people interact with their environment. In the case of general aviation, it is the study of how pilot performance is influenced by such issues as the design of cockpits, the function of the organs of the body, the effects of emotions, and the interaction and communication with other participants in the aviation community, such as other crew members and air traffic control personnel.

**Hunting.** Movement of a blade with respect to the other blades in the plane of rotation, sometimes called leading or "lagging."

**Inertia.** The property of matter by which it will remain at rest or in a state of uniform motion in the same direction unless acted upon by some external force.

**In-ground-effect (IGE) hover.** A hover close to the surface (usually less than one rotor diameter distance above the surface) under the influence of ground effect.

**Induced drag.** That part of the total drag that is created by the production of lift.

**Induced flow.** The component of air flowing vertically through the rotor system resulting from the production of lift.

**Isogonic line.** Lines on charts that connect points of equal magnetic variation.

**Knot.** A unit of speed equal to one nautical mile per hour.

**$L_{DMAX}$.** The maximum ratio between total lift (L) and total drag (D). This point provides the best glide speed. Any deviation from the best glide speed increases drag and reduces the distance you can glide.

**Lateral vibration.** A vibration in which the movement is in a lateral direction, such as imbalance of the main rotor.

**Lead and flag.** The fore (lead) and aft (lag) movement of the rotor blade in the plane of rotation.

**Licensed empty weight.** Basic empty weight plus only undrainable oil.

**Lift.** One of the four main forces acting on a rotorcraft. It acts perpendicular to the relative wind.

**Load factor.** The ratio of a specified load to the total weight of the aircraft.

**Loss of Tail Rotor Effectiveness (LTE).** A manifestation of the Vortex ring aerodynamics on a vertical rotating wing, anti-torque rotor in most instances. Usually inertia and winds combine with Vortex ring state aerodynamics to constitute the hazard. It is characterized by a loss of heading control and requires flight to gain airspeed to exit the phenomenon.

**Low-G Maneuvers.** A low-G condition is a phase of aerodynamic flight where the airframe is temporarily unloaded and the rotor is not supporting the weight of the helicopter. This usually occurs during low gravity or negative gravity maneuvers. This allows tail rotor thrust to tilt the airframe prompting the pilot to add lateral cyclic which is a fatal movement. Helicopter pilots experiencing less than one gravity of force should first apply aft cyclic to reload the rotor system with the weight of the helicopter.

**Married needles.** A term used when two hands of an instrument are superimposed over each other, as on the engine/rotor tachometer.

**Mast.** The component that supports the main rotor.

**Mast bumping.** Action of the rotor head striking the mast, occurring on only underslung rotors.

**Navigational aid (NAVAID).** Any visual or electronic device, airborne or on the surface, that provides point-to-point guidance information, or position data, to aircraft in flight.

**Negative transfer.** When previously learned procedures, techniques, and judgment may lead to negative outcomes or poor results in a different environment, or in some cases, a different type aircraft or different category of aircraft.

**Night.** The time between the end of evening civil twilight and the beginning of morning civil twilight, as published in the American Air Almanac.

**Normally aspirated engine.** An engine that does not compensate for decreases in atmospheric pressure through turbocharging or other means.

**One-to-one vibration.** A low-frequency vibration having one beat per revolution of the rotor. This vibration can be either lateral, vertical, or horizontal.

**Out-of-ground-effect (OGE) hover.** A hover greater than one rotor diameter distance above the surface. Because induced drag is greater while hovering out of ground effect, it takes more power to achieve a hover out of ground effect.

**Parasite drag.** The part of total drag created by the form or shape of helicopter parts.

**Payload.** The term used for passengers, baggage, and cargo.

**Pendular action.** The lateral or longitudinal oscillation of the fuselage due to its suspension from the rotor system.

**Pitch angle.** The angle between the chord line of the rotor blade and the reference plane of the main rotor hub or the rotor plane of rotation.

**Pressure altitude.** The height above the standard pressure level of 29.92 "Hg. It is obtained by setting 29.92 in the barometric pressure window and reading the altimeter.

**Profile drag.** Drag incurred from frictional or parasitic resistance of the blades passing through the air. It does not change significantly with the angle of attack of the airfoil section, but it increases moderately as airspeed increases.

**Resultant relative wind.** Airflow from rotation that is modified by induced flow.

**Retreating blade.** Any blade, located in a semicircular part of the rotor disk, in which the blade direction is opposite to the direction of flight.

**Retreating blade stall.** A stall that begins at or near the tip of a blade in a helicopter because of the high angles of attack required to compensate for dissymmetry of lift.

**Rigid rotor.** A rotor system permitting blades to feather but not flap or hunt.

**Rotational velocity.** The component of relative wind produced by the rotation of the rotor blades.

**Rotor.** A complete system of rotating airfoils creating lift for a helicopter.

**Rotor disk area.** See disk area.

**Rotor brake.** A device used to stop the rotor blades during shutdown.

**Rotor force.** The force produced by the rotor. It is composed of rotor lift and rotor drag.

**Semirigid rotor.** A rotor system in which the blades are fixed to the hub, but are free to flap and feather.

**Settling with power.** A condition of a rotor experiencing vortex ring state. The term describes how the helicopter keeps losing altitude, or settling, even though adequate engine power is available for flight. It is characterized by a rate of descent of more than 300 feet per minute, less than effective translational lift (around 15 knots), and 20 percent to 100 percent of engine power applied to the rotor system. See vortex ring state.

**Shaft turbine.** A turbine engine used to drive an output shaft, and commonly used in helicopters.

**Skid.** A flight condition in which the rate of turn is too great for the angle of bank.

**Skid shoes.** Plates attached to the bottom of skid landing gear, protecting the skid.

**Slip.** A flight condition in which the rate of turn is too slow for the angle of bank.

**Solidity ratio.** The ratio of the total rotor blade area to total rotor disk area.

**Span.** The dimension of a rotor blade or airfoil from root to tip.

**Split needles.** A term used to describe the position of the two needles on the engine/rotor tachometer when the two needles are not superimposed.

**Standard atmosphere.** A hypothetical atmosphere based on averages in which the surface temperature is 59 °F (15 °C), the surface pressure is 29.92 "Hg (1,013.2 mb) at sea level, and the temperature lapse rate is approximately 3.5 °F (2 °C) per 1,000 feet.

**Static stop.** A device used to limit the blade flap, or rotor flap, at low rpm or when the rotor is stopped.

**Steady-state flight.** Straight-and-level, unaccelerated flight, in which all forces are in balance.

**Symmetrical airfoil.** An airfoil having the same shape on the top and bottom.

**Tail rotor.** A rotor turning in a plane perpendicular to that of the main rotor and parallel to the longitudinal axis of the fuselage. It is used to control the torque of the main rotor and to provide movement about the yaw axis of the helicopter.

**Teetering hinge.** A hinge that permits the rotor blades of a semirigid rotor system to flap as a unit.

**Translational thrust.** As the tail rotor works in less turbulent air, it reaches a point of translational thrust. At this point the tail rotor becomes aerodynamically efficient and the improved efficiency produces more antitorque thrust. The pilot will be able to recognize when the tail rotor has reached translational thrust because once there is more antitorque thrust produced, the nose of the helicopter will yaw to the left (opposite direction of the tail rotor thrust) which will force the pilot correct with applying right pedal. This will in turn decrease the AOA in the tail rotor blades.

**Thrust.** The force developed by the rotor blades acting parallel to the relative wind and opposing the forces of drag and weight.

**Tip-path plane.** The imaginary circular plane outlined by the rotor blade tips as they make a cycle of rotation.

**Torque.** The tendency of helicopters with a single, main rotor system to turn in the opposite direction of the main rotor rotation.

**Torsion Load.** A type of load that causes objects to twist due to torque.

**Trailing edge.** The rearmost edge of an airfoil.

**Translating tendency.** The tendency of the single-rotor helicopter to move laterally during hovering flight. Also called tail rotor drift.

**Translational lift.** The additional lift obtained when entering forward flight, due to the increased efficiency of the rotor system.

**Transverse-flow effect.** A condition of increased drag and decreased lift in the aft portion of the rotor disk caused by the air having a greater induced velocity and angle in the aft portion of the disk.

**True altitude.** The actual height of an object above mean sea level.

**Turboshaft engine.** A turbine engine that transmits power through a shaft, as would be found in a turbine helicopter.

**Twist grip.** The power control on the end of the collective control.

**Underslung.** A rotor hub that rotates below the top of the mast, as on semirigid rotor systems.

**Unloaded rotor.** The state of a rotor when rotor force has been removed, or when the rotor is operating under a low or negative G condition.

**Useful load.** The difference between the gross weight and the basic empty weight. It includes the flight crew, usable fuel, drainable oil, if applicable, and payload.

**Variation.** The angular difference between true north and magnetic north, indicated on charts by isogonic lines.

**Vertical vibration.** A vibration in which the movement is up and down or vertical, as in an out-of-track condition.

**Vortex ring state.** A transient condition of downward flight (descending through air after just previously being accelerated downward by the rotor) during which an appreciable portion of the main rotor system is being forced to operate at angles of attack above maximum. Blade stall starts near the hub and progresses outward as the rate of descent increases.

**Weight.** One of the four main forces acting on a rotorcraft. Equivalent to the actual weight of the rotorcraft. It acts downward toward the center of the earth.

**Yaw.** The movement of a rotorcraft about its vertical, or yaw, axis.

# Index

# W

# Y